Y. Imayoshi · M. Taniguchi

An Introduction to Teichmüller Spaces

With 43 Illustrations

Springer-Verlag
Tokyo Berlin Heidelberg
New York London Paris
Hong Kong Barcelona

Yoichi Imayoshi
Department of Mathematics, College of General Education, Osaka University, Toyonaka, Osaka 560, Japan

Masahiko Taniguchi
Department of Mathematics, Faculty of Science, Kyoto University, Sakyo-ku, Kyoto 606, Japan

ISBN-13: 978-4-431-68176-2 e-ISBN-13: 978-4-431-68174-8
DOI: 10.1007/978-4-431-68174-8

Softcover reprint of the hardcover 1st edition 1992

Preface to the English Edition

This is the English translation of our introductory book on Teichmüller space written in Japanese.

We have taken advantage of the opportunity afforded by this translation to correct some minor errors in the original text, and to include several new related topics as additional sections or subsections.

Among other things, we add the construction of Thurston's compactification of the Teichmüller space in §4 of Chapter 3, and the Thurston and Bers theory on the classification of Teichmüller modular transformations in §5 of Chapter 6. We also include a sketch of the proof of celebrated theorems of Royden (§4 of Chapter 6), connection between Teichmüller theory and deformation theory of the complex structures of Riemann surfaces due to Kodaira and Spencer (§2.4 of Chapter 7), and a derivation of negativity of curvatures of the Teichmüller space with respect to the Weil-Petersson metric (§3.4 of Chapter 7). Further, we indicate how to verify that the compactified moduli space of a compact Riemann surface, constructed in Appendix B, is actually compact.

We include several new references chiefly related to added parts. Other than these, the main body of the text is unchanged.

Professors Tadashi Kuroda and Kôtaro Oikawa kindly read the manuscript of the English edition and gave much valuable advice. We also record our gratitude to all friends and colleagues, especially Sadayoshi Kojima, Makoto Masumoto, Toshiyuki Sugawa, and Harumi Tanigawa, who rendered us much help by reading various parts of the manuscript.

Finally, we would like to express our hearty thanks to Professor Kenji Ueno of Kyoto University, and to Springer-Verlag Tokyo for their kind and helpful support in achieving this English edition of our book.

Yoichi Imayoshi
Masahiko Taniguchi

August, 1991

Preface

The problem of how to parametrize the variations of complex structures on a fixed base surface originated with G. F. Bernhard Riemann. This problem has spurred extensive investigations, and progress has been considerable in the areas of the theory of Riemann surfaces, algebraic geometry, and differential geometry.

In this book, we present the theory of Teichmüller spaces which give a parametrization of all the complex structures on a given surface. This subject lies in the intersection of many important areas of mathematics. These include complex manifolds, Fuchsian groups, Kleinian groups, Lie groups, automorphic forms, complex analysis, algebraic geometry, differential geometry, topology in two and three dimensions, differential equations, complex dynamics, and ergodic theory. Recently, the theory of Teichmüller spaces has begun playing important roles in the string theory.

We have attempted to make the book as self-contained as possible. We begin from the most elementary aspects and primitive motivations. We also present subjects through typical examples and heuristic methods. We hope that these plans help the reader grasp the substance of Teichmüller spaces.

A Riemann surface is a connected one-dimensional complex manifold. Two Riemann surfaces R_1 and R_2 are biholomorphically equivalent if there exists a biholomorphic mapping from R_1 onto R_2. It is said that R_1 and R_2 have the same complex structure if they are biholomorphically equivalent.

A Riemann surface can be also regarded as a real two-dimensional oriented differentiable manifold. Even if there is an orientation-preserving diffeomorphism between two Riemann surfaces, they are not necessarily biholomorphically equivalent.

The question naturally arises as to how many distinct complex structures could be assigned on a given oriented two-dimensional differentiable manifold. This is called Riemann's moduli problem.

Tracing its evolution, let us examine this problem more closely for closed Riemann surfaces.

Let M_g be Riemann's moduli space of genus g, that is, the set of all biholomorphic equivalence classes of closed Riemann surfaces of genus g. Since every closed Riemann surface of genus zero is biholomorphically equivalent to the Riemann sphere, obviously M_0 consists of one point. As is well known classically, the theory of elliptic functions and elliptic curves shows that M_1 is identified with the complex plane. In 1857, Riemann asserted that M_g ($g \geq 2$) is parametrized by $3g-3$ complex parameters. He represented closed Riemann surfaces of genus

g as finite branched covering surfaces of the Riemann sphere, and determined the number of parameters of M_g by the number of degrees of freedom of the branch points.

In this book, we treat moduli spaces through Teichmüller spaces and Teichmüller modular groups as follows.

Let R be a closed Riemann surface of genus g, and let Σ be a marking on R, i.e., a canonical system of generators of a fundamental group of R. Two pairs (R, Σ) and (R', Σ') are defined to be equivalent if there exists a biholomorphic mapping $f\colon R \to R'$ such that $f_*(\Sigma)$ is equivalent to Σ'. Denote by $[R, \Sigma]$ the equivalence class of (R, Σ). Such an equivalence class $[R, \Sigma]$ is called a marked closed Riemann surface of genus g. The Teichmüller space T_g of genus g consists of all marked closed Riemann surfaces of genus g. It is verified that T_g has a canonical complex manifold structure, and it is a branched covering manifold of the moduli space M_g. Its covering transformation group is called the Teichmüller modular group Mod_g which corresponds to the change of markings. It turns out that M_g is identified with the quotient space T_g/Mod_g, which has a normal complex analytic space structure.

The Teichmüller space T_g has appeared implicitly in the continuity arguments of Felix Klein and Henri Poincaré, who studied Fuchsian groups and automorphic functions from the 1880s. Robert Fricke, Werner Fenchel and Jakob Nielsen constructed T_g ($g \geq 2$) as a real $(6g-6)$-dimensional manifold. Fricke also asserted that T_g is a cell. Their method was based on the uniformization theorem of Riemann surfaces due to Klein, Poincaré, and Paul Koebe: every closed Riemann surface of genus g (≥ 2) is identified with the quotient space H/Γ of the upper half-plane H by a Fuchsian group Γ which is isomorphic to a fundamental group of R. Then each point $[R, \Sigma]$ in T_g corresponds to a canonical system of generators of Γ. Hence we see that $[R, \Sigma]$ is represented by a point in $\mathbf{R}^{6g-6}$ which is called the Fricke coordinates of $[R, \Sigma]$. Moreover, the Poincaré metric on H induces the hyperbolic metric on R, and the conformal structure defined by this hyperbolic metric corresponds to the complex structure of R.

One of Oswald Teichmüller's great contributions to the moduli problem was to recognize that it becomes more accessible if we consider not only conformal mappings but also quasiconformal mappings. A quasiconformal mapping means a homeomorphism which satisfies the Beltrami equation $w_{\bar{z}} = \mu w_z$. A Beltrami coefficient μ measures the magnitude of deformation of a complex structure or a conformal structure. Around 1940 Teichmüller discovered an intimate relation between extremal quasiconformal mappings and holomorphic quadratic differentials, and asserted that T_g is homeomorphic to $\mathbf{R}^{6g-6}$. He also introduced the Teichmüller distance on T_g.

In the end of the 1950s, Lars V. Ahlfors and Lipman Bers developed the fundamentals of the theory of Teichmüller spaces, and they gave rigorous proofs for Teichmüller's results. They also showed that T_g ($g \geq 2$) has a natural complex structure of dimension $3g-3$, and can be embedded in $A_2(R)$ as a bounded domain, where $A_2(R)$ is the space of holomorphic quadratic differentials of a closed Riemann surface R of genus g. From the Riemann-Roch theorem, it is known that $A_2(R)$ is a complex $(3g-3)$-dimensional vector space. Furthermore, the action of the Teichmüller modular group Mod_g on T_g turns out to be a discrete group of biholomorphic automorphisms of T_g.

In these studies, at least the following three descriptions of deformations of Riemann surfaces were used:

(i) deformations of Fuchsian groups,
(ii) deformations of conformal structures induced by hyperbolic metrics, and
(iii) deformations of complex structures induced by quasiconformal mappings.

These deformations are closely related to each other. The essence of these studies is to investigate Teichmüller spaces from these various points of view, and to clarify their relationships.

The content of each chapter of this book is now briefly described.

The purpose of Chapter 1 is to give primitive motivations and backgrounds for the following chapters. First of all, we recall the moduli space and the Teichmüller space of closed Riemann surfaces of genus 1. After presenting the model of these spaces, we construct the Teichmüller space T_g of genus g as the set of marked closed Riemann surfaces of genus g. We also study T_g from the viewpoints of quasiconformal mappings, and of conformal structures induced by Riemannian metrics.

In Chapter 2, we construct the Fricke space F_g of genus g (≥ 2) which represents T_g as a subset of $\mathbf{R}^{6g-6}$. We show that every Riemann surface except for a few types is represented by a quotient space H/Γ of the upper half-plane H by a Fuchsian group Γ. In particular, each marked closed Riemann surface $[R, \Sigma]$ of genus g (≥ 2) is identified with the corresponding canonical systems of generators of Γ. This gives the identification of T_g with F_g. In order to show this fact, we also explain briefly the uniformization theorem of Riemann surfaces, Möbius transformations, and Fuchsian groups.

Chapter 3 deals with the construction of T_g $(g \geq 2)$ from the viewpoint of hyperbolic geometry induced by the Poincaré metric. The fundamental method is to decompose Riemann surfaces into a set of $2g-2$ pairs of pants by simple closed geodesics. Then the Fenchel-Nielsen coordinates on T_g are defined by geodesic length functions of $3g-3$ simple closed geodesics and twist parameters along these geodesics. We also study in the problem of what geodesic length functions of simple closed geodesics determine the points of T_g.

Chapter 4 is devoted to fundamentals of quasiconformal mappings. First of all, we define quasiconformal mappings, using two analytic procedures and a geometric method. Then we prove two fundamental theorems due to Ahlfors and Bers, i.e., the existence of a quasiconformal mapping satisfying a given Beltrami equation, and the holomorphic dependence of solutions on Beltrami coefficients.

In Chapter 5, the Teichmüller space $T(R)$ of a closed Riemann surface R of genus g (≥ 2) is constructed by using quasiconformal mappings. A Teichmüller mapping of R means a quasiconformal mapping which is "locally affine" in certain sense, and obtained from a holomorphic quadratic differential on R. The essence of Teichmüller's idea is that the extremal quasiconformal mapping in those which determine a given point of $T(R)$ is a Teichmüller mapping. Then it turns out that T_g $(g \geq 2)$ is homeomorphic to the space $A_2(R)$ of holomorphic quadratic differentials on R. Hence, T_g is homeomorphic to $\mathbf{R}^{6g-6}$. We also show that $T(R)$ is complete with respect to the Teichmüller distance.

In Chapter 6, using the Schwarzian derivative, we construct the Bers embedding of $T(R)$ into a bounded domain in $A_2(R^*)$, the space of holomorphic quadratic differentials on R^*. Here, R^* denotes the mirror image of R. By the Riemann-Roch theorem, $A_2(R^*)$ is also identified with the $(3g-3)$-dimensional complex Euclidean space $\mathbf{C}^{3g-3}$. Using this embedding, we see that $T(R)$ has a natural complex manifold structure of dimension $3g-3$. It is also proved that the Teichmüller modular group Mod_g is a discrete group of biholomorphic automorphisms of T_g, and acts properly discontinuously on T_g. This shows that the moduli space $M_g = T_g/Mod_g$ has a normal complex analytic space structure of dimension $3g-3$.

Chapter 7 treats the Weil-Petersson metric on T_g. The holomorphic tangent space of T_g at a point $[R, \Sigma]$ is identified with the dual space of $A_2(R)$. Then the Petersson scalar product on $A_2(R)$ induces the Weil-Petersson metric on T_g. We give two proofs for the fundamental fact that the Weil-Petersson metric is Kählerian. Both of them are due to Ahlfors.

In Chapter 8, we establish a beautiful formula due to S. Wolpert, which states that the Weil-Petersson Kähler form on T_g has a simple representation with respect to Fenchel-Nielsen coordinates.

We also give two appendixes. Appendix A deals with Schiffer's interior variation from the viewpoint of quasiconformal mappings. We explain Ahlfors' construction of the complex structure for T_g, which was the first construction of its natural complex structure. We also discuss variations with respect to degenerations of Riemann surfaces. In Appendix B, we explain briefly the compactification of moduli spaces.

At the end of each chapter, there are bibliographical notes of books and articles to which we referred in the text. The bibliography is not complete. There is a vast literature relating to the theory of Teichmüller spaces. We hope that this list helps the reader to begin to explore these research papers. Any omissions of references, or failure to attribute theorems, reflects only our ignorance.

The authors are extremely grateful to Professor Osamu Takenouchi who recommended that we write this book. They also gratefully acknowledge the generous contributions of our friends and colleagues Makoto Masumoto, Hiromi Ohtake, Hiroshige Shiga, and Toshiyuki Sugawa, who read the original manuscript, and made many helpful mathematical suggestions and improvements.

Yoichi Imayoshi
Masahiko Taniguchi

October, 1989

Contents

An Introduction to Teichmüller Spaces

Chapter 1

Teichmüller Space of Genus g

In this chapter, we construct the Teichmüller space T_g of genus g, and give motivations and backgrounds for the following chapters. Some proofs are rather sketchy, and some shall be omitted.

First, in Section 1, we give the definition of Riemann surfaces and some typical examples. Section 2 deals with the moduli space and the Teichmüller space of genus 1, which gives a model of the Teichmüller space T_g of arbitrary genus g. The Teichmüller space T_g is defined in Section 3 as the set of all marked closed Riemann surfaces of genus g. In Section 4, we study T_g from the viewpoint of quasiconformal mappings. Section 5 treats the relationship between the complex structure and the conformal structure induced by a Riemannian metric on an oriented surface.

1.1. Riemann Surfaces

1.1.1. Definition of Riemann Surfaces

A one-dimensional connected complex manifold is called a *Riemann surface*. Intuitively, it is a collection of domains in the complex plane which are patched by biholomorphic mappings. The standard definition of Riemann surface is as follows. Let R be a connected Hausdorff space with a family $\{(U_j, z_j)\}_{j\in J}$ satisfying the following three conditions (see Fig. 1.1):

(i) Every U_j is an open subset of R, and $R = \bigcup_{j\in J} U_j$.
(ii) Every z_j is a homeomorphism of U_j onto an open subset D_j in the complex plane.
(iii) If $U_j \cap U_k \neq \phi$, the transition mapping

$$z_{kj} = z_k \circ z_j^{-1} : z_j(U_j \cap U_k) \to z_k(U_j \cap U_k)$$

is a biholomorphic mapping, i.e., a holomorphic homeomorphism.

We say that $\{(U_j, z_j)\}_{j\in J}$ is a *system of coordinate neighborhoods* on R, and that it defines a *one-dimensional complex structure* on R. A *Riemann surface* or *one-dimensional complex manifold* means a connected Hausdorff space with one-dimensional complex structure.

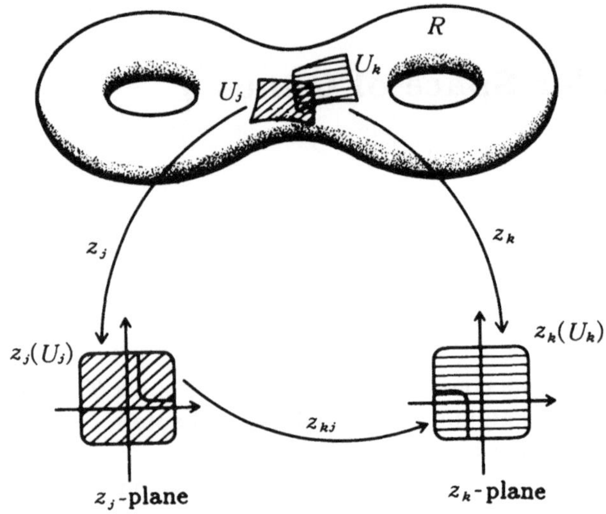

Fig. 1.1.

A *coordinate neighborhood* (U, z) of a Riemann surface R is a pair of an open set U in R and a homeomorphism z of U into the complex plane such that for any element (U_j, z_j) of a system of coordinate neighborhoods with $U \cap U_j \neq \phi$, the mapping

$$z \circ z_j^{-1} : z_j(U \cap U_j) \rightarrow z(U \cap U_j)$$

is biholomorphic. This U is also called a *coordinate neighborhood* of R. Such a homeomorphism z is said to be a *local coordinate* or a *local parameter* on U of R. A coordinate neighborhood (U, z) with $p \in U$ is called a *coordinate neighborhood around* p, and z is called a *local coordinate* or *local parameter around* p.

Local analysis on a Riemann surface R is reduced to analysis on domains in the complex plane via local parameters. For example, a *holomorphic function* on R is a function f on R such that $f \circ z^{-1}$ is holomorphic on $z(U)$ for any coordinate neighborhood (U, z) of R. A mapping f of R into a Riemann surface S is said to be a *holomorphic mapping* if $w \circ f \circ z^{-1}$ is holomorphic for all coordinate neighborhoods (U, z) of R and (V, w) of S with $f(U) \subset V$. A *biholomorphic mapping* $f : R \rightarrow S$ means a holomorphic mapping f of R onto S which has the holomorphic inverse mapping $f^{-1} : S \rightarrow R$. Two Riemann surfaces R and S are *biholomorphically equivalent* if there exists a biholomorphic mapping between R and S. In this case, we regard R and S as the same Riemann surface and write $R = S$. We say also that R and S have the *same complex structure*. Complex structures, biholomorphic mappings, and biholomorphic equivalence may be and are actually often said to be *conformal structures*, *conformal mappings*, and *conformal equivalence*, respectively (see §1.5).

Remark. A Riemann surface is a two-dimensional real-analytic manifold, and the Cauchy-Riemann equation implies that local coordinates determine its orientation. Hereafter, we assume that a Riemann surface is always equipped with this orientation. It is well-known that every Riemann surface has a countable basis of open sets and can be triangulated. For these facts and the general theory of Riemann surfaces, consult, for instance, Ahlfors and Sario [A-6], Bers [A-13], Cohn [A-22], Farkas and Kra [A-28], Forster [A-32], Gunning [A-40], Jones and Singerman [A-48], Siegel [A-98], and Springer [A-99].

1.1.2. Examples of Riemann Surfaces

First of all, every domain D in the complex z-plane is a Riemann surface. Namely, a complex structure on D is given by only one coordinate neighborhood (D, z). The *Riemann sphere* $\hat{\mathbf{C}} = \mathbf{C} \cup \{\infty\}$, which is the one point compactification of the complex plane $\mathbf{C}$, is also a Riemann surface. A complex structure on $\hat{\mathbf{C}}$ is defined by two coordinate neighborhoods $(\mathbf{C}, z)$ and $(\hat{\mathbf{C}} - \{0\}, 1/z)$.

Next, let us see the Riemann surface of the algebraic function $w = \sqrt{z}$. This is the Riemann surface on which the inverse function of the holomorphic function $z = f(w) = w^2$ is single-valued. Classically, it is constructed by the method of "cut and paste" or by analytic continuation. Here, we state the former one.

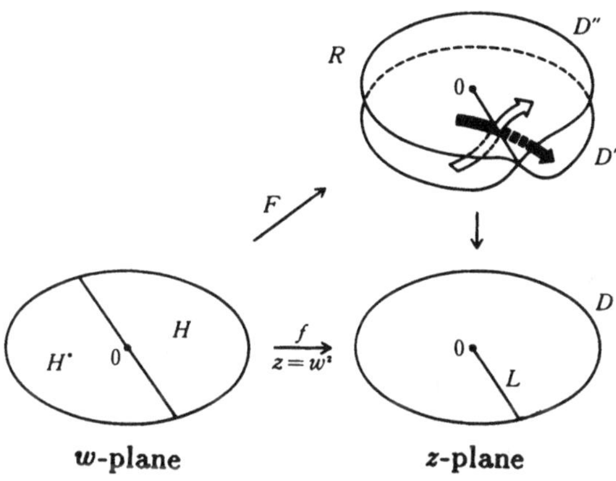

Fig. 1.2.

The holomorphic function $z = w^2$ maps biholomorphically both the upper half-plane $H = \{\, w \in \mathbf{C} \mid \operatorname{Im} w > 0 \,\}$ and the lower half-plane $H^* = \{\, w \in \mathbf{C} \mid \operatorname{Im} w < 0 \,\}$ onto the domain $D = \mathbf{C} - L$, where $L = \{\, x \in \mathbf{R} \mid x \geq 0 \,\}$ is a cut on the z-plane. Take two copies D', D'' of D and paste crosswise along their cuts L', L'' as in Fig. 1.2. Then we get a two-sheeted covering surface R over the z-sphere. Since the function $f(w) = w^2$ induces a homeomorphism F of the w-sphere onto R, we can define the complex structure of R from the complex structure of the w-sphere in such a way that $F\colon \hat{\mathbf{C}} \to R$ is a biholomorphic mapping. This R is the Riemann surface of $w = \sqrt{z}$. (See Ahlfors [A-4], Chap. 8; Jones and Singerman [A-48], Chap. 4; and Springer [A-99], Chap. 1.)

Note that the Riemann surface R of the algebraic function $w = \sqrt{z}$ is also regarded as the algebraic curve defined by the equation $w^2 = z$.

Finally, we see elliptic curves, i.e., tori from the viewpoint of algebraic curves. For any complex number $\lambda\,(\neq 0, 1)$, let R be the algebraic curve defined by the equation

$$w^2 = z(z-1)(z-\lambda). \tag{1.1}$$

In other words, R consists of all points $(z, w) \in \mathbf{C} \times \mathbf{C}$ satisfying algebraic equation (1.1) and the point $p_\infty = (\infty, \infty)$. We can define the complex structure of R by the complex structure of the z-sphere so that the projection $\pi \colon R \to \hat{\mathbf{C}}$, $\pi(z, w) = z$, is holomorphic. This R is a two-sheeted branched covering surface over the z-sphere with branch points 0, 1, λ, and ∞. The mapping $f \colon R \to \hat{\mathbf{C}}$, $f(z, w) = w$, is holomorphic. This function f is written as $w = \sqrt{z(z-1)(z-\lambda)}$ and R is a Riemann surface on which the algebraic function $w = \sqrt{z(z-1)(z-\lambda)}$ is single-valued.

The Riemann surface R defined by algebraic equation (1.1) is regarded topologically as a surface illustrated in Fig. 1.5. Take two copies of the Riemann spheres S_1, S_2 with cuts between 0 and 1, and between λ and ∞ (Fig. 1.3). Place them face to face (Fig. 1.4), and join along their cuts (Fig. 1.5). The resulting surface is homeomorphic to the Riemann surface R. Hence, R looks like the surface of a doughnut. We call such a Riemann surface a *torus*. A torus is also called an *elliptic curve*; this name comes from the elliptic integral (see §1.4).

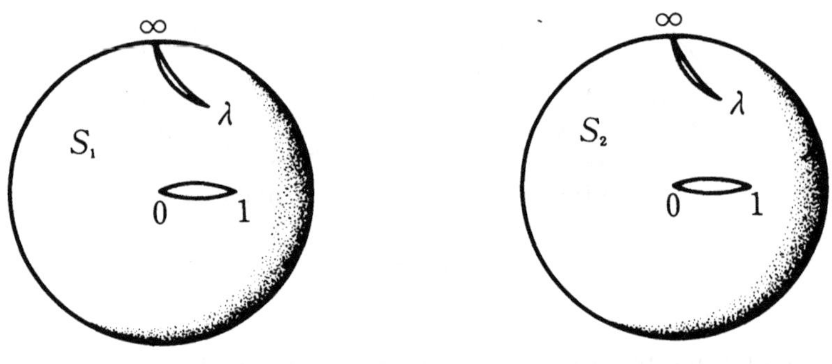

Fig. 1.3.

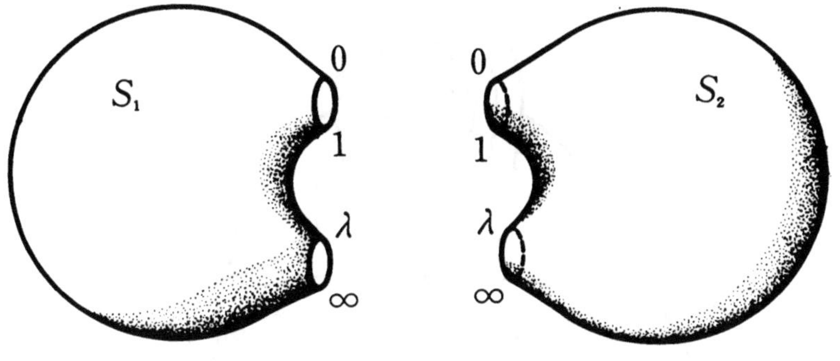

Fig. 1.4.

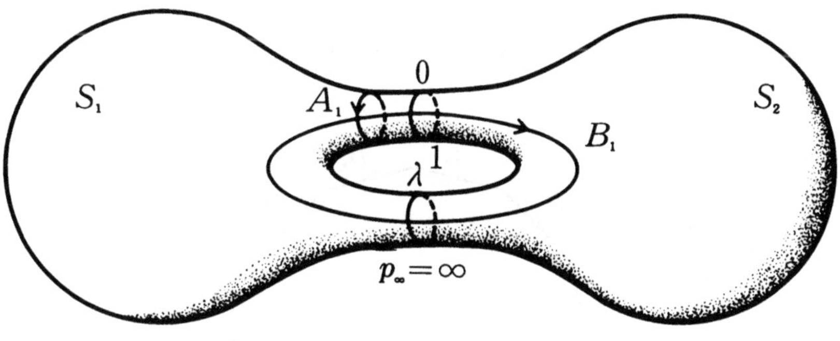

Fig. 1.5.

1.1.3. Closed Riemann Surfaces

A Riemann surface homeomorphic to a sphere with g handles as in Fig. 1.6 is called a *closed Riemann surface of genus g*. The Riemann sphere is of genus 0, and a torus is of genus 1. It is well-known that every compact Riemann surface is a closed Riemann surface of finite genus. A non-compact Riemann surface is called an *open Riemann surface*.

Take a point p_o on a closed Riemann surface R of genus g and cut R along simple closed curves A_1, B_1, ..., A_g, B_g with base point p_o as in Fig. 1.6. Then we get a domain homeomorphic to a convex polygon with $4g$ sides (Fig. 1.7).

The fundamental group $\pi_1(R, p_o)$ of R with base point p_o is generated by the homotopy classes $[A_1]$, $[B_1]$, ..., $[A_g]$, $[B_g]$ induced from A_1, B_1, ..., A_g, B_g and satisfies the fundamental relation

$$\prod_{j=1}^{g}[A_j][B_j][A_j]^{-1}[B_j]^{-1} = 1 \quad \text{(the unit)}.$$

We call $\{[A_j],[B_j]\}_{j=1}^{g}$ or $\{A_j, B_j\}_{j=1}^{g}$ a *canonical system of generators* of $\pi_1(R, p_o)$.

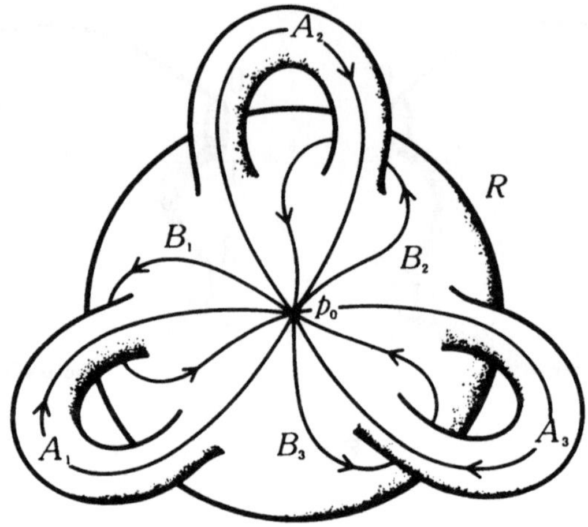

Fig. 1.6. ($g = 3$)

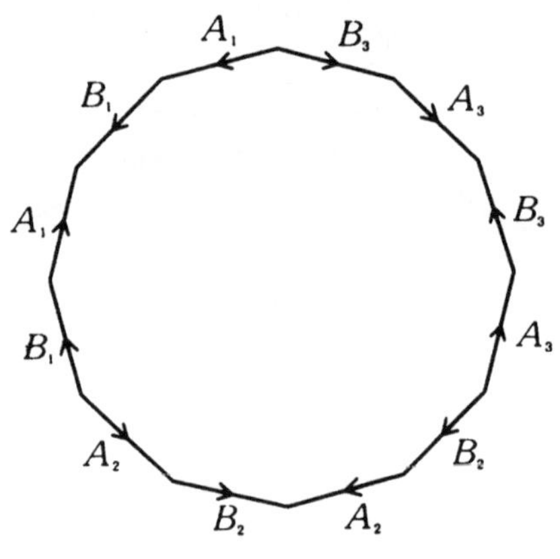

Fig. 1.7. ($g = 3$)

1.1.4. Lattice Group Representations of Tori

We shall represent a torus as the quotient space $\mathbf{C}/\Gamma$ of the complex plane $\mathbf{C}$ by a lattice group Γ. Since $w(z) = \sqrt{z(z-1)(z-\lambda)}$ is a single-valued meromorphic function on the torus R defined by equation (1.1), we can consider the complex integral of $1/w(z)$ along paths on R. For any point $p = (z(p), w(p))$ on R, the *elliptic integral* $\Phi(p)$ is defined by selecting a branch of algebraic function $w(z)$ and a path from ∞ to $z(p)$, and by setting

$$\Phi(p) = \int_{\infty}^{z(p)} \frac{dz}{\sqrt{z(z-1)(z-\lambda)}}, \tag{1.2}$$

where ∞ corresponds to the point p_∞ in Fig.1.5 and the value of $w(z)$ is determined by analytic continuation along the path of integral.

Remark. It is more adequate to regard (1.2) as a line integral, along a path, of the holomorphic differential

$$\frac{dz}{\sqrt{z(z-1)(z-\lambda)}}$$

of degree 1 on R.

The value of this elliptic integral Φ is not determined uniquely. It depends on a path joining p_∞ and p. The values of Φ along the simple closed curves A_1, B_1 in Fig. 1.5 are represented by

$$\pi_1 = 2\int_0^1 \frac{dz}{\sqrt{z(z-1)(z-\lambda)}} \quad \text{and} \quad \pi_2 = 2\int_0^\lambda \frac{dz}{\sqrt{z(z-1)(z-\lambda)}},$$

respectively. Setting

$$\Gamma = \{\, m\pi_1 + n\pi_2 \mid m, n \in \mathbf{Z}\,\},$$

we see that the function $\Phi(p)$ has infinitely many values which differ from each other by elements of Γ. Every element of Γ is called a *period* of Φ. Since the periods π_1 and π_2 satisfy $\operatorname{Im}(\pi_1/\pi_2) > 0$, they are linearly independent over the real number field $\mathbf{R}$. We call such Γ a *lattice group* for R.

A lattice group Γ for R is regarded as a subgroup of the analytic automorphism group $Aut(\mathbf{C})$ of $\mathbf{C}$. In fact, every $\gamma = m\pi_1 + n\pi_2 \in \Gamma$ is identified with a translation $\gamma(z) = z + m\pi_1 + n\pi_2$ of $\mathbf{C}$. We say that two points $z, z' \in \mathbf{C}$ are *equivalent* under Γ if there exists an element $\gamma \in \Gamma$ with $z' = \gamma(z)$. Denote by $[z]$ the equivalence class represented by z.

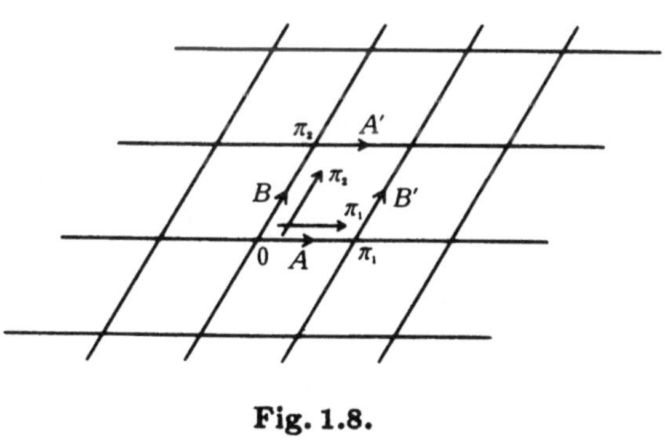

Fig. 1.8.

The quotient space $\mathbf{C}/\Gamma$ of $\mathbf{C}$ by Γ consists of all equivalence classes $[z]$. This quotient space $\mathbf{C}/\Gamma$ is realized as a surface obtained by identifying sides

A with A' and B with B' in the lattice of Fig. 1.8 by the translations π_1, π_2, respectively.

Now, we define a complex structure of $\mathbf{C}/\Gamma$. Let $\pi: \mathbf{C} \to \mathbf{C}/\Gamma$ be the projection, i.e., $\pi(z) = [z]$ for $z \in \mathbf{C}$. Introduce the quotient topology on $\mathbf{C}/\Gamma$, which is defined as follows: a subset U of $\mathbf{C}/\Gamma$ is open if the inverse image $\pi^{-1}(U)$ is open in $\mathbf{C}$. It is verified that $\mathbf{C}/\Gamma$ is a connected topological space.

For any two points $[a], [b] \in \mathbf{C}/\Gamma$, we can take neighborhoods V_a, V_b of a, b with $\pi(V_a) \cap \pi(V_b) = \phi$. Since π is an open mapping, this shows that $\mathbf{C}/\Gamma$ is a Hausdorff space. Moreover, for any point $[a] \in \mathbf{C}/\Gamma$, taking a sufficiently small neighborhood V_a of a, we see that π gives a homeomorphism of V_a into $\mathbf{C}/\Gamma$. Let $U_a = \pi(V_a)$ and $z_a: U_a \to V_a$ be a homeomorphism with $z_a([z]) = z$. Then (U_a, z_a) gives a coordinate neighborhood around $[a]$ in $\mathbf{C}/\Gamma$. Thus $\mathbf{C}/\Gamma$ becomes a torus, i.e., a closed Riemann surface of genus 1 such that the projection $\pi: \mathbf{C} \to \mathbf{C}/\Gamma$ is holomorphic. The triple $(\mathbf{C}, \pi, \mathbf{C}/\Gamma)$ gives an example of universal coverings, considered in §2.1 of Chapter 2.

As is known in the theory of elliptic functions, the mapping $[\Phi]: R \to \mathbf{C}/\Gamma$ sending a point $p \in R$ to a point $[\Phi(p)] \in \mathbf{C}/\Gamma$ is biholomorphic. Hence we see that a torus defined by equation (1.1) is represented by a Riemann surface $\mathbf{C}/\Gamma$ for a lattice group Γ. In Chapter 2, we shall show that every torus is represented by a lattice group Γ in $\mathbf{C}$ (see the Corollary to Theorem 2.13). Conversely, it is known that such a Riemann surface $\mathbf{C}/\Gamma$ is always biholomorphic to an elliptic curve defined by algebraic equation (1.1). For details, we refer to Ahlfors [A-4], Chap.7; Clemens [A-21], Chap. 2; Jones and Singerman [A-48], Chap. 3; Siegel [A-98], Chap. 1; or Springer [A-99], Chap.1.

1.2. Teichmüller Space of Genus 1

Let us construct the Teichmüller space of genus 1.

1.2.1. The Moduli Space of Tori

We use the fact that every torus is represented by a Riemann surface $\mathbf{C}/\Gamma$, where Γ is a lattice group on $\mathbf{C}$ as in §1.4 (see the Corollary to Theorem 2.13). On performing the transformation $z \mapsto z/\pi_1$, if necessary, we may assume from the beginning that the generators π_1 and π_2 for Γ are the canonical ones 1 and τ with $\operatorname{Im}\tau > 0$, respectively.

Now, consider a lattice group

$$\Gamma_\tau = \{\gamma = m + n\tau \mid m, n \in \mathbf{Z}\},$$

where $\tau \in H = \{\tau \in \mathbf{C} \mid \operatorname{Im}\tau > 0\}$. As was seen in §1.4, the lattice group Γ_τ corresponds to a subgroup of $Aut(\mathbf{C})$, and the Riemann surface $R_\tau = \mathbf{C}/\Gamma_\tau$ is a torus. Denote by π_τ the projection of $\mathbf{C}$ to $\mathbf{C}/\Gamma_\tau$. Notice that $\mathbf{C}/\Gamma_\tau$ has the structure of an additive group.

Theorem 1.1. *For any two points τ and τ' in the upper half-plane H, two tori R_τ and $R_{\tau'}$ are biholomorphically equivalent if and only if τ and τ' satisfy the relation*

$$\tau' = \frac{a\tau + b}{c\tau + d}, \tag{1.3}$$

where a, b, c, and d are integers with $ad - bc = 1$.

Proof. First, assume that there is a biholomorphic mapping f of $R_{\tau'}$ onto R_τ. Since $\mathbf{C}$ is simply connected, the monodromy theorem implies that there exists a lift $\tilde{f}$ of f, that is, a holomorphic mapping $\tilde{f}\colon \mathbf{C} \to \mathbf{C}$ such that $\pi_\tau \circ \tilde{f} = f \circ \pi_{\tau'}$ (cf. Theorem 2.4). Because f is biholomorphic, so is $\tilde{f}$. Then $\tilde{f}$ is written as $\tilde{f}(z) = \alpha z + \beta$, where α and β are complex numbers and $\alpha \neq 0$ (cf. Lemma 2.8). Moreover, we may assume that $\tilde{f}(0) = 0$, and hence $\beta = 0$. Thus both $\tilde{f}(\tau')$ and $\tilde{f}(1)$ are equivalent to $\tilde{f}(0) = 0$ under Γ_τ. Hence, we have

$$\tilde{f}(\tau') = \alpha\tau' = a\tau + b,$$
$$\tilde{f}(1) = \alpha = c\tau + d,$$

where a, b, c, and d are integers. Therefore, we obtain

$$\tau' = \frac{a\tau + b}{c\tau + d}.$$

Applying the same argument to $\tilde{f}^{-1}$, we get

$$\tau = \frac{a'\tau' + b'}{c'\tau' + d'},$$

where a', b', c', and d' are integers. Furthermore, from the relations $\tilde{f}^{-1} \circ \tilde{f}(1) = 1$ and $\tilde{f}^{-1} \circ \tilde{f}(\tau') = \tau'$, we see that $ad - bc = \pm 1$. Since

$$\operatorname{Im} \tau' = \frac{ad - bc}{|c\tau + d|^2} (\operatorname{Im} \tau) > 0,$$

we have $ad - bc = 1$.

Conversely, if (1.3) holds, then a biholomorphic mapping $f\colon R_{\tau'} \to R_\tau$ is given by $f([z]) = [(c\tau + d)z]$. ∎

Now, we call the group

$$PSL(2, \mathbf{Z}) = \left\{ \gamma(\tau) = \frac{a\tau + b}{c\tau + d} \;\middle|\; a, b, c, d \in \mathbf{Z} \quad \text{and} \quad ad - bc = 1 \right\}$$

the ***modular group***. Every $\gamma \in PSL(2, \mathbf{Z})$ is a biholomorphic automorphism of the upper half-plane H.

Let M_1 be the ***moduli space of tori***, i.e., the set of all biholomorphic equivalence classes of tori. Theorem 1.1 implies that M_1 is identified with the quotient space of H by $PSL(2, \mathbf{Z})$, that is,

$$M_1 \cong H/PSL(2, \mathbf{Z}).$$

It is known that the quotient space $H/PSL(2,\mathbf{Z})$ is a Riemann surface (cf. §2.4 of Chapter 2) and that a fundamental domain (cf. §4.2 of Chapter 2) for $PSL(2,\mathbf{Z})$ is the shaded area in Fig. 1.9. Intuitively, we get the Riemann surface $H/PSL(2,\mathbf{Z})$ by identifying the sides of this fundamental domain under the transformations $z \mapsto z+1$ and $z \mapsto -1/z$ as is illustrated in Fig. 1.9. Hence we see that the moduli space of tori is biholomorphic to the complex plane. For more details, see, for example, Ahlfors [A-4], Chap.7; and Jones and Singerman [A-48], Chap. 6.

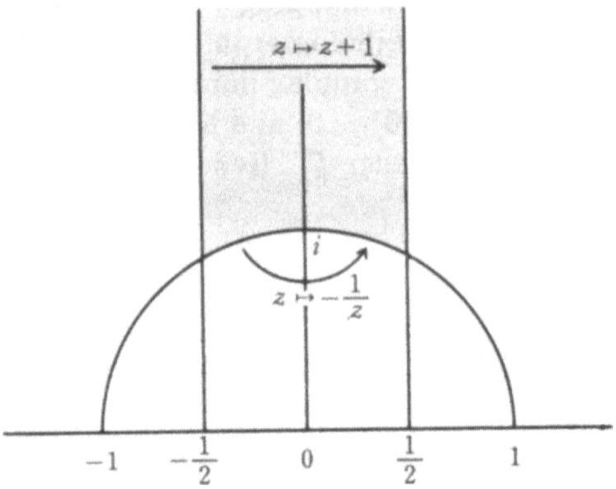

Fig. 1.9.

Remark. A torus given by equation (1.1) depends on a complex parameter $\lambda(\neq 0,1)$, which is denoted by S_λ. It is well known that two such tori S_λ and $S_{\lambda'}$ are biholomorphically equivalent if and only if there exists a linear fractional transformation which takes the set of branch points $\{0,1,\lambda,\infty\}$ of S_λ to the set of branch points $\{0,1,\lambda',\infty\}$ of $S_{\lambda'}$ (see, for example, Clemens [A-21], Chap. 2.7). Thus we see that S_λ and $S_{\lambda'}$ are biholomorphically equivalent if and only if λ' is equal to one of the following numbers:

$$\lambda, \quad \frac{1}{\lambda}, \quad 1-\lambda, \quad \frac{1}{1-\lambda}, \quad \frac{\lambda-1}{\lambda}, \quad \frac{\lambda}{\lambda-1}.$$

Now, let G be a finite group of order 6, generated by $g_1(\lambda) = 1/\lambda$ and $g_2(\lambda) = 1-\lambda$ which are analytic automorphisms of $D = \mathbf{C} - \{0,1\}$. This fact shows that $M_1 \cong D/G$, where D/G means the quotient space of D by G (cf. §2.4 of Chapter 2). Moreover, we find a biholomorphic mapping $F\colon D/G \to \mathbf{C}$, which is defined by $F([\lambda]) = f(\lambda)$ with

$$f(\lambda) = \frac{(\lambda^2-\lambda+1)^3}{\lambda^2(\lambda-1)^2}.$$

1.2.2. Teichmüller Space of Genus 1

First of all, we study the meaning of the difference between τ and τ' which represent biholomorphically equivalent tori $R_\tau = \mathbf{C}/\Gamma_\tau$ and $R_{\tau'} = \mathbf{C}/\Gamma_{\tau'}$, respectively. For a biholomorphic mapping $f\colon R_{\tau'} \to R_\tau$, take the lift $\tilde{f}(z) = (c\tau + d)z$ of f as in the proof of Theorem 1.1. Then $\tilde{f}$ maps the lattice group $\Gamma_{\tau'}$ to the lattice group Γ' generated by $\tilde{f}(1)$ and $\tilde{f}(\tau')$. From relation (1.3), we see that $\tilde{f}(1) = c\tau + d$, $\tilde{f}(\tau') = a\tau + b$, and $\Gamma' = \Gamma_\tau$. Thus this difference is considered as the one corresponding to the choice of generators of Γ_τ, i.e., $\{1, \tau\}$ and $\{c\tau + d, a\tau + b\}$. Note that the coordinate transformation $z \mapsto z/\tilde{f}(1)$ sends $\tilde{f}(1)$ and $\tilde{f}(\tau')$ to 1 and τ', respectively (see Fig. 1.10).

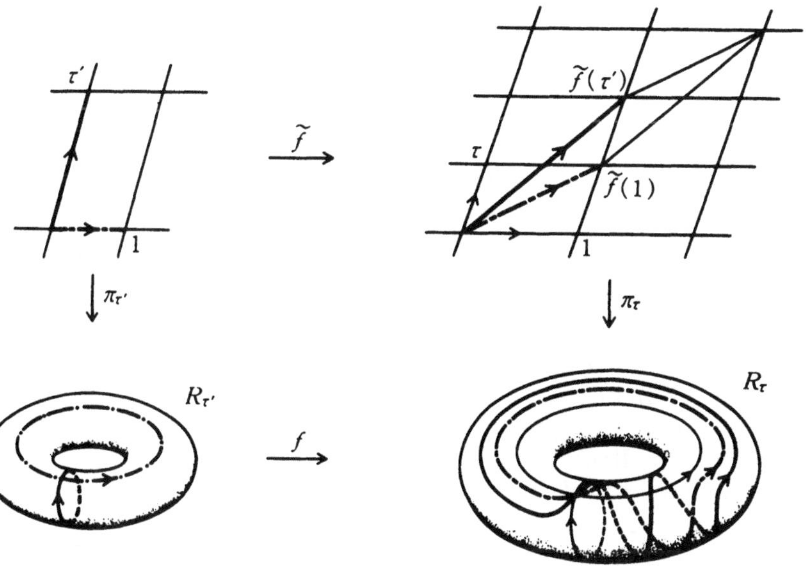

Fig. 1.10.

We consider the above statement from the viewpoint of the fundamental group. Take $p_o = [0]$ as a base point of the fundamental group $\pi_1(R_\tau, p_o)$ of R_τ. The segments between 0 and 1, and between 0 and τ in $\mathbf{C}$, respectively, determine simple closed curves $A_1(\tau)$ and $B_1(\tau)$ on R_τ with base point p_o. The homotopy classes $[A_\tau]$ and $[B_\tau]$ give a canonical system of generators of $\pi_1(R_\tau, p_o)$. Then Γ_τ is isomorphic to $\pi_1(R_\tau, p_o)$ under the correspondence which sends 1 and τ to $[A_1(\tau)]$ and $[B_1(\tau)]$, respectively. Similarly, $\Gamma_{\tau'}$ is identified with $\pi_1(R_{\tau'}, p_o)$ which has a canonical system of generators $\{[A_1(\tau')], [B_1(\tau')]\}$. Hence we see that the above biholomorphic mapping $f\colon R_{\tau'} \to R_\tau$ induces an isomorphism $f_*\colon \pi_1(R_{\tau'}, p_o) \to \pi_1(R_\tau, p_o)$ such that $f_*([A_1(\tau')]) = [f(A_1(\tau'))]$ and $f_*([B_1(\tau')]) = [f(B_1(\tau'))]$, which corresponds to the isomophism $\tilde{f}\colon \Gamma_{\tau'} \to \Gamma_\tau$ between lattice groups. Thus we can consider that the difference between τ

and τ' corresponds also to the different choices of generators of $\pi_1(R_\tau, p_o)$, i.e., $\{[A_1(\tau)], [B_1(\tau)]\}$ and $\{f_*([A_1(\tau')]), f_*([B_1(\tau')])\}$ (see Fig. 1.10).

Now, for any torus R, take a canonical system of generators $\Sigma_p = \{[A_1], [B_1]\}$ of the fundamental group $\pi_1(R_\tau, p)$ of R, and consider the pair (R, Σ_p). Such a Σ_p is called a ***marking*** on R. Two markings $\Sigma_p = \{[A_1], [B_1]\}$ and $\Sigma_{p'} = \{[A_1'], [B_1']\}$ are said to be ***equivalent*** when there exists a continuous curve C_o on R from p to p' which induces the isomorphism $T_{C_o}\colon \pi_1(R,p) \to \pi_1(R,p')$ with $[A_1'] = T_{C_o}([A_1])$ and $[B_1'] = T_{C_o}([B_1])$. Here, T_{C_o} sends an element $[C]$ of $\pi_1(R,p)$ to an element $[C_o^{-1} \cdot C \cdot C_o]$ of $\pi_1(R,p')$. For the definition of product of curves, see §2.2 of Chapter 2. Next, two pairs (R, Σ_p) and (S, Σ_q) as above are *equivalent* if and only if there exists a biholomorphic mapping $h\colon S \to R$ such that $h_*(\Sigma_q) = h_*(\{[A_1'], [B_1']\}) = \{h_*([A_1']), h_*([B_1'])\}$ is equivalent to $\Sigma_p = \{[A_1], [B_1]\}$. Denote by $[R, \Sigma_p]$ the equivalence class of (R, Σ_p). We call such a $[R, \Sigma_p]$ a ***marked torus***. The *Teichmüller space* T_1 *of genus* 1 consists of all marked tori.

Theorem 1.2. *For every point $\tau \in H$, let $\Sigma(\tau) = \{[A_1(\tau)], [B_1(\tau)]\}$ be the marking on $R_\tau = \mathbf{C}/\Gamma_\tau$ for which $[A_1(\tau)]$ and $[B_1(\tau)]$ correspond to 1 and τ in Γ_τ, respectively. Then $[R_\tau, \Sigma(\tau)] = [R_{\tau'}, \Sigma(\tau')]$ in T_1 if and only if $\tau = \tau'$.*

Proof. Assume that $[R_\tau, \Sigma(\tau)] = [R_{\tau'}, \Sigma(\tau')]$. Then there is a biholomorphic mapping $h\colon R_{\tau'} \to R_\tau$ such that $h_*(\Sigma(\tau')) = \{h_*([A_1(\tau')]), h_*([B_1(\tau')])\}$ is equivalent to $\Sigma(\tau) = \{[A_1(\tau)], [B_1(\tau)]\}$. We may assume that $h([0]) = [0]$ by replacing h with $h_1([z]) = h([z]) - h([0])$ if necessary. Then, the definition of equivalence of $\Sigma(\tau)$ and $h_*(\Sigma(\tau'))$ implies that $h_*([A_1(\tau')]) = [A_1(\tau)]$ and $h_*([B_1(\tau')]) = [B_1(\tau)]$. Take a lift $\tilde{h}$ of h with $\tilde{h}(0) = 0$. Then $\tilde{h}(z) = \alpha z$ for some complex number α. Hence we conclude that $\tilde{h}(1) = \alpha = 1$, and $\tilde{h}(\tau') = \alpha\tau' = \tau$. Therefore, we have $\tau = \tau'$. The converse is obvious. □

Since every marked torus $[R, \Sigma_p]$ is represented by $[R_\tau, \Sigma(\tau)]$ for some τ in H, this theorem shows that T_1 is identified with H.

Another method to mark tori is realized via orientation-preserving diffeomorphisms between tori instead of systems of generators of fundamental groups. For that purpose, fix a marking $\Sigma = \{[A_1], [B_1]\}$ on R. Then any pair (S, f) of a torus S and an orientation-preserving diffeomorphism $f\colon R \to S$ defines a marking $f_*(\Sigma) = \{f_*([A_1]), f_*([B_1])\}$ on S.

Theorem 1.3. *Let R, S, and S' be tori, and let $f\colon R \to S, g\colon R \to S'$ be orientation-preserving diffeomorphisms. Then $[S, f_*(\Sigma)] = [S', g_*(\Sigma)]$ in T_1 if and only if $g \circ f^{-1}\colon S \to S'$ is homotopic to a biholomorphic mapping $h\colon S \to S'$.*

Proof. Suppose that $[S, f_*(\Sigma)] = [S', g_*(\Sigma)]$. Take two points $\tau, \tau' \in H$ for which $[R, \Sigma] = [R_\tau, \Sigma(\tau)]$ and $[S, f_*(\Sigma)] = [R_{\tau'}, \Sigma(\tau')]$, where R_τ, $R_{\tau'}$, $\Sigma(\tau)$, and $\Sigma(\tau')$ are defined as above. Regard f and g as diffeomorphisms of R_τ to $R_{\tau'}$. We may assume that their lifts $\tilde{f}$ and $\tilde{g}$ send 0, 1, and τ to 0, 1, and τ', respectively. Thus we obtain a homotopy between $\tilde{f}$ and $\tilde{g}$ by setting

$$\tilde{F}_t(z) = (1-t)\tilde{f}(z) + t\tilde{g}(z), \quad z \in \mathbf{C},\ 0 \leqq t \leqq 1.$$

Putting $F_t([z]) = [\tilde{F}_t(z)]$, we have a homotopy F_t between f and g. Hence, $g \circ f^{-1}: R_{\tau'} \to R_{\tau'}$ is homotopic to the identity.

Conversely, suppose that $g \circ f^{-1}: S \to S'$ is homotopic to a biholomorphic mapping h. Two mappings $h \circ f$ and g from R to S are homotopic. Let $F_t: R \to S'$ $(0 \leqq t \leqq 1)$ be a homotopy between $h \circ f$ and g, and let p_o be a base point for the marking Σ. Let C_o be a continuous curve on S' from $h \circ f(p_o)$ to $g(p_o)$ which is given by $F_t(p_o), 0 \leqq t \leqq 1$. By the isomorphism $T_{C_o}: \pi_1(S', h \circ f(p_o)) \to \pi_1(S', g(p_o))$ induced by C_o, we see that markings $(h \circ f)_*(\Sigma)$ and $g_*(\Sigma)$ on S' are equivalent, which implies that $[S, f_*(\Sigma)] = [S', g_*(\Sigma)]$. □

For an arbitrary marked torus $[S, \Sigma_p]$, we find an orientation-preserving diffeomorphism $f: R \to S$ such that $[S, \Sigma_p] = [S, f_*(\Sigma)]$. Actually, taking two points $\tau_o, \tau \in H$ such that $[R, \Sigma] = [R_{\tau_o}, \Sigma(\tau_o)]$ and $[S, \Sigma_p] = [R_\tau, \Sigma(\tau)]$, we see that the linear mapping

$$\tilde{f}_\tau(z) = \frac{(\tau - \bar{\tau}_o)z - (\tau - \tau_o)\bar{z}}{\tau_o - \bar{\tau}_o}, \quad z \in \mathbf{C},$$

induces an orientation-preserving diffeomorphism $f_\tau: R_{\tau_o} \to R_\tau$ by setting $f_\tau([z]) = [\tilde{f}_\tau(z)]$. In particular, we get $(f_\tau)_*(\Sigma(\tau_o)) = \Sigma(\tau)$.

With this in mind, we say that two pairs (S, f) and (S', g) as above are *equivalent* if and only if, as in Theorem 1.3, $g \circ f^{-1}: S \to S'$ is homotopic to a biholomorphic mapping $h: S \to S'$. We call the set of all equivalence classes $[S, f]$ of such pairs (S, f), the *Teichmüller space* of R, which is denoted by $T(R)$. From the preceding remarks and Theorem 1.3, we can identify T_1 with $T(R)$ under the correspondence sending $[R_\tau, \Sigma(\tau)]$ to $[R_\tau, f_\tau]$. Therefore, it turns out that $T(R)$ also has a complex structure.

Now, using the identification $T_1 \cong T(R)$ we explain that the modular group $PSL(2, \mathbf{Z})$ acts on $T(R)$. Recall that every element $\gamma \in PSL(2, \mathbf{Z})$ is a linear fractional transformation in the form (1.3). The action of γ on $T(R)$ is primarily given by the correspondence sending $[R_\tau, f_\tau]$ to $[R_{\gamma(\tau)}, f_{\gamma(\tau)}]$. On the other hand, for any point $\tau \in H$, the element $\tilde{h}_\tau$ of $Aut(\mathbf{C})$ defined by $\tilde{h}_\tau(z) = (c\tau + d)z$ induces a biholomorphic mapping $h_\tau: R_{\gamma(\tau)} \to R_\tau$ given by $h_\tau([z]) = [\tilde{h}_\tau(z)]$. Since $h_\tau \circ f_{\gamma(\tau)} = f_\tau \circ h_{\tau_o} \circ f_{\gamma(\tau_o)}$, we obtain $[R_{\gamma(\tau)}, f_{\gamma(\tau)}] = [R_\tau, f_\tau \circ h_{\tau_o} \circ f_{\gamma(\tau_o)}]$ in $T(R)$. Put $\omega_\gamma = (h_{\tau_o} \circ f_{\gamma(\tau_o)})^{-1}$, which is a diffeomorphism of $R = R_{\tau_o}$ onto itself, and define its action on $T(R)$ by $[\omega_\gamma]_*([R_\tau, f_\tau]) = [R_\tau, f_\tau \circ \omega_\gamma^{-1}]$. Then $[\omega_\gamma]_*$ is the biholomorphic automorphism of $T(R)$ corresponding to $\gamma \in PSL(2, \mathbf{Z})$. By this correspondence, $PSL(2, \mathbf{Z})$ is considered to act on $T(R)$, and the moduli space M_1 of tori is identified with $T(R)/PSL(2, \mathbf{Z})$.

1.3. Teichmüller Space of Genus g

In accordance with observations in the previous section, let us construct the Teichmüller space of arbitrary genus g in two ways.

The first construction is given by considering marked Riemann surfaces. A system of canonical generators $\Sigma_p = \{[A_j],[B_j]\}_{j=1}^g$ of a fundamental group $\pi_1(R,p)$ of a closed Riemann surface R is called a *marking* on R. Two markings $\Sigma_p = \{[A_j],[B_j]\}_{j=1}^g$ and $\Sigma_{p'} = \{[A'_j],[B'_j]\}_{j=1}^g$ on R are *equivalent* if there exists a continuous curve C_o on R such that $[A'_j] = T_{C_o}([A_j])$and $[B'_j] = T_{C_o}([B_j])$ for $j = 1,\ldots,g$, where T_{C_o} is the isomorphism of $\pi_1(R,p)$ to $\pi_1(R,p')$ sending any $[C]$ to $[C_o^{-1}\cdot C\cdot C_o]$. Let Σ_p and Σ_q be markings on closed Riemann surfaces R and S of genus g, respectively. Two pairs (R,Σ_p) and (S,Σ_q) are said to be *equivalent* if there exists a biholomorphic mapping $h\colon S\to R$ such that the marking $h_*(\Sigma_q) = \{h_*([A'_j]), h_*([B'_j])\}_{j=1}^g$ is equivalent to $\Sigma_p = \{[A_j],[B_j]\}_{j=1}^g$. The equivalence class of (R,Σ_p) is denoted by $[R,\Sigma_p]$ and called a *marked closed Riemann surface of genus* g. The *Teichmüller space* T_g *of genus* g is the set of all marked closed Riemann surfaces of genus g.

The second construction is given by considering orientation-preserving diffeomorphisms. Fix a closed Riemann surface R of genus g. Consider an arbitrary pair (S,f) of a closed Riemann surface S and an orientation-preserving diffeomorphism $f\colon R\to S$. Two pairs (S,f) and (S',g) are said to be *equivalent* if $g\circ f^{-1}\colon S\to S'$ is homotopic to a biholomorphic mapping $h\colon S\to S'$. Let $[S,f]$ be the equivalence class of (S,f). The set of all these equivalence classes $[S,f]$ is denoted by $T(R)$ and is called the *Teichmüller space of* R.

As in the case of tori, we assert that the Teichmüller space T_g of genus $g(\geqq 2)$ is identified with the Teichmüller space $T(R)$ of a closed Riemann surface R of genus g.

To see this, first fix a marking $\Sigma = \{[A_j],[B_j]\}_{j=1}^g$ on R with base point p_o. Corresponding to a point $[S,f]$ in $T(R)$, a marking $f_*(\Sigma)$ on S determines a point $[S,f_*(\Sigma)]$ in T_g. It is noted that this point $[S,f_*(\Sigma)]$ in T_g does not depend on a representative of $[S,f]$ in $T(R)$, which is seen from Lemma 5.1. Hence we define a mapping $\Phi_\Sigma\colon T(R)\to T_g$ by setting

$$\Phi_\Sigma([S,f]) = [S,f_*(\Sigma)]$$

for any $[S,f]\in T(R)$.

Theorem 1.4. *The mapping* $\Phi_\Sigma\colon T(R)\to T_g$ *is bijective.*

Proof (an outline). The injectivity of Φ_Σ follows from the so-called Nielsen's theorem (Harvey [A-41], p.43). It can also be proved by Lemma 5.1. However, we shall give an intuitive explanation for its injectivity. Suppose that two points $[S,f],[S',g]\in T(R)$ satisfy $\Phi_\Sigma([S,f]) = \Phi_\Sigma([S',g])$, i.e., $[S,f_*(\Sigma)] = [S',g_*(\Sigma)]$ in T_g. Then we can take a biholomorphic mapping h of S' onto S and an orientation-preserving homeomorphism g_o of S onto itself, which is homotopic to the identity, so that $g_1 = g_o\circ h\circ g$ coincides with f on each A_j and B_j $(1\leqq j\leqq g)$.

Note that $[S', g] = [S, g_1]$ by the definition. Since the domain obtained by deleting all A_j and B_j from R is homeomorphic to the unit disk in the plane, we can prove easily that f and g_1 are homotopic. Thus we have $[S, f] = [S', g]$.

To prove the surjectivity, it is sufficient to show that for any $[S, \Sigma'] \in T_g$ there exists an orientation-preserving homeomorphism f of R onto S for which $[S, \Sigma'] = [S, f_*(\Sigma)]$. This is a well-known fact, but we shall give a construction of such an f.

First, we may assume that all A_j, B_j, A'_j, and B'_j in $\Sigma = \{[A_j], [B_j]\}_{j=1}^g$ and $\Sigma' = \{[A'_j], [B'_j]\}_{j=1}^g$ are simple closed smooth curves. Let p'_o be the base point for Σ'. Further, set

$$C = \bigcup_{j=1}^{g}(A_j \cup B_j), \quad C' = \bigcup_{j=1}^{g}(A'_j \cup B'_j), \quad R_o = R - C, \quad S_o = S - C'.$$

Then there exists an orientation-preserving diffeomorphism between R_o and a closed polygon P with $2g$ sides on the plane (cf. Fig. 1.7). Replacing differentiable structure around each vertex of P, we may suppose that P is diffeomorphic to the closed unit disk $\bar{\Delta}_R$. Denote by $\bar{\Delta}_S$ the closed unit disk obtained in the same way from (S, Σ').

Next, take an orientation-preserving diffeomorphism f_Σ of $C - \{p_0\}$ onto $C' - \{p'_0\}$ satisfying

$$f_\Sigma(A_j - \{p_o\}) = A'_j - \{p'_o\}, \quad f_\Sigma(B_j - \{p_o\}) = B'_j - \{p'_o\}$$

for all $j = 1, \ldots, g$. Here we assume that f_Σ preserves the orientation of all loops A_j, B_j, A'_j, and B'_j. Let $\tilde{f}$ be the restriction of f_Σ to the boundary $\partial\Delta_R$ of $\bar{\Delta}_R$. Take a sufficiently small neighborhood U of p_o in R which is diffeomorphic to a closed disk, and consider the set E on $\partial\Delta_R$ corresponding to $C \cap U$. Then we can construct a diffeomorphism $\tilde{f}_\partial \colon \partial\Delta_R \to \partial\Delta_S$ for which $\tilde{f}_\partial$ coincides with $\tilde{f}$ on $\partial\Delta_R - E$. It is shown that this $\tilde{f}_\partial$ extends to an orientation-preserving diffeomorphism $\tilde{F} \colon \bar{\Delta}_R \to \bar{\Delta}_S$ (Hirsh [A-42], Chap. 8, Theorem 3.3).

Now, projecting $\tilde{F}$ to $R - U$, we obtain an orientation-preserving homeomorphism $g \colon R - U \to S - g(U)$. This g is not necessarily smooth in a neighborhood of each A_j and B_j. Thus, using a smoothing theorem (Hirsh [A-42], Chap. 8, Theorem 1.9), deform g homotopically to an orientation-preserving diffeomorphism $g_1 \colon R - U \to S - g(U)$. Furthermore, take a neighborhood U_1 of p_o which contains U and is diffeomorphic to a closed disk. In the same way as for $\tilde{f}$, construct an orientation-preserving diffeomorphism g_o on U_1 from g_1 on ∂U_1. Define an orientation-preserving diffeomorphism $g_2 \colon R \to S$ so that $g_2 = g$ on $R - U_1$ and $g_2 = g_o$ on U_1. This g_2 is not necessarily smooth in a neighborhood of ∂U_1, and hence by using again the smoothing theorem, deform g_2 homotopically to an orientation-preserving diffeomorphism $f \colon R \to S$. The construction of f shows that $f_*(\Sigma)$ and Σ' are equivalent, and hence $[S, f_*(\Sigma)] = [S, \Sigma']$. □

Remark. The surjectivity in Theorem 1.4 can also be proved as follows. By Nielsen's theorem stated above, we know that there exists an orientation-preserving homeomorphism $f \colon R \to S$ satisfying $[S, f_*(\Sigma)] = [S, \Sigma']$. Then we find a quasiconformal mapping f_o homotopic to f (Bers [26] or Lehto [A-68], Chap.5, Theorem 1.5). This f_o is not necessarily smooth; however, there exists a real-analytic quasiconformal mapping homotopic to f_o (the Corollary to Theorem 6.9).

Finally, we define a canonical group action on the Teichmüller space $T(R)$. Let $Mod(R)$ be the set of all homotopy classes $[\omega]$ of orientation-preserving diffeomorphisms $\omega: R \to R$. We call $Mod(R)$ the ***Teichmüller modular group*** or the ***mapping class group*** of R. Every element $[\omega]$ acts on $T(R)$ by

$$[\omega]_*([S, f]) = [S, f \circ \omega^{-1}]$$

for any $[S, f] \in T(R)$. We call every $[\omega]_*$ a ***Teichmüller modular transformation.***

Let M_g be the ***moduli space of closed Riemann surfaces of genus*** g, i.e., the set of all biholomorphic equivalence classes $[S]$ of closed Riemann surfaces S of genus g. Since for an arbitrary closed Riemann surface S of genus g there exists an orientation-preserving diffeomorphism of R onto S, the moduli space M_g is identified with the quotient space $T(R)/Mod(R)$ of $T(R)$ by the action of $Mod(R)$. Therefore, we can study the moduli space M_g via the Teichmüller space $T(R)$ and the Teichmüller modular group $Mod(R)$. In Chapter 6, we shall see that $T(R)$ has a $(3g-3)$-dimensional complex manifold structure and that $Mod(R)$ acts properly discontinuously on $T(R)$ as a group of biholomorphic automorphisms. In particular, the moduli space M_g has a $(3g-3)$-dimensional normal complex analytic space structure.

1.4. Quasiconformal Mappings and Teichmüller Space

Let us review the Teichmüller space $T(R)$ constructed in the previous section from the view-point of the theory of quasiconformal mappings.

1.4.1. Deformation of Complex Structures and Beltrami Coefficients

For a point $[S, f] \in T(R)$, we want to compare the complex structures of R and S. Take a coordinate neighborhood (U, z) on R and a coordinate neighborhood (V, w) on S with $f(U) \subset V$, and set $F = w \circ f \circ z^{-1}$. Then

$$\mu = \frac{F_{\bar{z}}}{F_z}$$

is a smooth complex-valued function defind on an open set $z(U)$ in the complex plane. Note that it is independent of the choice of a local coordinate w. Since f is an orientation-preserving diffeomorphism, the Jacobian of F, i.e., $|F_z|^2 - |F_{\bar{z}}|^2$ is positive-definite on $z(U)$. Thus we have $|\mu| < 1$ on $z(U)$. Further, F is biholomorphic on $z(U)$ if and only if $\mu = 0$ on $z(U)$. We call μ the *Beltrami coefficient* of f with respect to (U, z).

It should be noted that a Beltrami coefficient of f depends on the choice of a local coordinate z on R. How it depends is shown as follows: take coordinate neighborhoods (U_j, z_j), (U_k, z_k) of R and coordinate neighborhoods (V_j, w_j), (V_k, w_k) of S such that $f(U_j) \subset V_j$ and $f(U_k) \subset V_k$. Let μ_j and μ_k be the Beltrami coefficients of f with respect to (U_j, z_j) and (U_k, z_k), respectively. When $U_j \cap U_k \neq \phi$, we have

$$\mu_j = (\mu_k \circ z_{kj}) \cdot \overline{\left(\frac{dz_{kj}}{dz_j}\right)} \Big/ \left(\frac{dz_{kj}}{dz_j}\right) \quad \text{on} \quad z_j(U_j \cap U_k), \tag{1.4}$$

where $z_{kj} = z_k \circ z_j^{-1}$. This shows that the set of Beltrami coefficients of f on coordinate neighborhoods of R induces a differential form of type $(-1, 1)$ on R. Thus we denote this differential form of type $(-1, 1)$ simply by

$$\mu_f = \mu \frac{d\bar{z}}{dz}, \tag{1.5}$$

which is called the *Beltrami coefficient* of f on R.

Now, for a system of coordinate neighborhoods $\{(V_\alpha, w_\alpha)\}_{\alpha \in A}$ on S and for an orientation-preserving diffeomorphism $f\colon R \to S$, a system of coordinate neighborhoods $\{(f^{-1}(V_\alpha), w_\alpha \circ f)\}_{\alpha \in A}$ defines a complex structure on R. In this way, we have a new Riemann surface R_f equipped with system of coordinate neighborhoods $\{(f^{-1}(V_\alpha), w_\alpha \circ f)\}_{\alpha \in A}$. Note that $R_f = R$ as sets, that the identity map $id\colon R \to R_f$ is an orientation-preserving diffeomorphism, and that $f\colon R_f \to S$ is a biholomorphic mapping. This shows that $[S, f] = [R_f, id]$ in $T(R)$. Hence it means that a point $[S, f]$ in $T(R)$ represents a deformation of the complex structure on R, and the Beltrami coefficient of f is used to measure the deviation of f from conformality.

1.4.2. Quasiconformal Mappings

First, we explain the geometric meaning of Beltrami coefficients. To see this, considering coordinate neighborhoods, we assume that f is an orientation-preserving diffeomorphism of a domain D containing the origin 0 in the complex z-plane onto a domain D' in the complex w-plane. Let $L(z) = f_z(0)z + f_{\bar{z}}(0)\bar{z}$ be the first order term of the Taylor expansion of f at $z = 0$. Since f is an orientation-preserving diffeomorphism, its Jacobian $J_f(0)$ at $z = 0$ satisfies

$$J_f(0) = |f_z(0)|^2 - |f_{\bar{z}}(0)|^2 > 0,$$

which implies that $f_z(0) \neq 0$ and $|\mu(0)| = |f_{\bar{z}}(0)/f_z(0)| < 1$. Moreover, the linear map L sends a circle with center 0 in the z-plane to an ellipse in the w-plane (Fig. 1.11). By the inequalities

$$|f_z(0)|\,(1 - |\mu(0)|)\,|z| \leqq |L(0)| \leqq |f_z(0)|\,(1 + |\mu(0)|)\,|z|,$$

the ratio of the major axis to the minor axis of this ellipse is

$$K(0) = \frac{1 + |\mu(0)|}{1 - |\mu(0)|}.$$

This shows that any infinitesimally small circle with center 0 is mapped by f to an ellipse whose ratio of the major axis to the minor axis is $K(0)$.

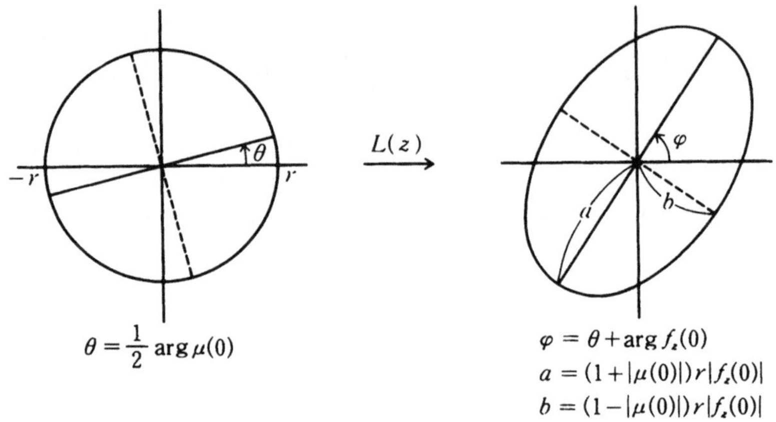

Fig. 1.11.

This statement holds at every point in D. Thus we also call the Beltrami coefficient

$$\mu_f(z) = \frac{f_{\bar{z}}(z)}{f_z(z)}, \quad z \in D,$$

the *complex dilatation* of f at z. As we saw before, $\mu_f = 0$ on D if and only if f is a biholomorphic mapping on D. We call f a *quasiconformal mapping* of D to D' if f satisfies

$$K_f = \sup_{z \in D} \frac{1 + |\mu_f(z)|}{1 - |\mu_f(z)|} < \infty.$$

Further, f is called a *quasiconformal mapping with Beltrami coefficient* μ_f. We call K_f the *maximal dilatation* of f.

In this chapter, we only consider smooth quasiconformal mappings. We shall study more general quasiconformal mappings in Chapter 4.

Transformation formula (1.4) implies that the absolute value $|\mu_f(z)|$ of the Beltrami coefficient $\mu_f = \mu_f(z)\, d\bar{z}/dz$ of an orientation-preserving diffeomorphism $f\colon R \to S$ does not depend on local coordinates on R. Thus $|\mu_f|$ is a continuous function and $|\mu_f| < 1$ on R. Since R is compact, we get

$$\|\mu_f\|_\infty = \sup_{z \in R} |\mu_f(z)| < 1. \tag{1.6}$$

In particular, we have

$$K_f = \sup_{z \in R} \frac{1 + |\mu_f(z)|}{1 - |\mu_f(z)|} = \frac{1 + \|\mu_f\|_\infty}{1 - \|\mu_f\|_\infty} < \infty. \tag{1.7}$$

Hence we may say that an orientation-preserving diffeomorphism $f: R \to S$ is a quasiconformal mapping of R.

The following formula for Beltrami coefficients of a composite mapping is easily proved by the chain rule.

Proposition 1.5. *For Riemann surfaces R, S, T, and orientation-preserving diffeomorphisms $f: R \to S$, $g: S \to T$, the relation*

$$\mu_g \circ f = \frac{f_z}{\overline{f_z}} \frac{\mu_{g\circ f} - \mu_f}{1 - \overline{\mu_f} \cdot \mu_{g\circ f}} . \tag{1.8}$$

holds.

In particular, for orientation-preserving diffeomorphisms $f_j: R \to S_j$ ($j = 1, 2$), the mapping $f_2 \circ f_1^{-1}: S_1 \to S_2$ is biholomorphic if and only if $\mu_{f_1} = \mu_{f_2}$.

Remark. Formula (1.8) has the same form as a biholomorphic automorphism

$$\gamma(z) = e^{i\theta} \frac{z - \alpha}{1 - \bar{\alpha} z}$$

of the unit disc Δ in the complex plane, where θ is a real number and α is a point in Δ (Lemma 2.8, formula (2.6)).

Moreover, formula (1.8) shows that for a fixed f, the Beltrami coefficient $\mu_{g\circ f}$ depends holomorphically on μ_g. This plays an important role in Chapter 6, where we develop a complex analytic theory of Teichmüller spaces.

1.4.3. Spaces of Beltrami Coefficients

Let us reconsider Teichmüller spaces by using the space of Beltrami coefficients on R. Let $B(R)_1$ be the set of Beltrami coefficients of all orientation-preserving diffeomorphisms of a fixed closed Riemann surface R onto closed Riemann surfaces. Define a topology on $B(R)_1$ by using the L^∞-norm given by (1.6).

Denote by $Diff_+(R)$ the group of all orientation-preserving diffeomorphisms on R onto itself. Let $Diff_o(R)$ be a group which consists of all elements in $Diff_+(R)$ homotopic to the identity map id. Clearly, $Diff_o(R)$ is a normal subgroup of $Diff_+(R)$.

The action of $Diff_+(R)$ on $B(R)_1$ is given by

$$\omega^*(\mu_f) = \mu_{f\circ\omega^{-1}} = \left(\frac{\omega_z}{\overline{\omega_z}} \frac{\mu_f - \mu_\omega}{1 - \overline{\mu_\omega}\, \mu_f}\right) \circ \omega^{-1}, \tag{1.9}$$

where ω is an element of $Diff_+(R)$ and μ_f is an element of $B(R)_1$.

Theorem 1.6. *For orientation-preserving diffeomorphisms $f: R \to S$ and $g: R \to S'$, there exists a biholomorphic mapping $h: S \to S'$ if and only if the relation $\mu_g = \omega^*(\mu_f)$ holds for some ω in $Diff_+(R)$. Furthermore, $g^{-1} \circ h \circ f$ is homotopic to the identity map of R if and only if ω belongs to $Diff_o(R)$.*

Proof. Suppose that there exists a biholomorphic mapping $h: S \to S'$. Setting $\omega = g^{-1} \circ h \circ f$, we see that formula (1.8) gives

$$\mu_g = \mu_{h \circ f \circ \omega^{-1}} = \mu_{f \circ \omega^{-1}} = \omega^*(\mu_f).$$

Conversely, if there exists an element $\omega \in Diff_+(R)$ with $\mu_g = \omega^*(\mu_f)$, then Proposition 1.5 shows that $h = g \circ \omega \circ f^{-1}: S \to S'$ is a biholomorphic mapping.

The second assertion is clear from the definition. □

Corollary. *The mapping of sending (S, f) to $\mu_f \in B(R)_1$ induces the following identifications:*

$$T(R) \cong B(R)_1 / Diff_o(R),$$
$$M_g \cong B(R)_1 / Diff_+(R).$$

1.5. Complex Structures and Conformal Structures

In this section, let us reconstruct the Teichmüller space $T(R)$ by means of conformal structures induced by Riemannian metrics on R.

1.5.1. Riemannian Metrics and Conformal Structures

Suppose that a Riemannian metric ds^2 is given on a real two-dimensional oriented smooth manifold M. This metric is represented as

$$ds^2 = E dx^2 + 2F dx dy + G dy^2$$

on a coordinate neighborhood $(U, (x, y))$ of M. Setting $z = x + iy$, we see that it is written in the form

$$ds^2 = \lambda \, |dz + \mu d\bar{z}|^2, \tag{1.10}$$

where λ is a positive smooth function on U and μ is a complex-valued smooth function with $|\mu| < 1$ on U. Actually, λ and μ are given by

$$\lambda = \frac{1}{4}\left(E + G + 2\sqrt{EG - F^2}\right),$$
$$\mu = \frac{E - G + 2iF}{E + G + 2\sqrt{EG - F^2}}.$$

Local coordinates (u, v) on U are said to be *isothermal coordinates* for ds^2 if ds^2 is represented as

$$ds^2 = \rho \, (du^2 + dv^2) \tag{1.11}$$

on U, where ρ is a positive smooth function on U. Here we assume that both of the orientations induced by the coordinates (x, y) and (u, v) on U coincide with the one on M. The complex coordinate $w = u + iv$ is also called an *isothermal coordinate* for ds^2.

Since an isothermal coordinate w for ds^2 satisfies

$$\rho\,|dw|^2 = \rho\,|w_z|^2 \left|dz + \frac{w_{\bar z}}{w_z}\,d\bar z\right|^2,$$

comparing with (1.10), we conclude that an isothermal coordinate w for ds^2 exists if the partial differential equation

$$\frac{\partial w}{\partial \bar z} = \mu \frac{\partial w}{\partial z} \tag{1.12}$$

has a diffeomorphic solution w. This equation is called the *Beltrami equation*. As is proved in §2 of Chapter 4, such a solution w always exists provided that $\|\mu\|_\infty < 1$. Hence, for a system of coordinate neighborhoods $\{(U_j,(x_j,y_j))\}_{j\in J}$ of M, there exists an isothermal coordinate w_j on each U_j. It is not diffcult to verify that $\{(U_j,w_j)\}_{j\in J}$ defines a complex structure on M. Denote by R the Riemann surface obtained in this way. The complex structure on R may be called the *conformal structure* induced by the Riemannian metric ds^2.

For oriented 2-dimensional Riemannian manifolds (M,ds^2) and (N,ds_1^2), an orientation-preserving diffeomorphism $f\colon M \to N$ is a *conformal mapping* if the pull back of ds_1^2 by f is equal to $\exp(\varphi)ds^2$ on M, where φ is a real-valued smooth function on M. Intuitively, it means that the angle, measured by ds^2, between any smooth curves C_1 and C_2 on M equals the angle, measured by ds_1^2, between $f(C_1)$ and $f(C_2)$ on N. We say that (M,ds^2) and (N,ds_1^2) are *conformally equivalent* or have the *same conformal structure* if there exists a conformal mapping between them.

In the case of dimension 2, it is easy to see that the uniqueness of the representation (1.10) leads to the following theorem.

Theorem 1.7. *Let R and S be Riemann surfaces induced by oriented 2-dimensional Riemannian manifolds (M,ds^2) and (N,ds_1^2), respectively. Then $f\colon (M,ds^2) \to (N,ds_1^2)$ is conformal if and only if $f\colon R \to S$ is biholomorphic.*

This theorem shows that, in the two-dimensional case, concepts of complex structure and of conformal structure are equivalent. This is a reason that a biholomorphic mapping is called a conformal mapping. This assertion is a remarkable property for 1-dimensional complex manifolds, i.e., 2-dimensional real manifolds, which is not true for higher dimensional cases. B. Riemann first recognized this relationship and founded the geometric function theory.

Example. Let M be a surface in the Euclidean space $\mathbf{R}^3$ which is obtained by revolving the circle $(y-a)^2 + z^2 = b^2$ on the (y,z)-plane around the z axis, where $0 < b < a$ (Fig. 1.12).

Then M has a canonical Riemannian metric ds^2 induced by the Euclidean metric on $\mathbf{R}^3$. Using the local coordinates (θ,φ) as illustrated in Fig. 1.12, a point $p = (x,y,z)$ on M is given by

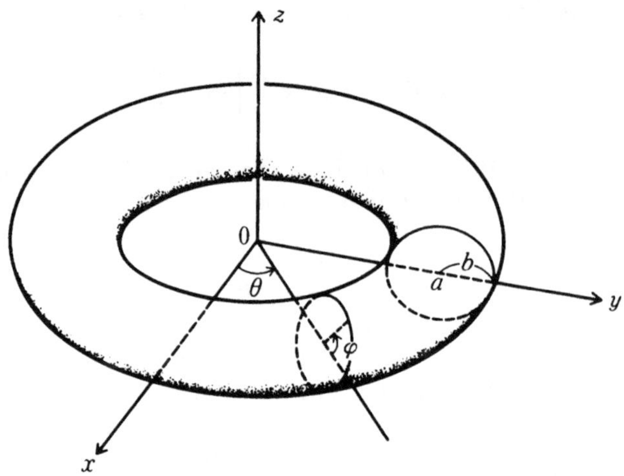

Fig. 1.12.

$$\begin{aligned} x &= (a + b\cos\varphi)\cos\theta, \\ y &= (a + b\cos\varphi)\sin\theta, \\ z &= b\sin\varphi. \end{aligned}$$

We assume that the orientation of M is induced from (θ, φ). If we set

$$\psi = \psi(\varphi) = \int_0^{\varphi} \frac{b}{a + b\cos\varphi}\, d\varphi,$$

then the metric $ds^2 = (a + b\cos\varphi)^2 d\theta^2 + b^2 d\varphi^2$ has the form

$$ds^2 = \lambda(\psi)(d\theta^2 + d\psi^2).$$

Thus $w = \theta + i\psi$ is an isothermal coordinate for ds^2 on M, which defines a complex structure on M. Hence R is a torus. A little more calculation shows that R is biholomorphic to $\mathbf{C}/\Gamma$, where Γ is a lattice group generated by 1 and $ib/\sqrt{a^2 - b^2}$.

1.5.2. Reconstruction of Teichmüller Spaces by Riemannian Metrics

Fix a closed Riemann surface R of genus $g \geq 1$. Take any local coordinate z on R.

For an arbitrary Riemannian metric ds^2 on R, from the uniqueness of the expression in (1.10), we obtain a globally defined Beltrami coefficient μ on R, being a differential form of type $(-1, 1)$ and $\|\mu\|_\infty < 1$. Such a μ is called the *Beltrami coefficient induced by a Riemannian metric.*

Let us observe the relationship between the Beltrami coefficient of an orientation-preserving diffeomorphism and the one induced by a Riemannian metric.

First, suppose that μ is the Beltrami coefficient induced by a Riemannian metric ds^2 on R. As has been seen in §5.1, solving the Beltrami equation

$$\frac{\partial w}{\partial \bar{z}} = \mu \frac{\partial w}{\partial z}$$

on each coordinate neighborhood (U, z), we obtain a Riemann surface R' whose complex structure is induced by the isothermal coordinate (U, w). Then the identity mapping $f_o: R \to R'$ is an orientation-preserving diffeomorphism and its Beltrami coefficient coincides with μ.

Conversely, let an orientation-preserving diffeomorphism $f: R \to S$ be given. We can take a Riemannian metric ds_1^2 on S so that the complex structure induced by ds_1^2 is equivalent to the original one of S. In fact, the universal covering surface of S is biholomorphic to the Riemann sphere, the complex plane, or the upper half-plane (Theorem 2.2 of Chapter 2). We take the metric on S which is induced by the spherical, Euclidean, or Poincaré metric, respectively (cf. §1.3 of Chapter 3). Then the pull back $f^*(ds_1^2)$ of ds_1^2 under f gives the same Beltrami coefficient as that of f. Note that this assertion does not depend on the choice of a metric ds_1^2 inducing the complex structure of S. Such a metric $f^*(ds_1^2)$ is said to be a *Riemannian metric on R corresponding to f*.

In this way, we see that the set $B(R)_1$ of Beltrami coefficients of orientation-preserving diffeomorphisms on R is equal to the set of Beltrami coefficients induced by Riemannian metrics on M.

Now, using the set $\mathcal{M}(R)$ of Riemannian metrics on R, we reconstruct the Teichmüller space $T(R)$ of R. Two elements ds^2 and ds_1^2 in $\mathcal{M}(R)$ are defined to be *equivalent* if there exists an element ω in $Diff_+(R)$ such that $\omega: (R, ds^2) \to (R, ds_1^2)$ is conformal. Further, ds^2 and ds_1^2 are defined to be *strongly equivalent* if this ω belongs to $Diff_o(R)$. Denote by $\mathcal{M}(R)/Diff_+(R)$ and $\mathcal{M}(R)/Diff_o(R)$ the set of all equivalence and all strong equivalence classes of $\mathcal{M}(R)$, respectively.

This observation and the Corollary to Theorem 1.6 lead to the following assertion.

Theorem 1.8. *The mappings which send an element $[S, f]$ in $T(R)$ to the equivalence and strong equivalence class, respectively, of a metric corresponding to f give the following identifications:*

$$T(R) \cong \mathcal{M}(R)/Diff_o(R),$$
$$M_g \cong \mathcal{M}(R)/Diff_+(R).$$

Notes

The geometric function theory originated with Riemann's 1851 Göttingen dissertation [181] and his 1857 paper [182]. In connection with multi-valued analytic functions such as algebraic functions, he introduced the concept of the Riemann surface as a branched covering surface over the Riemann sphere. He also recognized clearly the intimate relationship between holomorphic functions and conformal mappings on a domain in the complex plane. In [181], he proved Riemann's mapping theorem which asserts that any simply connected domain in the complex plane with more than one boundary point is biholomorphic to the unit disk. In [182], he obtained the Riemann-Roch theorem. By using this theorem, he determined the degree of freedom of finite branched coverings over the Riemann sphere which represent closed Riemann surfaces of genus g, and he obtained the complex dimension m_g of the moduli space of closed Riemann surfaces of genus g, that is, $m_g = 0$, 1, and $3g - 3$ for $g = 0$, 1, and $g \geqq 2$, respectively. For more complete exposition of Riemann's work, we refer to Ahlfors [4] and Klein [A-53].

The standard definition of a Riemann surface, that is, a one-dimensional complex manifold was introduced for the first time in Weyl's classic [A-111] "Die Idee der Riemannschen Fläche" in 1913.

The material of this chapter is classical. Some of the many celebrated books on Riemann surfaces are Ahlfors and Sario [A-6], Bers [A-13], Cohn [A-22], Farkas and Kra [A-28], Forster [A-32], Griffiths and Harris [A-39], Gunning [A-40], Jones and Singerman [A-48], Schlichenmaier [A-95], Siegel [A-98], and Springer [A-99]. For details of topology on surfaces, there are further books by Birman [A-18], Harvey [A-41], Chapters 1 and 6, Moise [A-75], Stillwell [A-101], and Zieschang, Vogt and Coldewey [A-114]. For algebraic curves, we refer to the books by Arbarello, Cornalba, Griffiths and Harris [A-9], Griffiths [A-38], Mumford [A-78], Namba [A-82], and Shafarevich [A-97]. The moduli space of tori considered in §2 is studied in the context of elliptic curves, elliptic integrals, and theta functions. For an interesting exposition on this subject, see Clemens [A-21], and Jones and Singerman [A-48].

For textbooks on Teichmüller spaces, there are Abikoff [A-1], Ahlfors [A-2], Gardiner [A-34], Harvey [A-41], Krushkal' [A-60], Lehto [A-68], and Nag [A-80]. For expository papers on this subject, consult articles by Ahlfors [8] and [11], and Bers [22], [29], and [40]. The approaches to Teichmller spaces as in §4 and §5 are found in Earle and Eells [62], [63], and Fischer and Tromba [71], respectively.

Chapter 2

Fricke Space

The purpose of the present chapter is to show that the Teichmüller space of genus $g(\geqq 2)$ is realized as a subset in $\mathbf{R}^{6g-6}$, which is called the Fricke space of genus g.

For this purpose, we provide the uniformization theorem of Riemann surfaces, basic facts on universal covering surfaces, Möbius transformations, and Fuchsian groups. In Section 1, we explain the uniformization theorem due to Klein, Poincaré, and Koebe, which asserts that every simply connected Riemann surface is biholomorphic to one of the three Riemann surfaces: the Riemann sphere $\hat{\mathbf{C}}$, the complex plane $\mathbf{C}$, or the upper half-plane H. In order to apply the uniformization theorem to an arbitary Riemann surface R, we consider the universal covering surface $\tilde{R}$ of R. Its covering transformation group Γ is constructed in Section 2. By the uniformization theorem, $\tilde{R}$ is biholomorphically equivalent to $\hat{\mathbf{C}}$, $\mathbf{C}$, or H, and Γ acts properly discontinuously on $\tilde{R}$ as a group consisting of Möbius transformations. In particular, when $\tilde{R} = H$, we call Γ a Fuchsian group. In this way, we conclude that every closed Riemann surface of genus $g(\geqq 2)$ is represented by a quotient space H/Γ of H by a Fuchsian group Γ.

In Sections 3 and 4, we prove some elementary properties of Fuchsian groups, which are used later.

Finally, in Section 5, using canonical systems of generators of Fuchsian groups representing closed Riemann surfaces of genus $g(\geqq 2)$, we define the Fricke space F_g as a subset in $\mathbf{R}^{6g-6}$, and prove that it is identified with the Teichmüller space T_g of genus g.

2.1. Uniformization Theorem

It is well known that a remarkable fact which is called the uniformization theorem holds for Riemann surfaces. Especially, for domains in the complex plane $\mathbf{C}$, this theorem is reduced to *Riemann's mapping theorem* which asserts that every simply connected domain in $\mathbf{C}$ with more than one boundary point is biholomorphic to the unit disk (see Ahlfors [A-4], p.247).

Uniformization Theorem. (Klein, Poincaré and Koebe) *Every simply connected Riemann surface is biholomorphically equivalent to one of the three Riemann surfaces $\hat{\mathbf{C}}$, $\mathbf{C}$, or H.*

Remark. These Riemann surfaces $\hat{\mathbf{C}}$, $\mathbf{C}$, and H are not mutually biholomorphically equivalent. The Möbius transformation $w = (z - i)/(z + i)$ maps biholomorphically H onto the unit disk Δ, and hence we often use the unit disk Δ instead of the upper half-plane H.

Corollary. *A closed Riemann surface of genus* 0 *is biholomorphic equivalent to the Riemann sphere* $\hat{\mathbf{C}}$. *Thus the moduli space* M_0 *of closed Riemann surfaces of genus* 0 *consists of one point.*

Proof. Since a closed Riemann surface R of genus 0 is simply connected, the uniformization theorem implies that R is biholomorphic to one of the three Riemann surfaces $\hat{\mathbf{C}}$, $\mathbf{C}$, and H. Since R is compact, it should be biholomorphically equivalent to $\hat{\mathbf{C}}$. □

For proofs of the uniformization theorem, we refer to books on Riemann surfaces listed in the notes of Chapter 1. See also Ahlfors [A-3]. For historical and expository accounts, see Abikoff [2], and Bers [29] and [36].

Among standard proofs for the uniformization theorem, there is a method in which a mapping function is constructed by using Green functions. Let us elucidate the concept of Green functions by using an intuitive example from electromagnetism. We regard a *Green function* on a Riemann surface R as the electric potential on R where a positive charge is given at a point p and whose boundary is earthed. Mathematically, when z is a local coordinate around p on R, we define the Green function $g(\cdot, p)$ on R with pole at p as the minimal function in the family of positive superharmonic functions which are harmonic in $R - \{p\}$ and have the singularity $-\log|z - z(p)|$ at p. The existence of a Green function depends on "capacity" of the boundary of R and is indepent of the choice of a point p. For example, the Green function on the unit disk Δ with pole at z_o is given by $\log|(1 - \overline{z_o}\, z)/(z - z_o)|$. On the other hand, there are no Green functions on $\hat{\mathbf{C}}$ or $\mathbf{C}$.

Now, assume that there exists a Green function $g = g(\cdot, p)$ on R. Then we obtain a biholomorphic mapping $f\colon R \to \Delta$;

$$f(q) = \exp\left(-g(q) + ig^*(q)\right), \tag{2.1}$$

where g^* is the conjugate harmonic function of g on $R - \{p\}$. Note that g^* is a multi-valued function whose periods are $2n\pi$ ($n \in \mathbf{Z}$) because of the singularity of g and simple connectedness of R. Hence, f itself is a single-valued holomorphic function on R. By the argument principle we see that f is univalent, i.e., a biholomorphic mapping of R onto Δ (see Fig. 2.1).

Next, we deal with the case where there exist no Green functions on R. We take a sequence $\{R_n\}_{n=1}^{\infty}$ of simply connected subdomains of R such that R_n is a relatively compact subset of R_{n+1} for each n, that $\bigcup_{n=1}^{\infty} R_n$ covers R except for at most one point, and that every R_n has the Green function g_n with the common pole p. Then, in the same way as before, we can construct a biholomorphic mapping $f_n : R_n \to \Delta$ for every n. By multiplying each f_n by a suitable constant, we get a sequence of biholomorphic mappings $F_n : R_n \to \{ w \in \mathbf{C} \mid |w| < r_n \}$ so

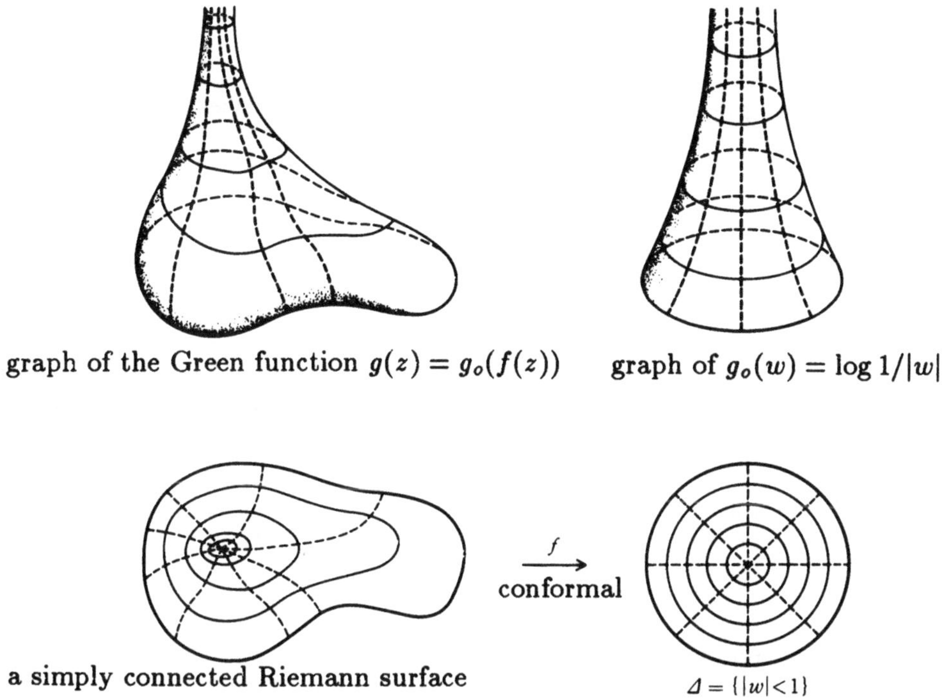

Fig. 2.1.

that the limit function on $\bigcup_{n=1}^{\infty} R_n$ induces a desired biholomorphic mapping of R onto $\hat{\mathbf{C}}$ or $\mathbf{C}$.

2.2. Universal Coverings

In order to apply the uniformization theorem, we construct a simply connected covering surface of an arbitrary Riemann surface.

2.2.1. Definitions of Covering Surfaces and Covering Transformation Groups

Let R and $\tilde{R}$ be Riemann surfaces. A surjective holomorphic mapping $\pi: \tilde{R} \to R$ is said to be a *covering map* if every point p of R has a neighborhood U such that for each connected component V of the inverse image $\pi^{-1}(U)$ of U, the restricted map $\pi: V \to U$ is biholomorphic. In this book, we call $(\tilde{R}, \pi, R)$ a *covering* of R, and $\tilde{R}$ a *covering surface* of R. The covering map π is also called the *projection* of $\tilde{R}$ onto R. Further, when $\tilde{R}$ is simply connected, we call $(\tilde{R}, \pi, R)$ a *universal covering* of R, and $\tilde{R}$ a *universal covering surface* of R. A universal covering of R means that it is the "highest" covering surface of all coverings of R (cf. Theorems 2.2 and 2.4, and the remark in §2.2).

Example 1. We give a few simple examples of covering surfaces.

(i) Let $\pi : \mathbf{C} \to \mathbf{C} - \{0\}$ be given by $\pi(z) = e^z$. Then $\mathbf{C}$ is a universal covering surface of $\mathbf{C} - \{0\}$.
(ii) Let $\pi : H \to \Delta - \{0\}$ be given by $\pi(z) = e^{2\pi i z}$. Then H is a universal covering surface of $\Delta - \{0\}$.
(iii) Let $\pi : \mathbf{C} - \{0\} \to \mathbf{C} - \{0\}$ be given by $\pi(z) = z^n$, where n is a positive integer. Then $\mathbf{C} - \{0\}$ is a covering surface of itself, but it is not a universal covering surface.
(iv) For a given $\lambda\,(> 1)$, set $r = \exp(-2\pi^2/\log\lambda)$ and $A = \{w \in \mathbf{C} \mid r < |w| < 1\}$. Define $\pi : H \to A$ by $\pi(z) = \exp(2\pi i \log z/\log\lambda)$, where $\log z$ denotes its principal branch. Then H becomes a universal covering surface of the annulus A.
(v) Let Γ_τ be a lattice group generated by 1 and a point $\tau \in H$, and let π be the projection of $\mathbf{C}$ onto the quotient space $\mathbf{C}/\Gamma_\tau$. Then $\mathbf{C}$ is a universal covering surface of the torus $\mathbf{C}/\Gamma_\tau$.

Any biholomorphic mapping $\gamma : \tilde{R} \to \tilde{R}$ with $\pi \circ \gamma = \pi$ is called a *covering transformation* of a covering $(\tilde{R}, \pi, R)$. For a given covering $(\tilde{R}, \pi, R)$, denote by Γ the set of all its covering transformations. By the composition of mappings, Γ forms a group, which is called the *covering transformation group* of $(\tilde{R}, \pi, R)$. In particular, we call Γ the ***universal covering transformation group*** of $(\tilde{R}, \pi, R)$ if $\tilde{R}$ is a universal covering surface of R.

Example 2. We give covering transformation groups (i)$'$, ..., (v)$'$ of coverings (i), ..., (v) in Example 1. The notation $\langle \gamma_1, \ldots, \gamma_n \rangle$ expresses the group generated by $\gamma_1, \ldots, \gamma_n$.

(i)$'$ $\Gamma = \langle \gamma_1 \rangle$ with $\gamma_1(z) = z + 2\pi i$.
(ii)$'$ $\Gamma = \langle \gamma_1 \rangle$ with $\gamma_1(z) = z + 1$.
(iii)$'$ $\Gamma = \langle \gamma_1 \rangle$ with $\gamma_1(z) = z \exp(2\pi i/n)$, which is a finite group of order n.
(iv)$'$ $\Gamma = \langle \gamma_1 \rangle$ with $\gamma_1(z) = \lambda z$.
(v)$'$ $\Gamma = \langle \gamma_1, \gamma_2 \rangle = \Gamma_\tau$, where $\gamma_1(z) = z + 1$ and $\gamma_2(z) = z + \tau$.

2.2.2. Construction of Universal Covering Surfaces

First of all, we need several definitions. A ***path*** on a Riemann surface R means a continuous curve $C : I \to R$, where I is the interval $[0, 1]$. The points $C(0)$ and $C(1)$ are said to be the ***initial*** and ***terminal points*** of C, respectively. We also say that C is a path from $C(0)$ to $C(1)$. Throughout the book, if no confusion is possible, its image $C(I)$ is also denoted by the same letter C.

For two paths C and C' on R such that $C(1) = C'(0)$, by connecting the terminal point of C with the initial point of C', we get a path $C \cdot C'$ on R with the initial point $C(0)$ and the terminal point $C'(1)$.

Let $(\tilde{R}, \pi, R)$ be a covering of a Riemann surface R. A point $\tilde{p}$ in $\tilde{R}$ is said to be *lie over a point* p in R if $\pi(\tilde{p}) = p$. A *lift* of a path C on R is a path $\tilde{C}$ on $\tilde{R}$ with $\pi \circ \tilde{C} = C$.

Now, by using paths we shall construct concretely a universal covering surface of a Riemann surface. Fix a base point p_o on a given Riemann surface R. Let (C, p) be a pair of any point p on R and any path C on R from p_o to p. These two pairs (C, p) and (C', p') are *equivalent* if $p = p'$ and C is homotopic to C' on R. Denote by $[C, p]$ the equivalence class of (C, p). Let $\tilde{R}$ be the set of all these equivalence classes $[C, p]$.

We see that this $\tilde{R}$ becomes a universal covering surface of R as follows. First, we need to introduce a topology on $\tilde{R}$. For any point $\tilde{p} = [C, p]$ of $\tilde{R}$, take a neighborhood U_p of p which is a simply connected domain in R. Denote by $U_{\tilde{p}}$ the set of all points $[C \cdot C_q, q]$ in $\tilde{R}$ in such a way that q is a point in U_p and C_q is an arbitrary path contained in U_p from p to q. Since U_p is a simply connected domain, we have a canonical one-to-one correspondence between U_p and $U_{\tilde{p}}$. By these $U_{\tilde{p}}$, we define a system of fundamental neighborhoods of $\tilde{p}$ in $\tilde{R}$. Then $\tilde{R}$ becomes a Hausdorff topological space. Let $\pi \colon \tilde{R} \to R$ be the projection given by $\pi([C, p]) = p$. By the construction, it is easy to see that π is a continuous mapping of $\tilde{R}$ onto R and satisfies the condition of a covering map.

Next, we define a complex structure on $\tilde{R}$ so that $\pi \colon \tilde{R} \to R$ becomes a holomorphic mapping. In fact, for any point $\tilde{p} = [C, p]$ of $\tilde{R}$, take a coordinate neighborhood (U_p, z_p) such that U_p is a simply connected domain in R. Setting $z_{\tilde{p}} = z_p \circ \pi$, we see that the family $\{ (U_{\tilde{p}}, z_{\tilde{p}}) \}$ yields the required complex structure on $\tilde{R}$.

Lemma 2.1. *The surface $\tilde{R}$, constructed as explained above, is connected and simply connected.*

Proof. Let I_o be the path on R defined by $I_o(t) = p_o$ for any $t \in I = [0, 1]$. To prove the connectedness of $\tilde{R}$, it suffices to show that every point $[C, p] \in \tilde{R}$ is connected with $[I_o, p_o]$ by a path on $\tilde{R}$. For each $s \in I$, define a path C_s on R by $C_s(t) = C(st)$ for $t \in I$. Setting $\tilde{C}(s) = [C_s, C(s)]$ for every $s \in I$, we have a path $\tilde{C}$ on $\tilde{R}$ from $[I_o, p_o]$ to $[C, p]$, which implies that $\tilde{R}$ is connected.

Now, we prove that $\tilde{R}$ is simply connected. It is sufficient to see that every closed path $\tilde{C}$ on $\tilde{R}$ with base point $[I_o, p_o]$ is homotopic to $[I_o, p_o]$. Put $C = \pi \circ \tilde{C}$. Then C is a closed path on R with base point p_o, and the terminal point of $\tilde{C}$ is represented by $[C, p_o]$. Since $\tilde{C}$ is a closed path, we conclude that $[I_o, p_o] = [C, p_o]$, which means that C is homotopic to I_o. Let $F \colon I \times I \to R$ be a homotopy between them such that $\{ F_s = F(\cdot, s) \}_{s \in I}$ satisfies $F_o = I_o$, $F_1 = C$, and $F_s(0) = F_s(1) = p_o$ for any $s \in I$. Then, for any $(t, s) \in I \times I$, we define a path $C_{(t,s)}$ on R by $C_{(t,s)}(u) = F_s(tu)$ for $u \in I$. Finally, we have a homotopy

$\tilde{F}(t,s)\colon I \times I \to \tilde{R}$ between $\tilde{C}$ and $[I_o, p_o]$ by setting $\tilde{F}(t,s) = [C_{(t,s)}, F_s(t)]$ for $(t,s) \in I \times I$. Therefore, we conclude that $\tilde{R}$ is simply connected. □

On putting these observations together with the uniformization theorem, we obtain the following theorem.

Theorem 2.2. *For every Riemann surface R, there exists a universal covering surface $\tilde{R}$ of R, which is biholomorphic to one of the three Riemann surfaces $\hat{\mathbf{C}}$, $\mathbf{C}$, or H.*

Throughout this section, as a universal covering surface $\tilde{R}$ of a Riemann surface R we always take the one constructed above.

From the construction of such a universal covering, it is easy to get the following lemma by an argument similar to that used in the case of analytic continuation (Ahlfors [A-4], Chapter 8).

Lemma 2.3. (Existence and uniqueness of a lift of a path) *For any path C on R with initial point p, and for any point $\tilde{p}$ of $\tilde{R}$ over p, there exists a unique lift $\tilde{C}$ of C with initial point $\tilde{p}$.*

Theorem 2.4. (Lift of a mapping) *For Riemann surfaces R and S, let $(\tilde{R}, \pi_R, R)$ and $(\tilde{S}, \pi_S, S)$ be their universal coverings constructed as explained previously, respectively. Then given an arbitrary continuous mapping $f\colon R \to S$, there exists a continuous mapping $\tilde{f}\colon \tilde{R} \to \tilde{S}$ with $f \circ \pi_R = \pi_S \circ \tilde{f}$. This mapping $\tilde{f}$ is uniquely determined under the condition that $\tilde{f}(\tilde{p}_1) = \tilde{q}_1$, where $\tilde{p}_1 \in \tilde{R}$ and $\tilde{q}_1 \in \tilde{S}$ are such that $\pi_S(\tilde{q}_1) = f(\pi_R(\tilde{p}_1))$.*

Moreover, if f is differentiable or holomorphic, then $\tilde{f}$ is also differentiable or holomorphic.

This mapping $\tilde{f}\colon \tilde{R} \to \tilde{S}$ is called a *lift* of $f\colon R \to S$.

Proof of Theorem 2.4. Setting $\tilde{p}_1 = [C_1, p_1]$ and $\tilde{q}_1 = [D_1, f(p_1)]$, we get a mapping defined by $\tilde{f}([C,p]) = [D_1 \cdot f(C_1)^{-1} \cdot f(C), f(p)]$ for all points $[C,p]$ in $\tilde{R}$. Then it is obvious that $\tilde{f}(\tilde{p}_1) = \tilde{q}_1$ and $f \circ \pi_R = \pi_S \circ \tilde{f}$. Since π_R and π_S are locally biholomorphic and f is continuous, $\tilde{f}$ must be continuous. It is also trivial that if f is differentiable or holomorphic, then so is $\tilde{f}$. The uniqueness assertion follows from Lemma 2.3. □

Remark. **(Uniqueness of universal covering)** For any two universal coverings $(\tilde{R}, \pi, R)$ and $(\tilde{R}_1, \pi_1, R)$ of a Riemann surface R, there exists a biholomorphic mapping φ of $\tilde{R}$ to $\tilde{R}_1$ with $\pi_1 \circ \varphi = \pi$. See, for example, Ahlfors and Sario [A-6], Theorem 18A of Chapter I.

2.2.3. Universal Covering Transformation Groups

For a given universal covering surface $(\tilde{R}, \pi, R)$ of a Riemann surface R, we shall see that its universal covering group Γ is isomorphic to the fundamental group $\pi_1(R, p_o)$ of R.

For any element $[C_o] \in \pi_1(R, p_o)$, we define the action $[C_o]_*$ on $\tilde{R}$ by

$$[C_o]_*([C,p]) = [C_o \cdot C, p], \quad [C,p] \in \tilde{R}.$$

Clearly, by the definition, this $[C_o]_*$ belongs to Γ, that is, it is a covering transformation of $(\tilde{R}, \pi, R)$.

Theorem 2.5. *The above correspondence $[C_o] \mapsto [C_o]_*$ yields an isomorphism of the fundamental group $\pi_1(R, p_o)$ of R onto the universal covering transformation group Γ of a universal covering $(\tilde{R}, \pi, R)$.*

Proof. It is trivial that this correspondence is a homomorphism of $\pi_1(R, p_o)$ to Γ. To prove that it is injective, suppose that $[C_o]_*$ is the unit element of Γ. Then we have

$$[C_o]_*([I_o, p_o]) = [C_o, p_o] = [I_o, p_o],$$

where I_o is the path on R such that $I_o(t) = p_o$ for any $t \in I = [0,1]$. Thus, C_o is homotopic to I_o, and hence $[C_o]$ is the unit element of $\pi_1(R, p_o)$. It follows that this correspondence is injective.

To prove that this correspondence is surjective, take any element $\gamma \in \Gamma$. Let $\tilde{C}$ be a path on $\tilde{R}$ from $[I_o, p_o]$ to $\gamma([I_o, p_o])$. Then the relation $\pi \circ \gamma = \pi$ implies that $C = \pi \circ \tilde{C}$ is a closed path on R with base point p_o. Hence $\tilde{C}$ is a lift of C, and $[I_o, p_o]$ and $[C, p_o]$ are the initial and terminal points of $\tilde{C}$, respectively. Thus Lemma 2.3 shows that $\gamma([I_o, p_o]) = [C, p_o] = [C]_*([I_o, p_o])$. Since $[C]$ is an element of $\pi_1(R, p_o)$, Theorem 2.4 implies that $\gamma = [C]_*$, and hence this correspondence is surjective. □

Lemma 2.6. *The universal covering transformation group Γ of $(\tilde{R}, \pi, R)$ of a Riemann surface R satisfies the following properties:*

(i) *For any $\tilde{p}, \tilde{q} \in \tilde{R}$ with $\pi(\tilde{p}) = \pi(\tilde{q})$, there exists an element $\gamma \in \Gamma$ with $\tilde{q} = \gamma(\tilde{p})$.*
(ii) *For every $\tilde{p} \in \tilde{R}$, there is a suitable neighborhood $\tilde{U}$ of $\tilde{p}$ in $\tilde{R}$ such that $\gamma(\tilde{U}) \cap \tilde{U} = \phi$ for every $\gamma \in \Gamma - \{id\}$. In particular, each element of Γ except for the identity has no fixed points.*
(iii) *Γ acts properly discontinuously on $\tilde{R}$; that is, for any compact subset K of $\tilde{R}$, there are at most finitely many elements $\gamma \in \Gamma$ such that $\gamma(K) \cap K \neq \phi$.*

Proof. To prove (i), suppose that $\pi(\tilde{p}) = \pi(\tilde{q}) = p$. Then we have $\tilde{p} = [C_1, p]$ and $\tilde{q} = [C_2, p]$ for some paths C_1 and C_2 on R. Putting $C_o = C_2 \cdot C_1^{-1}$, we see that $\gamma = [C_o]_*$ satisfies $\tilde{q} = \gamma(\tilde{p})$.

To see (ii), take a point $\tilde{p} \in \tilde{R}$, and set $p = \pi(\tilde{p}), \tilde{p} = [C, p]$. Choose a neighborhood U of p in R which satisfies the condition in the definition of a covering map in §2.1, and denote by $\tilde{U}$ the connected component of $\pi^{-1}(U)$ containing $\tilde{p}$. Actually, it is sufficient to take a simply connected domain U containing p. If $\gamma(\tilde{U}) \cap \tilde{U} \neq \phi$ for some $\gamma \in \Gamma$, then there are points $\tilde{p}_1, \tilde{q}_1 \in \tilde{U}$ with $\tilde{q}_1 = \gamma(\tilde{p}_1)$. Since $\pi \circ \gamma = \pi$, we get $\pi(\tilde{p}_1) = \pi(\tilde{q}_1)$, and hence $\tilde{q}_1 = \tilde{p}_1$, for π is biholomorphic on $\tilde{U}$. Thus we have $\gamma(\tilde{p}_1) = id(\tilde{p}_1)$, where id is the identity. By Theorem 2.4, we conclude that $\gamma = id$.

Finally, to verify (iii), assume that there exists a sequence $\{ \gamma_n \}_{n=1}^{\infty}$ consisting of mutually distinct elements of Γ such that $\gamma_n(K) \cap K \neq \phi$ for all n. Then for each n, we can take two points $\tilde{q}_n, \tilde{r}_n \in K$ with $\tilde{r}_n = \gamma_n(\tilde{q}_n)$. Since K is compact, taking a subsequence if necessary, we may assume that $\{ \tilde{q}_n \}_{n=1}^{\infty}$, $\{ \tilde{r}_n \}_{n=1}^{\infty}$ converge to $\tilde{q}_o, \tilde{r}_o \in K$, respectively, as $n \to \infty$. Since $\pi \circ \gamma_n = \pi$, we obtain $\pi(\tilde{q}_n) = \pi(\tilde{r}_n)$ and $\pi(\tilde{q}_o) = \pi(\tilde{r}_o)$. Take a neighborhood U of $\pi(\tilde{q}_o)$ in R satisfying the condition of the definition of a covering map in §2.1, and denote by $\tilde{U}$ and $\tilde{V}$ the connected components of $\pi^{-1}(U)$ containing $\tilde{q}_o$ and $\tilde{r}_o$, respectively. Since $\{ \gamma_n(\tilde{q}_n) \}_{n=1}^{\infty}$ converges to $\tilde{r}_o$, we have $\gamma_n(\tilde{U}) \cap \tilde{U} \neq \phi$ for a sufficiently large n. Since $\pi \circ \gamma_n(\tilde{U}) = U$, it follows that $\gamma_n(\tilde{U}) = \tilde{V}$, namely, $(\gamma_{n+1})^{-1} \circ \gamma_n(\tilde{U}) = \tilde{U}$. By the assertion (ii), we conclude that $\gamma_{n+1} = \gamma_n$. This is a contradiction. □

Example 3. Here is an example of a group which does not act properly discontinuously. Let α be a real number not equal to 2π multiplied by a rational number. Then the group generated by $\gamma(z) = e^{i\alpha} z$ does not act properly discontinuously on $\mathbf{C} - \{ 0 \}$.

2.2.4. Representation of Riemann Surfaces as Quotient Spaces

We shall explain a way to construct a Riemann surface $\tilde{R}/\Gamma$ from a Riemann surface $\tilde{R}$ and a subgroup Γ of the biholomorphic automorphism group $Aut(\tilde{R})$, where Γ is assumed to satisfy the properties (ii) and (iii) in Lemma 2.6, that is, every element of Γ except for the unit element has no fixed points in $\tilde{R}$, and acts properly discontinuously on $\tilde{R}$.

Two points $\tilde{p}, \tilde{q} \in \tilde{R}$ are said to be *Γ-equivalent* or *equivalent under Γ* if there exists an element $\gamma \in \Gamma$ satisfying $\tilde{q} = \gamma(\tilde{p})$. Denote by $[\tilde{p}]$ the equivalence class of $\tilde{p}$. Let $\tilde{R}/\Gamma$ be the set of all these equivalence classes $[\tilde{p}]$, which is called the *quotient space* of $\tilde{R}$ by Γ. Define the projection $\pi\colon \tilde{R} \to \tilde{R}/\Gamma$ by $\pi(\tilde{p}) = [\tilde{p}]$.

We introduce the quotient topology on $\tilde{R}/\Gamma$. A subset U of $\tilde{R}/\Gamma$ is said to be open if and only if the inverse image $\pi^{-1}(U)$ of U is an open subset of $\tilde{R}$. The projection π is readily seen to be a continuous mapping of $\tilde{R}$ onto $\tilde{R}/\Gamma$. Since $\tilde{R}$ is connected, so is $\tilde{R}/\Gamma$. Moreover, we see that $\tilde{R}/\Gamma$ is a Hausdorff space, for Γ acts properly discontinuously on $\tilde{R}$.

Now, we define a complex structure on $\tilde{R}/\Gamma$ as follows: for any point $\tilde{p} \in \tilde{R}$, take a neighborhood $\tilde{U}_{\tilde{p}}$ of $\tilde{p}$ satisfying the property (ii) in Lemma 2.6. We may assume that there exists a local coordinate $z_{\tilde{p}}$ on $\tilde{U}_{\tilde{p}}$. Then, putting $p = \pi(\tilde{p})$, $U_p = \pi(\tilde{U}_{\tilde{p}})$, we see that $\pi\colon \tilde{U}_{\tilde{p}} \to U_p$ is homeomorphic. Hence, setting $z_p = z_{\tilde{p}} \circ \pi^{-1}$, we conclude that $\{ (U_p, z_p) \}_{p \in \tilde{R}/\Gamma}$ defines a complex structure so that

$(\tilde{R}, \pi, \tilde{R}/\Gamma)$ is a covering of $\tilde{R}/\Gamma$. We call this Riemann surface $\tilde{R}/\Gamma$ the *quotient Riemann surface* of $\tilde{R}$ by Γ. Then we have immediately the following assertion.

Theorem 2.7. *Let $(\tilde{R}, \pi, R)$ be a universal covering of a Riemann surface R with universal covering transformation group Γ. Then the quotient Riemann surface $\tilde{R}/\Gamma$ of $\tilde{R}$ by Γ is biholomorphically equivalent to R under the correspondence $[\tilde{p}] \mapsto \pi(\tilde{p})$.*

2.3. Möbius Transformations

In the preceding section, we have seen that every Riemann surface R is represented by the quotient Riemann surface $\tilde{R}/\Gamma$ of a universal covering surface $\tilde{R}$ by its universal covering transformation group Γ. Here, by virtue of the uniformization theorem, $\tilde{R}$ is biholomorphic to one of the three Riemann surfaces $\hat{\mathbf{C}}$, $\mathbf{C}$, or H. We denote by $Aut(\tilde{R})$ the group of biholomorphic automorphisms of $\tilde{R}$. From Lemma 2.6, Γ is a subgroup of $Aut(\tilde{R})$, consists of elements without fixed points in $\tilde{R}$ except for the identity, and acts properly discontinuously on $\tilde{R}$.

With this in mind, let us study the biholomorphic automorphism groups $Aut(\hat{\mathbf{C}})$, $Aut(\mathbf{C})$, $Aut(\Delta)$, and $Aut(H)$.

2.3.1. Biholomorphic Automorphism Groups of Canonical Domains

Biholomorphic automorphisms of canonical domains $\hat{\mathbf{C}}$, $\mathbf{C}$, Δ, and H have the following forms:

Lemma 2.8

(i) *Every element of $Aut(\hat{\mathbf{C}})$ has a form*

$$\gamma(z) = \frac{az+b}{cz+d}, \tag{2.2}$$

where $a, b, c, d \in \mathbf{C}$ with $ad - bc = 1$.

(ii) *Every element of $Aut(\mathbf{C})$ has a form*

$$\gamma(z) = az + b, \tag{2.3}$$

where $a, b \in \mathbf{C}$ with $a \neq 0$.

(iii) *Every element of $Aut(\Delta)$ has a form*

$$\gamma(z) = \frac{az+b}{\bar{b}z+\bar{a}}, \tag{2.4}$$

where $a, b \in \mathbf{C}$ with $|a|^2 - |b|^2 = 1$. This is also written as

$$\gamma(z) = e^{i\theta}\,\frac{z-\alpha}{1-\bar{\alpha}\,z}, \tag{2.5}$$

where $\theta \in \mathbf{R}$ *and* $\alpha \in \Delta$.

(iv) *Every element of* $Aut(H)$ *has a form*

$$\gamma(z) = \frac{az+b}{cd+d}, \tag{2.6}$$

where $a, b, c, d \in \mathbf{R}$ *with* $ad - bc = 1$.

In (2.2), it is sufficient that complex numbers a, b, c, and d satisfy the condition $ad - bc \neq 0$. However, γ does not change when a, b, c, and d are multiplied by a common constant. Hence, we may normalize the expression of γ by $ad - bc = 1$. Every element of $Aut(\hat{\mathbf{C}})$ is called a *Möbius transformation* or a *linear fractional transformation*. In particular, an element of $Aut(H)$ is called a *real Möbius transformation* or a *real linear fractional transformation*.

Proof of Lemma 2.8. First of all, let us determine the form of an element $\gamma \in Aut(\hat{\mathbf{C}})$. If $\gamma(\infty) = \infty$, then in a neighborhood of ∞, γ has the Laurent expansion

$$\gamma(z) = az + \sum_{n=0}^{\infty} b_n z^{-n},$$

where $a \neq 0$. Then $\gamma(z) - az$ is holomorphic on $\hat{\mathbf{C}}$, and hence the maximum principle shows that $\gamma(z) - az$ must be a constant function, say b. Thus we have $\gamma(z) = az + b$ with $a \neq 0$. If $\gamma(\infty) = z_o \neq \infty$, then setting $\gamma_1(z) = 1/(z - z_o)$, we see that both γ_1 and $\gamma_1 \circ \gamma$ are elements of $Aut(\hat{\mathbf{C}})$, and $\gamma_1 \circ \gamma(\infty) = \infty$. Thus we have $\gamma_1 \circ \gamma(z) = 1/(\gamma(z) - z_o) = a_1 z + b_1$, where $a_1, b_1 \in \mathbf{C}$ with $a_1 \neq 0$. Therefore, γ is expressed in the form (2.2).

Next, every element $\gamma \in Aut(\mathbf{C})$ is extended to an element of $Aut(\hat{\mathbf{C}})$ if we put $\gamma(\infty) = \infty$. By the above argument, it is obvious that γ is represented in the form (2.3).

Let γ be an element in $Aut(\Delta)$. Set $\gamma(0) = \beta$. Then the Möbius transformation $\gamma_1(z) = (z - \beta)/(1 - \bar{\beta} z)$ belongs to $Aut(\Delta)$. Hence $\gamma_2 = \gamma_1 \circ \gamma$ also belongs to $Aut(\Delta)$ and $g_2(0) = 0$. Schwarz' lemma implies that γ_2 is a rotation $\gamma_2(z) = e^{i\theta} z$, with real number θ. Hence, γ is expressed in the form (2.5). It is easy to see that γ is written in the form (2.4).

Finally, for any element $\gamma \in Aut(H)$, taking a biholomorphic mapping $T(z) = (z-i)/(z+i)$ of H onto Δ, we have an element $\gamma_1 = T \circ \gamma \circ T^{-1} \in Aut(\Delta)$. Thus γ_1 is a Möbius transformation and is represented in the form (2.2). Since γ sends H onto itself, we may assume that a, b, c, and d are real numbers, and $ad - bc > 0$. Therefore, this γ is written in the form (2.5). □

For more on the fundamental properties of Möbius transformations, such as transformation of circles into circles, and the invariance of the cross ratio under them, we refer, for instance, to Ahlfors [A-4], §3 of Chapter 3; and Jones and Singerman [A-48], Chapter 2.

Now, for every $\gamma \in Aut(\hat{\mathbf{C}})$ given by

$$\gamma(z) = \frac{az+b}{cz+d}, \quad a,b,c,d \in \mathbf{C} \text{ with } ad-bc=1,$$

we have an element

$$A = \begin{bmatrix} a & b \\ c & d \end{bmatrix}$$

of the special linear group $SL(2,\mathbf{C})$. Conversely, we have a homomorphism M of $SL(2,\mathbf{C})$ onto $Aut(\hat{\mathbf{C}})$ defined by $M(A)(z) = (az+b)/(cz+d)$ for $A \in SL(2,\mathbf{C})$, where a, b, c, and d are the entries of A as above. Then the kernel of M is $\{\pm I\}$, where I is the unit matrix. Hence, by the homomorphism theorem, M induces an isomorphism

$$Aut(\hat{\mathbf{C}}) \cong SL(2,\mathbf{C})/\{\pm I\}.$$

We set $PSL(2,\mathbf{C}) = SL(2,\mathbf{C})/\{\pm I\}$ and call it the *projective special linear group of degree* 2. An element A of $M^{-1}(\gamma)$ for an element $\gamma \in Aut(\hat{\mathbf{C}})$ is called a *matrix representation* of γ. Note that γ is represented by two elements $\pm A$ in $SL(2,\mathbf{C})$.

Remark. When the Riemann sphere $\hat{\mathbf{C}}$ is identified with the one-dimensional complex projective space $\mathbf{P}^1$, an element of $Aut(\hat{\mathbf{C}})$ corresponds to a projective transformation of $\mathbf{P}^1$. Indeed, define a biholomorphic mapping $F\colon \mathbf{P}^1 \to \hat{\mathbf{C}}$ by $F([z_0 : z_1]) = z_0/z_1$, where $[z_0 : z_1]$ is a homogeneous coordinate of $\mathbf{P}^1$. Then we see that an element $\gamma(z) = (az+b)/(cz+d)$ of $Aut(\hat{\mathbf{C}})$ corresponds to a projective transformation

$$\begin{bmatrix} z_1 \\ z_0 \end{bmatrix} \mapsto A \begin{bmatrix} z_1 \\ z_0 \end{bmatrix} = \begin{bmatrix} az_1 + bz_0 \\ cz_1 + dz_0 \end{bmatrix},$$

and that $Aut(\mathbf{P}^1) = \{ F^{-1}\circ\gamma\circ F \mid \gamma \in Aut(\hat{\mathbf{C}}) \}$.

Similarly, we have

$$Aut(H) \cong PSL(2,\mathbf{R}) = SL(2,\mathbf{R})/\{\pm I\},$$

and

$$Aut(\Delta) \cong PSU(1,1) = SU(1,1)/\{\pm I\},$$

where $SL(2,\mathbf{R})$ and $SU(1,1)$ are the *real special linear group of degree* 2 and the *special unitary group of signature* $(1,1)$, respectively.

2.3.2. Canonical Forms of Möbius Transformations

Let γ be a Möbius transformation given by

$$\gamma(z) = \frac{az+b}{cz+d}, \qquad a,b,c,d \in \mathbf{C},\ ad-bc=1,$$

which is not the identity. Let $\mathrm{Fix}(\gamma)$ be the set of fixed points z_o of γ, that is, the set of all $z_o \in \hat{\mathbf{C}}$ satisfying $\gamma(z_o) = z_o$.

By a simple calculation, we have the following statements.

(i) The case where $\infty \in \mathrm{Fix}(\gamma)$, i.e., $c = 0$.

If $a/d = 1$, that is, $a = d = \pm 1$, then γ has a sole fixed point ∞ and it is written in the form

$$\gamma(z) = z + b,$$

where b is a non-zero complex number. On the other hand, if $a/d \neq 1$, then γ has another fixed point z_o, and is represented as

$$w - z_o = \lambda(z - z_o),$$

where $w = \gamma(z)$, and λ is a complex number equal neither to 0 nor to 1.

(ii) The case where $\infty \notin \mathrm{Fix}(\gamma)$, i.e., $c \neq 0$.

If $(a+d)^2 = 4$, then γ has a sole fixed point z_o and it is written as

$$\frac{1}{w - z_o} = \frac{1}{z - z_o} + \alpha,$$

where $w = \gamma(z)$, and α is a non-zero complex number. If $(a+d)^2 \neq 4$, then γ has two fixed points z_1 and z_2, and it is represented in the form

$$\frac{w - z_1}{w - z_2} = \lambda \frac{z - z_1}{z - z_2},$$

where $w = \gamma(z)$, and λ is a complex number equal neither to 0 nor to 1.

Now, two elements $\gamma_1, \gamma_2 \in Aut(\hat{\mathbf{C}})$ are said to be *$Aut(X)$-conjugate* or *conjugate in $Aut(X)$* if there exists an element $\delta \in Aut(X)$ such that $\gamma_2 = \delta \circ \gamma_1 \circ \delta^{-1}$, where X is one of $\hat{\mathbf{C}}$, $\mathbf{C}$, H, or Δ.

This leads us to the following lemma.

Lemma 2.9. *Every Möbius transformation $\gamma(\neq id)$ has one or two fixed points on $\hat{\mathbf{C}}$, and is $Aut(\hat{\mathbf{C}})$-conjugate to the following Möbius transformation γ_o:*

(i) *If γ has a sole fixed point, then $\gamma_o(z) = z + \alpha$ for some $\alpha \in \mathbf{C}, \alpha \neq 0$.*
(ii) *If γ has two fixed points, then $\gamma_o(z) = \lambda z$ for some $\lambda \in \mathbf{C}, \lambda \neq 0, 1$.*

We call this γ_o a *canonical form* of γ. Matrix representations of canonical forms in (i) and (ii) correspond to

$$\begin{bmatrix} 1 & \alpha \\ 0 & 1 \end{bmatrix}, \quad \begin{bmatrix} \sqrt{\lambda} & 0 \\ 0 & 1/\sqrt{\lambda} \end{bmatrix},$$

respectively.

A real Möbius transformation $\gamma(\neq id)$ whose fixed points are in $\hat{\mathbf{R}} = \mathbf{R} \cup \{\infty\}$ is $Aut(H)$-conjugate to a canonical form γ_o such that the entry α or λ of a matrix representation γ_o is a real number.

2.3.3. Classification of Möbius Transformations

We classify Möbius transformations into three types. Let γ be a Möbius transformation which is not the identity.

(i) γ is *parabolic* if it is conjugate to a translation $\gamma_o(z) = z + \alpha$ for some $\alpha \in \mathbf{C}, \alpha \neq 0$.
(ii) γ is *elliptic* if it is conjugate to a rotation $\gamma_o(z) = e^{i\theta} z$ for some $\theta \in \mathbf{R}, \theta \neq 2n\pi$ $(n \in \mathbf{Z})$.
(iii) γ is *hyperbolic* if it is conjugate to a dilation $\gamma_o(z) = \lambda z$ for some $\lambda > 0, \lambda \neq 1$.

Note that there are Möbius transformations which are neither parabolic, elliptic, nor hyperbolic. An example is given by $\gamma(z) = \lambda z$ $(\lambda \in \mathbf{C}, |\lambda| \neq 1$ and $\lambda \notin [0, \infty))$. A Möbius transformation which is not the identity is said to be *loxodromic* if it is neither parabolic nor elliptic. In this book, we assume that the loxodromic transformations include hyperbolic ones.

Let z_1 and z_2 be the fixed points of a loxodromic Möbius transformation γ. Suppose that z_1 and z_2, respectively, correspond to the fixed points 0 and ∞ of a canonical form $\gamma_o(z) = \lambda z$ for some λ with $|\lambda| > 1$. Then z_1 and z_2 are called the *repelling fixed point* and the *attractive fixed point* of γ, respectively. Denote by r_γ and a_γ the repelling fixed point and the attractive fixed point of γ, respectively.

It is easy to show that every element of $Aut(H)$ or $Aut(\Delta)$, which is not the identity, is either parabolic, or elliptic, or hyperbolic.

Combining these results, we obtain easily the following assertion.

Lemma 2.10. *Let γ be a real Möbius transformation which is not the identity. Then the following hold:*

(i) *γ is parabolic if and only if γ has a sole fixed point on $\hat{\mathbf{R}}$.*
(ii) *γ is elliptic if and only if γ has two fixed points z_1, z_2 such that $z_1 \in H, z_2 \in H^*$, and $\overline{z_1} = z_2$.*
(iii) *γ is hyperbolic if and only if γ has two fixed points on $\hat{\mathbf{R}}$.*

Now, a Möbius transformation γ which is not parabolic is $Aut(\hat{\mathbf{C}})$-conjugate to a canonical form $\gamma_o(z) = \lambda z$ $(\lambda \in \mathbf{C}, \lambda \neq 0, 1)$. This λ is called a *multiplier* of γ. We have two choices for the multiplier of γ, i.e., λ and $1/\lambda$. It depends on the choice of a fixed point of γ corresponding to the attractive fixed point of γ_o. Its multiplier λ satisfies the equation $(a+d)^2 = \lambda + 1/\lambda + 2$. Note that $a + d$ is the trace of a matrix representation of γ:

$$A = \begin{bmatrix} a & b \\ c & d \end{bmatrix}, \quad a, b, c, d \in \mathbf{C}, \ ad - bc = 1.$$

We put $\mathrm{tr}(\gamma) = a + d$, which is called a *trace* of a Möbius transformation γ. If A is replaced by $-A$, then its trace is altered into $-(a + d)$. Thus $\mathrm{tr}(\gamma)$ is not determined uniquely by γ, but its square $\mathrm{tr}^2(\gamma)$ is well-defined.

By a simple calculation, we see that Möbius transformations are classified by trace.

Lemma 2.11. *Let γ be a Möbius transformation which is not the identity. Then the following hold:*
(i) *γ is parabolic if and only if $tr^2(\gamma) = 4$.*
(ii) *γ is elliptic if and only if $0 \leqq tr^2(\gamma) < 4$.*
(iii) *γ is hyperbolic if and only if $tr^2(\gamma) > 4$.*
(iv) *γ is loxodromic if and only if $tr^2(\gamma) \in \mathbf{C} - [0, 4]$.*

Finally, we define the axis of a hyperbolic real Möbius transformation γ. Suppose that γ is conjugate to a canonical form $\gamma_o(z) = \lambda z$ with $\lambda > 1$, by a real Möbius transformation δ. Namely, suppose that $\gamma = \delta \circ \gamma_o \circ \delta^{-1}$. The half-line $L = \{\, iy \mid 0 < y < \infty \,\}$ in the upper half-plane H is the geodesic, joining 0 and ∞, with respect to the Poincaré metric $|dz|^2/(\operatorname{Im} z)^2$ on H (see §3 of Chapter 3). The image $\delta(L)$ of L under δ is called the *axis* of γ and is denoted by A_γ. Then A_γ is the geodesic joining the fixed points r_γ and a_γ of γ, which is characterized as a semi-circle which joins r_γ and a_γ and is orthogonal to the real axis. Similarly, we define the axis A_γ of a hyperbolic transformation γ in $Aut(\Delta)$.

2.4. Fuchsian Models

First, we show that a Riemann surface whose universal covering surface is not biholomorphic to the upper half-plane H is biholomorphic to one of $\hat{\mathbf{C}}$, $\mathbf{C}$, $\mathbf{C} - \{\, 0 \,\}$, or tori. Next, we study some fundamental properties of discrete subgroups of $Aut(H)$, i.e., Fuchsian groups.

2.4.1. Riemann Surfaces of Exceptional Type

Let us determine Riemann surfaces whose universal covering surfaces are biholomorphic to either $\hat{\mathbf{C}}$ or $\mathbf{C}$.

Theorem 2.12. *A Riemann surface R has a universal covering surface $\tilde{R}$ biholomorphic to the Riemann sphere $\hat{\mathbf{C}}$ if and only if R itself is biholomorphic to $\hat{\mathbf{C}}$.*

Proof. Assume that $\tilde{R} = \hat{\mathbf{C}}$. Since every element γ of its covering transformation group Γ is a Möbius transformation, it should have fixed points. From Lemma 2.6, any $\gamma \in \Gamma - \{\, id \,\}$ has no fixed points on $\hat{\mathbf{C}}$. Hence we have $\Gamma = \{\, id \,\}$, which implies that $R = \tilde{R}/\Gamma = \hat{\mathbf{C}}$.

Conversely, suppose that $R = \hat{\mathbf{C}}$. Since $\hat{\mathbf{C}}$ is simply connected, by the construction of $\tilde{R}$ in §2.2 it follows that $\tilde{R} = \hat{\mathbf{C}}$. □

Theorem 2.13. *A Riemann surface R has a universal covering surface biholomorphic to the complex plane $\mathbf{C}$ if and only if R is biholomorphic to one of $\mathbf{C}$, $\mathbf{C} - \{\, 0 \,\}$, or tori.*

Proof. Assume that $\tilde{R} = \mathbf{C}$. Let Γ be its universal covering transformation group. Since Γ is a subgroup of $Aut(\mathbf{C})$, by Lemma 2.8 every $\gamma \in \Gamma$ is written in the form $\gamma(z) = az + b$ $(a, b \in \mathbf{C}, a \neq 0)$. Further, $a = 1$, because any $\gamma \in \Gamma - \{id\}$ has no fixed points on $\mathbf{C}$. Moreover, we recall that Γ acts properly discontinuously on $\mathbf{C}$ (cf. Lemma 2.6). Then we can prove that there occur the following three cases (cf. Ahlfors [A-4], §2 of Chap.7):

(i) $\Gamma = \{id\}$.
(ii) $\Gamma = \langle \gamma_o \rangle$, that is, Γ is generated by a translation $\gamma_o(z) = z + b_o$ for some $b_o \in \mathbf{C} - \{0\}$.
(iii) $\Gamma = \langle \gamma_o, \gamma_1 \rangle$, where $\gamma_o(z) = z + b_o$ and $\gamma_1(z) = z + b_1$ for some $b_o, b_1 \in \mathbf{C}$ which are linearly independent over $\mathbf{R}$.

Therefore, in cases (i), (ii), and (iii), the Riemann surface $\tilde{R} = \mathbf{C}/\Gamma$ is biholomorphic to $\mathbf{C}$, $\mathbf{C} - \{0\}$, and a torus, respectively.

To show the converse, first assume that $R = \mathbf{C}$. Since $\mathbf{C}$ is simply connected, we get $\tilde{R} = \mathbf{C}$. Next, let $R = \mathbf{C} - \{0\}$. As was seen in Example 1 of §2.1, we know that $\tilde{R} = \mathbf{C}$. Finally, suppose that R is a torus. If $\tilde{R}$ is biholomorphic to H, then Γ must be a free abelian group of rank 2, for Γ is isomorphic to the fundamental group of R. This contradicts the following lemma (Lemma 2.14) which asserts that such a Γ should be a cyclic group. Hence a universal covering surface of a torus is biholomorphic to $\mathbf{C}$. □

Corollary. *For every torus R, there exists a lattice group Γ such that R is biholomorphic to $\mathbf{C}/\Gamma$.*

A Riemann surface which is biholomorphic to one of $\hat{\mathbf{C}}$, $\mathbf{C}$, $\mathbf{C} - \{0\}$, or tori is said to be of *exceptional type*.

Lemma 2.14. *Let Γ be a subgroup of $Aut(H)$ such that every element of $\Gamma - \{id\}$ has no fixed points on H and such that the action of Γ is properly discontinuous on H. If Γ is abelian, then it is cyclic.*

Proof. We may assume that $\Gamma \neq \{id\}$. Take an element $\gamma_o \in \Gamma$ with $\gamma_o \neq id$. Since γ_o has no fixed points on H, Lemma 2.10 implies that γ_o is parabolic or hyperbolic.

If γ_o is parabolic, then we may assume that $\gamma_o(z) = z + b_o$ $(b_o \in \mathbf{R}, b_o \neq 0)$ by conjugation in $Aut(H)$. It is easy to see that an element $\gamma \in Aut(H)$ is commutative with γ_o if and only if γ is written in the form $\gamma(z) = z + b$ for some $b \in \mathbf{R}$. Since Γ acts properly discontinuously on H, there exists a positive number b_1 such that Γ is generated by $\gamma_1(z) = z + b_1$.

Now, suppose that γ_o is hyperbolic. Then we may assume that $\gamma_o(z) = \lambda_o z$ $(\lambda_o > 0)$ by conjugation in $Aut(H)$. It is easy to show that an element $\gamma \in Aut(H)$ is commutative with γ_o if and only if $\gamma(z) = \lambda z$ for some positive number λ. Since Γ acts properly discontinuously on H, there exists a positive number λ_1 such that Γ is generated by $\gamma_1(z) = \lambda_1 z$. □

2.4.2. Fuchsian Models and Fundamental Domains

The following is an immediate consequence of Theorems 2.12 and 2.13.

Theorem 2.15. *A Riemann surface R has a universal covering surface $\tilde{R}$ biholomorphic to H if and only if R is not of exceptional type; that is, if and only if R is not biholomorphic to any one of $\hat{\mathbf{C}}$, $\mathbf{C}$, $\mathbf{C}-\{0\}$, or tori.*

If a universal covering surface $\tilde{R}$ of a Riemann surface R is the upper half-plane H, we call its universal covering transformation group Γ a *Fuchsian model* of R. In this case, Γ is a subgroup of $Aut(H)$. However, identifying H with Δ, we sometimes consider a Fuchsian model Γ as a subgroup of $Aut(\Delta)$.

Remark 1. By an argument similar to that in the proofs of Theorem 2.13 and Lemma 2.14, we see that the fundamental group of a Riemann surface R is commutative if and only if R is biholomorphic to one of $\hat{\mathbf{C}}$, $\mathbf{C}$, $\mathbf{C}-\{0\}$, tori, the unit disk Δ, $\Delta-\{0\}$, or annuli $\{z \in \mathbf{C} \mid 1 < |z| < r\}$.

In order to obtain a geometric image of correspondence between a Riemann surface R and its Fuchsian model Γ, we use a fundamental domain for Γ. An open set F of the upper half-plane H is a *fundamental domain* for Γ if F satisfies the following three conditions:

(i) $\gamma(F) \cap F = \phi$ for every $\gamma \in \Gamma$ with $\gamma \neq id$.
(ii) If $\bar{F}$ is the closure of F in H, then

$$H = \bigcup_{\gamma \in \Gamma} \gamma(\bar{F}).$$

(iii) The relative boundary ∂F of F in H has measure zero with respect to the two-dimensional Lebesgue measure.

These conditions tell us that the Riemann surface $R = H/\Gamma$ is considered as $\bar{F}$ with points on ∂F identified under the covering group Γ.

Example 4. For each covering group Γ in Example 2 in §2.1, we define similarly its fundamental domain. The following (i)″, ..., (v)″ give examples of fundamental domains for covering groups of (i)′, ..., (v)′ in Example 2, respectively.

(i)″ $F = \{z \in \mathbf{C} \mid 0 < \operatorname{Im} z < 2\pi\}$.
(ii)″ $F = \{z \in H \mid 0 < \operatorname{Re} z < 1\}$.
(iii)″ $F = \{z \in \mathbf{C}-\{0\} \mid 0 < \arg z < 2\pi/n\}$.
(iv)″ $F = \{z \in H \mid 1 < |z| < \lambda\}$.
(v)″ $F = \{z \in \mathbf{C} \mid z = a + b\tau,\quad 0 < a < 1,\quad 0 < b < 1\}$.

There is a simple way to construct canonically a fundamental domain for a Fuchsian model of a Riemann surface R. First, cut R along suitable smooth paths on R to get a simply connected domain R_o. Let F be a connected component of the inverse image $\pi^{-1}(R_o)$ of R_o under the covering map π. By the construction of a universal covering $(\tilde{R}, \pi, R)$ in §2.2, we see easily that this F is a fundamental domain for Γ.

In this way, we obtain the following three examples.

Example 5. Let $\{A_j, B_j\}_{j=1}^{g}$ be a canonical system of generators of $\pi_1(R, p_o)$ for a closed Riemann surface R of genus $g\,(\geqq 2)$ such that A_j and B_j are all smooth simple closed curves with base point p_o. Cutting R along all A_j and B_j, we get a simply connected domain R_o. A connected component F of $\pi^{-1}(R_o)$ is a fundamental domain for Γ as illustrated in Fig. 2.2. In particular, it follows that Γ has a fundamental domain which is relatively compact in H.

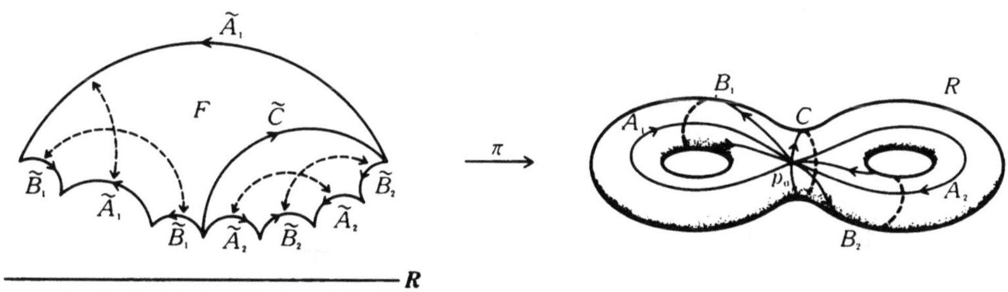

Fig. 2.2.

Example 6.

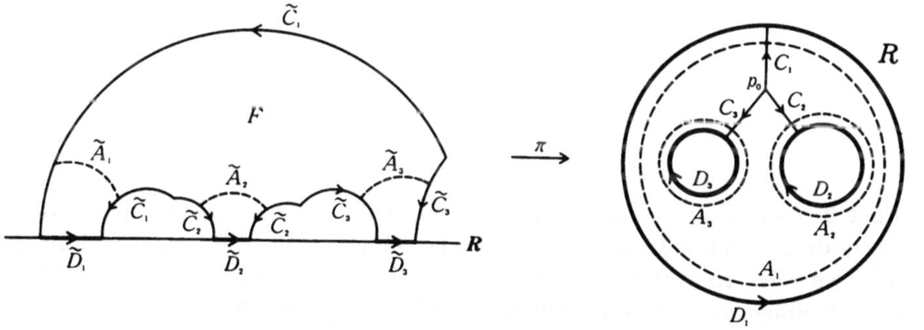

Fig. 2.3.

Consider, as illustrated in Fig. 2.3, a domain R in the complex plane, bounded by three circles D_1, D_2, and D_3. Take a point $p_o \in R$ and cut R along three smooth curves C_1, C_2, and C_3. By the same argument as that in example 6, we have a fundamental domain F for R as is shown in Fig. 2.3. The elements $\gamma_1, \gamma_2 \in \Gamma$ corresponding to the elements $[A_1], [A_2] \in \pi_1(R, p_o)$, respectively, give a canonical system of generators of Γ. In §1.5 of Chapter 3, we shall describe another way of cutting R to get a fundamental domain for this group.

Example 7. As a limiting case of Example 6, let each circle D_i degenerate to a single point p_i to obtain a Riemann surface R biholomorphic to $\hat{\mathbf{C}} - \{p_1, p_2, p_3\}$. A Fuchsian model of R is conjugate to the ***principal congruence subgroup*** $\Gamma(2)$ ***of level*** 2, which consists of all elements $\gamma(z) = (az+b)/(cz+d)$ such that $a, b, c, d \in \mathbf{Z}, ad - bc = 1$, and $a \equiv d \equiv 1, b \equiv c \equiv 0 \bmod 2$. As a system of generators of $\Gamma(2)$, we have $\gamma_1(z) = z + 2$ and $\gamma_1(z) = z/(2z+1)$. The picture on the left hand side of Fig. 2.4 shows an example of a fundamental domain for $\Gamma(2)$. The picture on the right hand side of this figure illustrates a fundamental domain for a subgroup of $Aut(\Delta)$ conjugate to $\Gamma(2)$. For details, see Ahlfols [A-4], §2 of Chapter 7; and Jones and Singerman [A-48], Chpater 6.

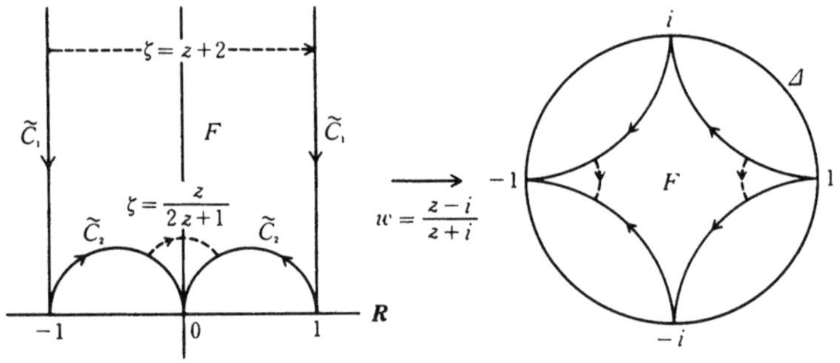

a fundamental domain in H for $\Gamma(2)$ a fundamental domain in Δ for $\Gamma(2)$

Fig. 2.4.

Remark 2. As canonical fundamental domains for a subgroup of $Aut(H)$ acting properly discontinuously on H, we have Dirichlet regions and Ford regions. For details, we refer to standard text books such as Beardon [A-11], Ford [A-31], Jones and Singerman [A-48], Lehner [A-66], [A-67], and Maskit [A-71].

2.4.3. Discrete Subgroups of $Aut(H)$

We begin with defining a natural topology on $Aut(H)$, i.e., the compact-open topology. This means that a sequence $\{\gamma_n\}_{n=1}^{\infty}$ of $Aut(H)$ converges to $\gamma \in Aut(H)$ if γ_n converges uniformly to γ on compact subsets of H as n tends to ∞. This topology is equivalent to the one of the Lie group $PSL(2,\mathbf{R})$ under the identification as is shown in §3.1. The topology of $PSL(2,\mathbf{R})$ is induced by the topology of $SL(2,\mathbf{R})$. Here, the sequence $\{A_n\}_{n=1}^{\infty}$ of $SL(2,\mathbf{R})$ with

$$A_n = \begin{bmatrix} a_n & b_n \\ c_n & d_n \end{bmatrix}$$

converges to

$$A = \begin{bmatrix} a & b \\ c & d \end{bmatrix}$$

in $SL(2,\mathbf{R})$ if and only if a_n, b_n, c_n, and d_n converge to a, b, c, and d, respectively, as n tends to ∞.

A subgroup Γ of $Aut(H)$ is said to be *discrete* if Γ is a discrete subset of $Aut(H)$, i.e., Γ consists of isolated points with respect to the relative topology on Γ induced from that of $Aut(H)$. A discrete subgroup of $Aut(H)$ is called a *Fuchsian group*. Since $SL(2,\mathbf{R})$ satisfies the second axiom of countability, a Fuchsian group consists of at most a countable number of elements. Similarly, we define a discrete subgroup of $Aut(\Delta)$, $Aut(\mathbf{C})$, or $Aut(\hat{\mathbf{C}})$. A discrete subgroup of $Aut(\Delta)$ is also referred to as a *Fuchsian group*.

Lemma 2.16. *For a subgroup Γ of $Aut(H)$ the following are equivalent:*

(i) *Γ is a Fuchsian group.*
(ii) *There exist no sequences of mutually distinct elements of Γ which converge in $Aut(H)$.*

Proof. Clearly, if Γ is not Fuchsian, then the assertion (ii) does not hold, and hence (ii) implies (i). Now, assume that there exists a sequence $\{\gamma_n\}_{n=1}^{\infty}$ of distinct elements of Γ which converges to an element $\gamma \in Aut(H)$. Considering their matrix representations, we see that $\{(\gamma_n)^{-1}\}_{n=1}^{\infty}$ converges to γ^{-1} in $Aut(H)$. Since $(\gamma_n)^{-1}\circ\gamma_{n+1} \in \Gamma$ and $(\gamma_n)^{-1}\circ\gamma_{n+1} \neq id$ for any n, and since the unit element id of Γ is not an isolated point of Γ, Γ is not discrete. Hence (i) implies (ii). The proof of Lemma 2.16 is hereby complete. □

Theorem 2.17. *For a subgroup Γ of $Aut(H)$ the following are equivalent:*

(i) *Γ is Fuchsian.*
(ii) *Γ acts properly discontinuously on H.*

Proof. That (ii) implies (i) is obvious by the definition. Conversely, assume that Γ does not act properly discontinuously on H. Then we have a point $z_o \in H$ and a sequence $\{\gamma_n\}_{n=1}^{\infty}$ of distinct elements of Γ such that $\gamma_n(z_o) \to w_o \in H$

as $n \to \infty$. Since $\{\gamma_n\}_{n=1}^{\infty}$ is a normal family, taking a subsequence, if necessary, we may assume that $\{\gamma_n\}_{n=1}^{\infty}$ converges uniformly on compact subsets of H to a holomorphic function γ defined in H. From the following lemma (Lemma 2.18), this γ must be an element of $Aut(H)$. Hence by Lemma 2.16, Γ is not Fuchsian, and hence (i) implies (ii). □

Remark 1. For a subgroup Γ of $Aut(\hat{\mathbf{C}})$, the discreteness of Γ does not always imply that it acts properly discontinuously on $\hat{\mathbf{C}}$. A typical example is given by

$$\Gamma = \left\{ \gamma(z) = \frac{az+b}{cz+d} \;\middle|\; a,b,c,d \in \mathbf{Z} + i\mathbf{Z} \right\}.$$

Lemma 2.18. *Let $\{\gamma_n\}_{n=1}^{\infty}$ be a sequence of $Aut(H)$ which converges uniformly on compact subsets of H to a holomorphic function f defined in H. Here, f admits a constant function with value ∞. Then either one of the following holds:*

(i) *f is an element of $Aut(H)$.*
(ii) *f is a constant function c with $c \in \hat{\mathbf{R}}$.*

Proof. We consider the unit disk Δ instead of H. Clearly, we have $|f| \leqq 1$ on Δ. If $|f(z_o)| = 1$ for some point $z_o \in \Delta$, the maximum principle implies that f is a constant function. Thus either $|f| < 1$ on Δ, or f is a constant function c with $|c| = 1$. If $|f| < 1$ on Δ, then f belongs to $Aut(\Delta)$. In fact, $\{(\gamma_n)^{-1}\}_{n=1}^{\infty}$ being a normal family, taking a subsequence, if necessary, we may assume that it converges uniformly on compact subsets of Δ to a holomorphic function g defined in Δ. In particular, we have $g \circ f = id$. By the same argument, we see that $|g| < 1$ on Δ and $f \circ g = id$. Hence, f belongs to $Aut(H)$. □

Proposition 2.19. *Let Γ be a Fuchsian model of a closed Riemann surface of genus $g \geqq 2$. For an arbitrary point $\zeta \in \hat{\mathbf{R}} = \mathbf{R} \cup \{\infty\}$, there exists a sequence $\{\gamma_n\}_{n=1}^{\infty}$ of Γ such that $\{\gamma_n(z_o)\}_{n=1}^{\infty}$ converges to ζ for any point $z_o \in H$.*

Proof. Take a fundamental domain F for Γ such that the closure $\bar{F}$ of F is compact in H (Example 5 in §4.2). By the definition of a fundamental domain, we can pick a sequence $\{\gamma_n\}_{n=1}^{\infty}$ of Γ and a sequence $\{z_n\}_{n=1}^{\infty}$ of points in $\bar{F}$ such that $\{\gamma_n(z_n)\}_{n=1}^{\infty}$ converges to ζ. Since $\{\gamma_n\}_{n=1}^{\infty}$ is a normal family, we may suppose that $\{\gamma_n\}_{n=1}^{\infty}$ converges uniformly on compacts subsets of H to a holomorphic function f defined in H. Thus Lemma 2.18 shows that f must be a constant ζ. □

Remark 2. Let $L(\Gamma)$ be the set of all accumulation points of the set $\{\gamma(z_o) \mid \gamma \in \Gamma\}$, where z_o is a point on H. Lemma 2.18 implies that $L(\Gamma)$ is independent of the choice of z_o. It is well known that $L(\Gamma)$ is a closed subset of $\hat{\mathbf{R}}$. We call $L(\Gamma)$ the *limit set* of a Fuchsian group Γ. Proposition 2.19 tells us $L(\Gamma) = \hat{\mathbf{R}}$ provided that $R = H/\Gamma$ is compact.

2.4.4. Further Properties of Fuchsian Groups

We present some properties of Fuchsian groups, which are used later.

Lemma 2.20. *Let γ and δ be two elements of a Fuchsian group Γ. If γ is hyperbolic and $\delta \neq id$, then one of the following holds:*

(i) $\mathrm{Fix}(\gamma) = \mathrm{Fix}(\delta)$.
(ii) $\mathrm{Fix}(\gamma) \cap \mathrm{Fix}(\delta) = \phi$.

Proof. Assume that neither (i) nor (ii) holds. By $Aut(H)$-conjugation, we may assume that $\mathrm{Fix}(\gamma) = \{0, \infty\}$ and $\mathrm{Fix}(\gamma) \cap \mathrm{Fix}(\delta) = \{\infty\}$. Then matrix representations A, B of γ, δ are given by

$$A = \begin{bmatrix} \lambda & 0 \\ 0 & \lambda^{-1} \end{bmatrix}, \qquad \lambda > 0, \quad \lambda \neq 1,$$

$$B = \begin{bmatrix} a & b \\ 0 & a^{-1} \end{bmatrix}, \qquad a, b \in \mathbf{R}, \quad ab \neq 0,$$

respectively. Now, setting $C_n = BA^nB^{-1}A^{-n}$ for any positive integer n, we get

$$C_n = \begin{bmatrix} 1 & ab(1 - \lambda^{2n}) \\ 0 & 1 \end{bmatrix}.$$

Hence, if $0 < \lambda < 1$, then C_n converges to

$$C = \begin{bmatrix} 1 & ab \\ 0 & 1 \end{bmatrix}$$

as $n \to +\infty$. If $\lambda > 1$, then $C_n \to C$ as $n \to -\infty$. Since $ab \neq 0$, $\lambda > 0$, and $\lambda \neq 1$, we see that $\{C_n\}_{n=-\infty}^{\infty}$ consists of distinct elements. Therefore, by Lemma 2.16, Γ is not discrete, which leads to a contradiction. □

Lemma 2.21. (Shimizu [203]) *Let Γ be a Fuchsian group containing a translation $\gamma_o(z) = z + 1$. Then every $\gamma \in \Gamma$ having a matrix representation*

$$A = \begin{bmatrix} a & b \\ c & d \end{bmatrix}, \qquad a, b, c, d \in \mathbf{R}, \quad ad - bc = 1$$

satisfies $|c| \geqq 1$ provided that $c \neq 0$.

Proof. Assume that there exists an element $\gamma \in \Gamma$ such that γ has a matrix representation $A \in SL(2, \mathbf{R})$ with $0 < |c| < 1$. We set

$$A_o = \begin{bmatrix} 1 & 1 \\ 0 & 1 \end{bmatrix},$$

which is a matrix representation of γ_o. Putting $A_1 = A$ and $A_{n+1} = A_nA_o(A_n)^{-1}$ for any positive integer n, we see that A_n is written in the form

$$A_n = \begin{bmatrix} a_n & b_n \\ c_n & d_n \end{bmatrix},$$

where $a_n = 1 - a_{n-1}c_{n-1}$, $b_n = (a_{n-1})^2$, $c_n = -(c_{n-1})^2$, and $d_n = 1 + a_{n-1}c_{n-1}$. Thus it follows that $c_n = -c^{2^{n-1}} \to 0$ as $n \to \infty$. Next, setting $M = \max\{\,|a|, 1/(1-|c|)\,\}$, we obtain inductively $|a_n| \leqq M$ for any n. Thus each a_n, b_n, and d_n converges to 1 as $n \to \infty$. Hence, A_n converges to A_o, which contradicts the discreteness of Γ. □

Theorem 2.22. *Every element of a Fuchsian model of a closed Riemann surface of genus g ($\geqq 2$) consists only of the identity and hyperbolic elements.*

Proof. Since every element $\gamma \in \Gamma - \{\,id\,\}$ has no fixed points on H, it is parabolic or hyperbolic. Assume that Γ contains a parabolic element γ_o. By $Aut(H)$-conjugation, we may suppose that $\gamma_o(z) = z+1$. From Lemma 2.20, any element γ ($\neq id$) of Γ with $\gamma(\infty) = \infty$ is parabolic, which is written in the form $\gamma(z) = z + b$ for some real number b. Hence, $\Gamma_\infty = \{\,\gamma \in \Gamma \mid \gamma(\infty) = \infty\,\}$ is a cyclic group. Replacing γ_o with another element, if necessary, we may assume that γ_o is a generator for Γ_∞. Since every $\gamma(z) = (az+b)/(cz+d)$, $ad - bc = 1$, belonging to $\Gamma - \Gamma_\infty$ satisfies $c \neq 0$, Lemma 2.21 shows that $|c| \geqq 1$. Thus we obtain

$$\operatorname{Im}\gamma(z) \leqq \frac{1}{(\operatorname{Im} z)|c|^2} \leqq 1$$

for all z with $\operatorname{Im} z > 1$. Set $U_o = \{\,z \in H \mid \operatorname{Im} z > 2\,\}$. Then any two distinct points on U_o are not equivalent under any element of $\Gamma - \Gamma_\infty$. Thus the quotient space $D_o = U_o/\Gamma_\infty$ is biholomorphic to a domain R_o in R. Since γ_o corresponds to a non-trivial element of the fundamental group of R, the closure $\overline{R_o}$ of R_o in R is not simply connected. Since D_o is biholomorphic to the punctured disk $\{\,z \in \mathbf{C} \mid 0 < |z| < 1\,\}$, we infer that $\overline{R_o}$ must be homeomorphic to $\{\,z \in \mathbf{C} \mid 0 < |z| \leqq 1\,\}$. This contradicts that R is compact. □

Remark. This theorem is also obtained by using the hyperbolic geometry discussed in §1 of Chapter 3. We present its outline. Let $ds^2 = |dz|^2/(\operatorname{Im} z)^2$ be the Poincaré metric on H, which induces the hyperbolic metric on $R = H/\Gamma$. Assume that Γ has a translation $\gamma_o(z) = z + 1$. For any positive number a, denote by C_a a closed path on R which is the image of the segment L_a joining ia and $\gamma_o(ia)$ by the projection $\pi\colon H \to R$. Let $\ell(C_a)$ be the hyperbolic length of C_a, i.e., the length of L_a with respect to the Poincaré metric. Then we see that $\ell(C_a) \to 0$ as $n \to \infty$. On the other hand, R being compact, we have a sequence $\{\,a_n\,\}_{n=1}^{\infty}$ of positive numbers such that $a_n \to \infty$ and $\pi(ia_n) \to p_o$ as $n \to \infty$, where p_o is a point on R. Hence, if we take a simply connected domain U which contains p_o, then the closed path C_{a_n} is included in U for sufficiently large n. This implies $\gamma_o = id$, a contradiction.

2.5. Fricke Space

By using a canonical system of generators of Fuchsian models, we shall define so-called Fricke coordinates on the Teichmüller space T_g of closed Riemann surfaces of genus $g\,(\geqq 2)$.

As was defined in §3 of Chapter 1, the Teichmüller space T_g of genus $g\,(\geqq 2)$ consists of all marked closed Riemann surfaces $[R, \Sigma]$ of genus g, where $\Sigma = \{[A_j], [B_j]\}_{j=1}^{g}$ is a marking on R, i.e., a canonical system of generators of the fundamental group $\pi_1(R, p_o)$ of a closed Riemann surface R of genus g. Under the isomorphism between $\pi_1(R, p_o)$ and a Fuchsian model Γ of R stated in Theorem 2.5, denote by α_j and β_j the elements of Γ corresponding to $[A_j]$ and $[B_j]$ in $\pi_1(R, p_o)$, respectively, for each $j = 1, 2, \ldots, g$.

Now, a Fuchsian model Γ of R has the ambiguity caused by inner automorphisms of $Aut(H)$; that is, for any $\delta \in Aut(H)$, the group $\Gamma' = \delta\Gamma\delta^{-1}$ is a Fuchsian model of the same R as well. In order to assign uniquely a Fuchsian model Γ to a given marking Σ on R, we impose the normalization conditions:

(i) β_g has its repelling and attractive fixed points at 0 and ∞, respectively.
(ii) α_g has its attractive fixed point at 1.

Remark. For a given marking Σ on a closed Riemann surface of genus g, there always exists a Fuchsian model of R which satisfies the normalization conditions. In fact, by Theorem 2.22, both α_g and β_g are hyperbolic. Since α_g and β_g are not commutative, Lemma 2.20 implies $\mathrm{Fix}(\alpha_g) \cap \mathrm{Fix}(\beta_g) = \phi$. Further, conjugating Γ in $Aut(H)$, if necessary, we may assume that α_g and β_g satisfy the normalization conditions (i) and (ii).

Proposition 2.23. *For a given marking Σ on a closed Riemann surface R of genus $g\,(\geqq 2)$, a canonical system of generators $\{\alpha_j, \beta_j\}_{j=1}^{g}$ of a Fuchsian model Γ of R which satisfies the normalization conditions with respect to Σ is uniquely determined by the point $[R, \Sigma]$ in T_g.*

We call this Fuchsian group Γ the *normalized Fuchsian model* of a marked closed Riemann surface $[R, \Sigma]$. The system of generators $\{\alpha_j, \beta_j\}_{j=1}^{g}$ is referred to as its *canonical system of generators*, which satisfies the sole fundamental relation

$$\prod_{j=1}^{g} [\alpha_j, \beta_j] = id,$$

where $[\alpha_j, \beta_j] = \alpha_j \circ \beta_j \circ \alpha_j^{-1} \circ \beta_j^{-1}$.

Proof of Proposition 2.23. Take another closed Riemann surface R' of genus g and a marking Σ' on it such that $[R, \Sigma] = [R', \Sigma']$ in T_g. Then there exists a biholomorphic mapping $f\colon R \to R'$ such that $f_*(\Sigma)$ is equivalent to Σ'. A lift $\tilde{f}$ of f to H, which is an element of $Aut(H)$, is taken to satisfy

$$\alpha_j' = \tilde{f} \circ \alpha_j \circ \tilde{f}^{-1} \quad \text{and} \quad \beta_j' = \tilde{f} \circ \beta_j \circ \tilde{f}^{-1},$$

where $\{\alpha'_j, \beta'_j\}_{j=1}^g$ is the canonical system of generators of a Fuchsian model of R' which satisfies the normalization conditions with respect to Σ'. From condition (i), we have $\tilde{f}(z) = \lambda z$ for some $\lambda > 0$. Further, by condition (ii), α_g and α'_g have the common fixed point at 1, and hence $\lambda = 1$, i.e, $\tilde{f} = id$. Thus we get $\alpha_j = \alpha'_j$ and $\beta_j = \beta'_j$. □

Lemma 2.24. *Let $\{\alpha_j, \beta_j\}_{j=1}^g$ be the canonical system of generators of the normalized Fuchsian model Γ for a point $[R, \Sigma]$ in T_g. If an element $\gamma(z) = (az+b)/(cz+d)$ of $\{\alpha_j, \beta_j\}_{j=1}^g$ does not coincide with β_g, then $bc \neq 0$.*

Proof. In the case where $b = c = 0$, we have $\mathrm{Fix}(\gamma) = \mathrm{Fix}(\beta_g) = \{0, \infty\}$, and hence γ and β_g are commutative, a contradiction. Next, in the case where $b = 0$ and $c \neq 0$, we get $\mathrm{Fix}(\gamma) = \mathrm{Fix}(\beta_g) = \{0\}$. Thus, γ and β_g being non-commutative, Lemma 2.20 implies that Γ is not Fuchsian. Hence we have a contradiction. By the same argument, in the case where $b \neq 0$ and $c = 0$, we obtain a contradiction. □

By this lemma, the canonical system $\{\alpha_j, \beta_j\}_{j=1}^g$ of generators of the normalized Fuchsian model Γ for a point $[R, \Sigma]$ in T_g is written uniquely in the form

$$\alpha_j = \frac{a_j z + b_j}{c_j z + d_j}, \qquad a_j, b_j, c_j \in \mathbf{R}, \quad c_j > 0, \quad a_j d_j - b_j c_j = 1,$$

$$\beta_j = \frac{a'_j z + b'_j}{c'_j z + d'_j}, \qquad a'_j, b'_j, c'_j \in \mathbf{R}, \quad c'_j > 0, \quad a'_j d'_j - b'_j c'_j = 1$$

for each $j = 1, 2, \ldots, g-1$.

Now, we define the *Fricke coordinates* $\mathcal{F}_g : T_g \to \mathbf{R}^{6g-6}$ by

$$\mathcal{F}_g([R, \Sigma]) = (a_1, c_1, d_1, a'_1, c'_1, d'_1, \ldots, a_{g-1}, c_{g-1}, d_{g-1}, a'_{g-1}, c'_{g-1}, d'_{g-1}).$$

The image $F_g = \mathcal{F}_g(T_g)$ is called the *Fricke space* of closed Riemann surfaces of genus g. The topology of F_g is introduced by the relative topology of F_g in $\mathbf{R}^{6g-6}$. In §2 of Chapter 5, we shall verify that F_g is a simply connected domain in $\mathbf{R}^{6g-6}$. By the following theorem (Theorem 2.25), $\mathcal{F}_g$ is a bijective mapping of T_g to F_g. Hence we define a topology on T_g by identifying T_g with F_g under $\mathcal{F}_g$. Therefore, a topology of the Teichmüller space $T(R)$ of a closed Riemann surface R of genus g is induced from that of T_g. In the rest of this book, we assume that T_g and $T(R)$ are equipped with these topologies.

Theorem 2.25. *The Fricke coordinates $\mathcal{F}_g : T_g \to \mathbf{R}^{6g-6}$ is injective.*

Proof. We need to show that every point $\mathcal{F}_g([R, \Sigma]) = (a_1, c_1, d_1, \ldots, a'_{g-1}, c'_{g-1}, d'_{g-1})$ in F_g determines uniquely the canonical system $\{\alpha_j, \beta_j\}$ of generators of the normalized Fuchsian model Γ for the point $[R, \Sigma] \in T_g$.

For each j $(j = 1, 2, \ldots, g-1)$, b_j is obtained from the relation $a_j d_j - b_j c_j = 1$ with $c_j > 0$, and hence α_j is determined uniquely by $\mathcal{F}_g([R, \Sigma])$. By the same argument, β_j $(j = 1, 2, \ldots, g-1)$ is also determined.

What remains to show is that both α_g and β_g are determined by $\mathcal{F}_g([R,\Sigma])$. By the normalization condition (i) for Γ, we have $\beta_g(z) = \lambda z$ with $\lambda > 1$. By the normalization condition (ii) for Γ, α_g has its attractive fixed point at 1, and hence

$$a_g + b_g = c_g + d_g. \tag{2.7}$$

Further, from the fundamental relation $\prod_{j=1}^{g}[\alpha_j, \beta_j] = id$, putting $\gamma = \prod_{j=1}^{g-1}[\alpha_j, \beta_j]$, we have $\gamma \circ \alpha_g = \beta_g \circ \alpha_g \circ \beta_g^{-1}$. Set

$$\gamma(z) = \frac{az+b}{cz+d}, \qquad a,b,c,d \in \mathbf{R}, \quad ad - bc = 1.$$

Replacing a, b, c, and d by $-a$, $-b$, $-c$, and $-d$, respectively, if necessary, we may assume that the following equations hold:

$$(a-1)a_g + bc_g = 0, \tag{2.8}$$
$$ca_g + (d - \lambda^{-1})c_g = 0, \tag{2.9}$$
$$cb_g + (d-1)d_g = 0. \tag{2.10}$$

Since at least one of a_g or c_g does not vanish, from (2.8) and (2.9) we have $a - 1 = \lambda(1-d)$. If $a = 1$, then $d = 1$, and hence $\mathrm{tr}^2(\gamma) = 4$, which implies that γ is parabolic. This contradicts Theorem 2.22. Thus it follows that $a \neq 1$, $d \neq 1$, and $\lambda = (a-1)/(1-d)$. Hence we have determined β_g.

From (2.8) and (2.10), we get

$$a_g = \frac{bc_g}{1-a}, \tag{2.11}$$
$$d_g = \frac{cb_g}{1-d}. \tag{2.12}$$

Substitution of (2.11) and (2.12) into (2.7) gives

$$\frac{a+b-1}{1-a} c_g = \frac{c+d-1}{1-d} b_g. \tag{2.13}$$

Here, if $c + d = 1$, then we have $a + b = 1$, because $c_g \neq 0$ by Lemma 2.24. Thus, from the relation $ad - bc = 1$, we find that $a + d = 2$, and hence γ is parabolic. Again this contradicts Theorem 2.22. Therefore, we have determined α_g by $\mathcal{F}_g([R,\Sigma])$. □

Notes

For historical and expository accounts of the uniformization theorem, we refer to Abikoff [2], and Bers [29] and [36]. The original idea of using universal covering surfaces is due to H. A. Schwarz (cf. Bers [29], pp.264-265). Complete details of covering surfaces are contained in the books on Riemann surfaces listed in the notes of Chapter 1.

The notion of a Fuchsian group was first introduced by Fuchs in the study of analytic continuation of solutions of certain ordinary differential equations of the second order (cf. Ford [A-31], Chapter XI). See also Yoshida [A-113]. For more details on Fuchsian groups, we refer to Jones and Singerman [A-48], and Lehner [A-66] and [A-67]. Discrete subgroups of $PSL(2, \mathbf{C})$ are called *Kleinian groups*, which are intimately related to the theory of Teichmüller spaces. It is most regrettable that this interesting subject cannot be covered. Concerning Kleinian groups, see Beardon [A-11], Bers et al. [A-15], Ford [A-31], Krushkal', Apanasov and Gusevskiĭ [A-61], Lehner [A-66], Magnus [A-70], and Maskit [A-71]. For relation between Kleinian groups and 3-manifolds, we also refer to Epstein [A-25] and [A-26], Fathi, Laudenbach and Poénaru [A-29], Morgan and Bass [A-76], McMullen [154], and Thurston [231]. Poincaré [A-90] is his collected works on Fuchsian groups and automorphic functions.

For the interaction between ergodic theory and discrete groups, see Nicholls [A-86], Bowen and Series [47], Morosawa [158], Series [195], and Velling and Matsuzaki [241].

Fricke spaces first appeared in Fricke and Klein [A-33]. For modern treatments, see Abikoff [A-1], Goldman and Magid [A-36], Bers and Gardiner [42], Keen [110], Saito [186], and Weil [243].

For a representation of a Riemann surface, we can use a *Schottky group* instead of a Fuchsian group, and we obtain a *Schottky space* instead of a Teichmüller space. This topic is discussed in Bers [35], Hejhal [98], and Sato [188] and [189].

Chapter 3

Hyperbolic Geometry and Fenchel-Nielsen Coordinates

In this chapter, we shall discuss some aspects of the hyperbolic geometry on Riemann surfaces which is induced by the Poincaré metric on the unit disk.

First, in Section 1, we define the Poincaré metric and study basic properties, especially those concerning geodesics. Using hyperbolic geometry, in Section 2 we define a system of coordinates, called Fenchel-Nielsen coordinates, on the Teichmüller space of a closed Riemann surface. In Section 3, we discuss an embedding of the Teichmüller space into an Euclidean space by means of geodesic lengths, which has its origin in classical investigations of Fricke and Klein.

Finally, in Section 4, we give a sketch of the construction of a notable compactification of the Teichmüller space, which was recently proposed by W. Thurston.

3.1. Poincaré Metric and Hyperbolic Geometry

3.1.1. Poincaré Metric

The unit disk $\Delta = \{z \in \mathbf{C} \mid |z| < 1\}$ has several "natural" metrics. One of them is the Euclidean metric $ds^2 = |dz|^2 = dx^2 + dy^2$, and another important one is the *Poincaré metric*

$$ds^2 = \frac{4|dz|^2}{(1-|z|^2)^2}.$$

H. Poincaré used this metric to construct a model of a non-Euclidean geometry.

For any two points $z_1, z_2 \in \Delta$, we set

$$\rho(z_1, z_2) = \inf_C \int_C \frac{2|dz|}{1-|z|^2}.$$

Here, C moves all rectifiable curves in Δ which connect z_1 and z_2. It is shown that ρ satisfies the axioms of distance. We call $\rho(z_1, z_2)$ the *Poincaré distance* between z_1 and z_2.

Proposition 3.1. (Schwarz-Pick's lemma) *Every holomorphic mapping $f : \Delta \longrightarrow \Delta$ satisfies*

$$\frac{|f'(z)|}{1-|f(z)|^2} \leq \frac{1}{1-|z|^2}, \quad z \in \Delta.$$

Moreover, if the equality holds at one point in Δ, then f is a biholomorphic automorphism of Δ, and the equality holds at any point of Δ.

Proof. Fix a point z in Δ arbitrarily, and set

$$\gamma_1(w) = \frac{w+z}{1+\bar{z}w},$$
$$\gamma_2(w) = \frac{w-f(z)}{1-\overline{f(z)}w}.$$

Then γ_1 and γ_2 belong to $Aut(\Delta)$, and $F(w) = \gamma_2 \circ f \circ \gamma_1(w)$ is a holomorphic mapping of Δ into Δ. Since $F(0) = 0$ and

$$F'(0) = \frac{1-|z|^2}{1-|f(z)|^2} f'(z),$$

we have the assertion by Schwarz' lemma. □

When we denote by $f^*(ds^2)$ the pull-back of the Poincaré metric $ds^2 = 4|dz|^2/(1-|z|^2)^2$ by f, Proposition 3.1 implies that

$$f^*(ds^2) \leq ds^2$$

and that $f^*(ds^2) = ds^2$ if and only if f belongs to $Aut(\Delta)$.

Corollary. *Every holomorphic mapping $f : \Delta \longrightarrow \Delta$ satisfies*

$$\rho(f(z_1), f(z_2)) \leq \rho(z_1, z_2), \quad z_1, z_2 \in \Delta.$$

Remark. In general, the *Gaussian curvature* $K(h)$ of a Riemannian metric $h(z)^2|dz|^2$ ($h(z) > 0$) is given by

$$K(h) = -\frac{4}{h^2}\frac{\partial^2 \log h}{\partial z \partial \bar{z}}.$$

A simple computation shows that the Gaussian curvature of the Poincaré metric is identically equal to -1 on Δ.

Moreover, we can see that, when a metric $h(z)^2|dz|^2$ is invariant under the action by $Aut(\Delta)$, it is coincident with the Poincaré metric, up to a constant factor.

3.1.2. Geodesics

For every rectifiable closed arc C in Δ, we call $\int_C ds$ the *hyperbolic length* of C, and denote it by $\ell(C)$. For any two points $z_1, z_2 \in \Delta$, we call a rectifiable closed arc C, connecting z_1 and z_2 in Δ, a *geodesic* (with respect to the Poincaré metric) connecting z_1 and z_2 in Δ, if we have

$$\rho(z_1, z_2) = \ell(C).$$

Proposition 3.2. *For arbitrary $z_1, z_2 \in \Delta$, there exists a geodesic connecting z_1 and z_2 in Δ. Moreover, it is unique and is a subarc of the circle or the line segment which passes through z_1 and z_2 and is orthogonal to the boundary $\partial\Delta$ of Δ.*

Proof. Since the Poincaré metric is invariant under the action by $Aut(\Delta)$, we may assume that $z_1 = 0$ and $z_2 > 0$, by transforming them by an element

$$\gamma(z) = e^{i\theta}\frac{z - z_1}{1 - \overline{z_1}z}$$

of $Aut(\Delta)$ with suitable $\theta \in \mathbf{R}$.

Then, for every closed arc C connecting 0 and z_2, we have

$$\int_C \frac{2|dz|}{1 - |z|^2} \geq \int_0^{z_2} \frac{2dx}{1 - x^2}.$$

Hence $\rho(0, z_2) = \ell(C)$ if and only if C is coincident with the line segment $L = [0, z_2]$. □

Recall that, when $\gamma \in Aut(\Delta)$ is hyperbolic, γ has two distinct fixed points r_γ and a_γ on $\partial\Delta$. Recall that the part in Δ of the circle or the line segment which passes through these points and is orthogonal to $\partial\Delta$ is called the axis A_γ of γ. The axis A_γ is invariant under the action by γ. Here note that, by Proposition 3.2, every geodesic connecting any two points on A_γ is a subarc of A_γ.

Remark. The Poincaré metric ds_H^2 on the upper half-plane H is defined by setting

$$ds_H^2 = \frac{|dz|^2}{(\mathrm{Im}z)^2},$$

which is nothing but the pull-back of the Poincaré metric ds^2 on Δ by the Möbius transformation $\gamma(z) = (z - i)/(z + i)$ of H onto Δ.

3.1.3. Hyperbolic Metric on a Riemann Surface

Let R be a Riemann surface whose universal covering surface is biholomorphically equivalent to Δ. Consider a Fuchsian model Γ of R acting on Δ. Let $\pi : \Delta \longrightarrow R$ be the projection of Δ onto $R = \Delta/\Gamma$. Since the Poincaré metric ds^2 is invariant under the action by Γ, we obtain a Riemannian metric ds_R^2 on R which satisfies

$$\pi^*(ds_R^2) = ds^2.$$

We call this ds_R^2 the Poincaré metric, or the *hyperbolic metric* on R.

Now, every $\gamma \in \Gamma$ corresponds to an element $[C_\gamma]$ of the fundamental group $\pi_1(R, p_0)$ of R (Theorem 2.5). In particular, γ determines the free homotopy class of C_γ, where C_γ is a representative of the class $[C_\gamma]$. We say that γ *covers the closed curve* C_γ.

When $\gamma \in \Gamma$ is hyperbolic, it is seen that the closed curve $L_\gamma = A_\gamma / < \gamma >$, the image on R of the axis A_γ by π, is the unique geodesic (with respect to the hyperbolic metric on R) belonging to the free homotopy class of C_γ. We call L_γ the *closed geodesic corresponding to* γ, or *to* C_γ.

Proposition 3.3. *Let R be a Riemann surface with universal covering surface H, and Γ_1 be a Fuchsian model of R acting on H. Let*

$$\gamma(z) = \frac{az+b}{cz+d}, \quad a, b, c, d \in \mathbf{R}, \quad ad - bc = 1,$$

be a hyperbolic element of Γ_1, and L_γ be the closed geodesic on R corresponding to γ. Then the hyperbolic length $\ell(L_\gamma)$ of L_γ satisfies

$$\mathrm{tr}^2(\gamma) = (a+d)^2 = 4\cosh^2\left(\frac{\ell(L_\gamma)}{2}\right).$$

Proof. Since $\ell(L_\gamma)$ and $\mathrm{tr}^2(\gamma)$ are invariant under the conjugation of γ by an element of $Aut(H)$, we may assume that $\gamma(z) = \lambda z$ $(\lambda > 1)$. We may also assume that $a = \sqrt{\lambda}$, $b = c = 0$, and $d = 1/\sqrt{\lambda}$. In this case, we have

$$\ell(L_\gamma) = \int_1^\lambda \frac{dy}{y} = \log \lambda = 2 \log a.$$

Hence we have the assertion. □

3.1.4. Pants

Consider cutting a Riemann surface R which admits the hyperbolic metric by a family of mutually disjoint simple closed geodesics on R. Let P be a relatively compact connected component of the resulting union of subsurfaces. If P contains no more simple closed geodesic of R, then P should be triply connected, i.e., homeomorphic to a planar region, say

$$P_0 = \{|z| < 2\} - \left(\left\{|z+1| \le \frac{1}{2}\right\} \cup \left\{|z-1| \le \frac{1}{2}\right\}\right).$$

Such a subsurface P can be considered as one of the smallest pieces for rebuilding R. Hereafter, we call a relatively compact subsurface P of R a pair of ***pants*** of R if P is triply connected and if every connected component of the relative boundary of P in R is a simple closed geodesic on R.

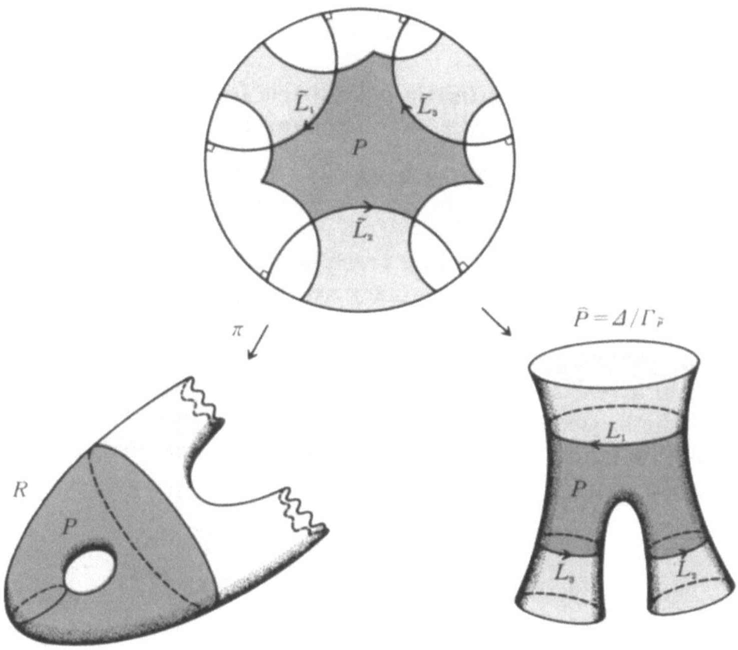

Fig. 3.1.

Fix a pair of pants P of R arbitrarily. Let Γ be a Fuchsian model of R acting on Δ, and $\pi : \Delta \longrightarrow R = \Delta/\Gamma$ be the projection. Let $\tilde{P}$ be a connected component of $\pi^{-1}(P)$. Denote by $\Gamma_{\tilde{P}}$ the subgroup of Γ consisting of all elements γ of Γ such that $\gamma(\tilde{P}) = \tilde{P}$. Then $\Gamma_{\tilde{P}}$ is a free group generated by two hyperbolic transformations, and $P = \tilde{P}/\Gamma_{\tilde{P}}$.

Set $\hat{P} = \Delta/\Gamma_{\tilde{P}}$. Then $\hat{P}$ is a surface obtained from P by attaching a suitable doubly connected region along each boundary component. Hence, $\hat{P}$ is again triply connected, and $\Gamma_{\tilde{P}}$ is a Fuchsian model of $\hat{P}$ (see Fig. 3.1).

Clearly, P is considered as a subsurface of $\hat{P}$, which is the unique pair of pants of $\hat{P}$. In other words, P is uniquely determined by $\Gamma_{\tilde{P}}$. Habitually, P is called the *Nielsen kernel* of $\hat{P}$, and $\hat{P}$ is called the *Nielsen extension* of P.

3.1.5. Existence and Uniqueness of Pants

We shall discuss the relationship between the complex structure of a triply connected domain Ω and the hyperbolic structure of P, the unique pair of pants of Ω, induced by the hyperbolic metric on Ω.

Let L_1, L_2, and L_3 be the boundary components, which are simple closed geodesics, of the pair of pants P. Let Γ_0 be a Fuchsian model of the domain Ω acting on Δ. Then Γ_0 is a free group generated by two hyperbolic transformations, say, γ_1 and γ_2. We may assume that γ_1 and γ_2 cover L_1 and L_2, respectively.

Theorem 3.4. *For an arbitrarily given triple* (a_1, a_2, a_3) *of positive numbers, there exists a triply connected planar Riemann surface* Ω *such that*

$$\ell(L_j) = a_j, \quad j = 1, 2, 3.$$

Proof. We prove it by constructing Ω explicitly.

Let C_1 be the part of the imaginary axis in Δ. Fix another geodesic, say C_2, on Δ such that the Poincaré distance between C_1 and C_2 is equal to $a_1/2$. On the other hand, geodesics on Δ from which the Poincaré distance to C_1 are equal to $a_3/2$ form a real one-parameter family (i.e., the family of circular arcs C_3' tangent to the broken circular arc in Fig. 3.2-i)). Hence there exists a geodesic, say C_3, in this family such that the Poincaré distance between C_2 and C_3 is equal to $a_2/2$.

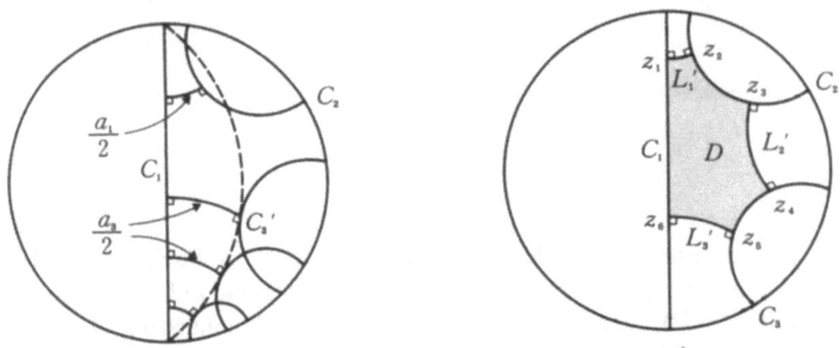

i) ii)

Fig. 3.2.

Next, let z_1 and z_2 be the points in Δ uniquely determined by the condition

$$\rho(z_1, z_2) = \frac{a_1}{2}, \quad z_1 \in C_1, z_2 \in C_2.$$

Let L_1' be the geodesic connecting z_1 and z_2. Similarly, let $\{z_3, z_4\}$ and $\{z_5, z_6\}$ be the pairs of points uniquely determined by the conditions

$$\rho(z_3, z_4) = \frac{a_2}{2}, \quad z_3 \in C_2, z_4 \in C_3,$$
$$\rho(z_5, z_6) = \frac{a_3}{2}, \quad z_5 \in C_3, z_6 \in C_1,$$

respectively. Denote by L_2' and L_3', respectively, the geodesics connecting z_3 and z_4, and z_5 and z_6. (See Fig. 3.2-ii).)

Let D be the closed hyperbolic hexagon bounded by $\{C_j, L_j'\}_{j=1}^3$. Let η_j ($j = 1, 2, 3$) be the reflection with respect to C_j, i.e., the anti-holomorphic automorphism of $\hat{\mathbf{C}}$ preserving C_j pointwise. Set

$$\gamma_1 = \eta_1 \circ \eta_2, \quad \gamma_2 = \eta_3 \circ \eta_1.$$

Then γ_1 and γ_2 are hyperbolic elements of $Aut(\Delta)$. Let Γ_0 be the group generated by these γ_1 and γ_2.

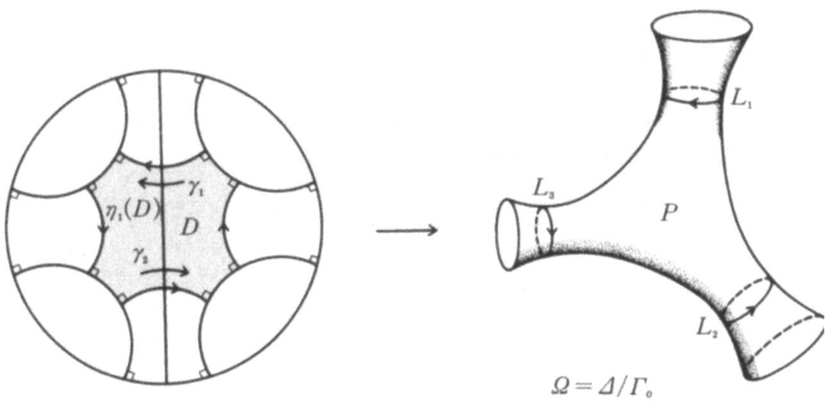

Fig. 3.3.

It is clear that $\Omega = \Delta/\Gamma_0$ is triply connected, and that the unique pair of pants P of Ω is the interior of the set obtained by identifying the boundary of $D \cup \eta_1(D)$ under the action by Γ_0. Thus Ω is a desired surface (see Fig. 3.3). □

Theorem 3.5. *The complex structure of a pair of pants P is uniquely determined by the hyperbolic lengths of the ordered boundary components of P.*

Proof. (Cf. Keen [110], III.) Suppose that a pair of pants P is given arbitrarily. Let a_j be the hyperbolic length of the boundary component L_j ($j = 1, 2, 3$) of P. Let $\hat{P}$ be the Nielsen extension of P, and Γ_0 be a Fuchsian model of $\hat{P}$ acting on the upper half-plane H. Let $\{\gamma_1, \gamma_2\}$ be a system of generators of Γ_0. Here, we may assume that γ_k covers L_k ($k = 1, 2$) and that $\gamma_3 = (\gamma_2 \circ \gamma_1)^{-1}$ covers L_3.

It suffices to show that γ_1 and γ_2 are uniquely determined by $\{a_j\}_{j=1}^3$. For this purpose, taking an $Aut(H)$-conjugation if necessary, we may assume that

$$\begin{aligned}\gamma_1(z) =&\lambda^2 z, \quad 0<\lambda<1,\\ \gamma_2(z) =&\frac{az+b}{cz+d}, \quad ad-bc=1, c>0,\end{aligned}$$

and that 1 is the attractive fixed point of γ_2, or equivalently,

$$a+b=c+d, \quad 0<-\frac{b}{c}<1.$$

Then we see that $\gamma_2(\infty) = a/c > 0$ and $a+d>0$, since the middle-point $(a-d)/(2c)$ of two fixed points of γ_2 has a value less than $\gamma_2(\infty)$.

Next, write

$$(\gamma_3)^{-1}(z) = \frac{\tilde{a}z+\tilde{b}}{\tilde{c}z+\tilde{d}}, \quad \tilde{a}\tilde{d}-\tilde{b}\tilde{c}=1.$$

Since $(\gamma_3)^{-1} = \gamma_2 \circ \gamma_1$, we may assume that

$$\tilde{a}=a\lambda, \quad \tilde{b}=b/\lambda, \quad \tilde{c}=c\lambda, \quad \tilde{d}=d/\lambda.$$

In particular, $\tilde{c}>0$. Moreover, the middle-point $(\tilde{a}-\tilde{d})/(2\tilde{c})$ of the fixed points of $(\gamma_3)^{-1}$ has a value greater than $(\gamma_3)^{-1}(\infty)=\tilde{a}/\tilde{c}$. Hence, $\tilde{a}+\tilde{d}<0$.

On the other hand, by Proposition 3.3, we have

$$\begin{aligned}(\lambda+1/\lambda)^2 =&4\cosh^2\left(\frac{a_1}{2}\right),\\ (a+d)^2 =&4\cosh^2\left(\frac{a_2}{2}\right),\\ (\tilde{a}+\tilde{d})^2 =&4\cosh^2\left(\frac{a_3}{2}\right).\end{aligned}$$

Therefore, γ_1 and γ_2 are uniquely determined by $\{a_1, a_2, a_3\}$. □

We have proved that, for any triple of positive numbers, there exists a pair of pants admitting a reflection (induced, for example, by η_1) such that the hyperbolic lengths of the ordered boundary components are the given triple (Theorem 3.4), and that it is uniquely determined by the given triple (Theorem 3.5). Thus we have the following corollary (see also Fig. 3.4).

Corollary. *Every pair of pants P has an anti-holomorphic automorphism J_P of order two.*

Moreover, the set $F_{J_P} = \{z \in P \mid J_P(z)=z\}$ of all fixed points of J_P consists of three geodesics $\{D_j\}_{j=1}^3$ in P satisfying the following condition:

For every j ($j=1,2,3$), D_j has the endpoints on, and is orthogonal to, both L_j and L_{j+1}, where $L_4=L_1$.

We call J_P described in this corollary the *reflection* of P.

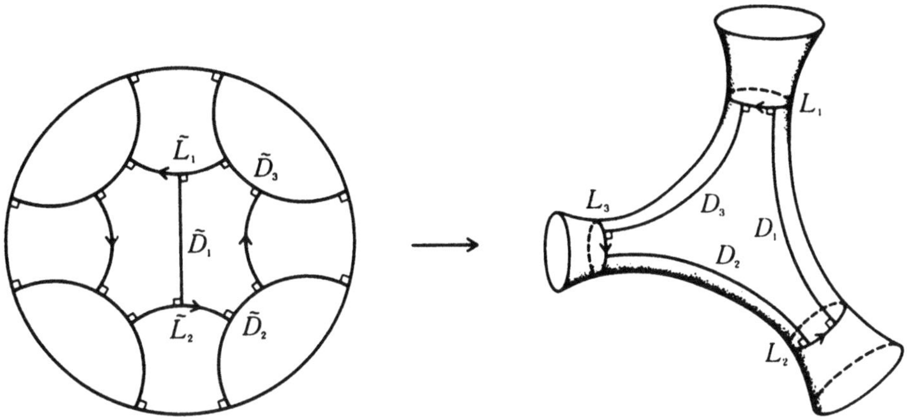

Fig. 3.4.

3.2. Fenchel-Nielsen Coordinates

In Chapter 2, we represented the Teichmüller space T_g of genus g (≥ 2) as a subset F_g (named the Fricke space) of real $(6g-6)$-dimensional Euclidean space, by using Fuchsian models of surfaces. In this section, we introduce another type of coordinates to T_g by using hyperbolic geometry.

As stated previously, the Fricke space F_g is a subset of $\mathbf{R}^{6g-6}$. Moreover, the following theorem was conceived a long time ago, but rather intuitively.

Teichmüller's theorem. (Theorem 5.15) *The Fricke space F_g is a domain in $\mathbf{R}^{6g-6}$ and homeomorphic to $\mathbf{R}^{6g-6}$.*

We postpone giving a proof of this theorem until Chapter 5, though we shall use the assertions freely in the remainder of this chapter.

3.2.1. Pants Decomposition

Let R be a closed Riemann surface of genus g (≥ 2). As before, consider cutting R along mutually disjoint simple closed geodesics with respect to the hyperbolic metric ds_R^2 on R. When there are no more simple closed geodesics of R contained in the remaining open set, then every piece should be a pair of pants of R.

Recall that the complex structure of each pair of pants of R is uniquely determined by the triple of the hyperbolic lengths of boundary geodesics of it by Theorem 3.5. It is clear that R is reconstructed by gluing all resulting pieces suitably. Hence, we can consider, as a system of coordinates for the Teichmüller space T_g, the pair of the set of lengths of all geodesics used in the above decomposition into pants and the set of the so-called twisting parameters used to glue the pieces. Such a system of coordinates is called *Fenchel-Nielsen coordinates* on T_g.

Now, we give a precise definition of these coordinates, and verify that they give a system of global coordinates on T_g.

For this purpose, first fix a point $[R, \Sigma]$ of T_g. A set $\mathcal{L}$ of mutually disjoint simple closed geodesics on R is termed *maximal* if there is no set $\mathcal{L}'$ which includes $\mathcal{L}$ properly. We call a maximal set $\mathcal{L} = \{L_j\}_{j=1}^N$ of mutually disjoint simple closed geodesics on R a *system of decomposing curves*, and the family $\mathcal{P} = \{P_k\}_{k=1}^M$ consisting of all connected components of $R - \cup_{j=1}^N L_j$ the *pants decomposition* of R corresponding to $\mathcal{L}$.

Example. When $g = 2$, there are the two kinds of pants decompositions shown in Fig. 3.5.

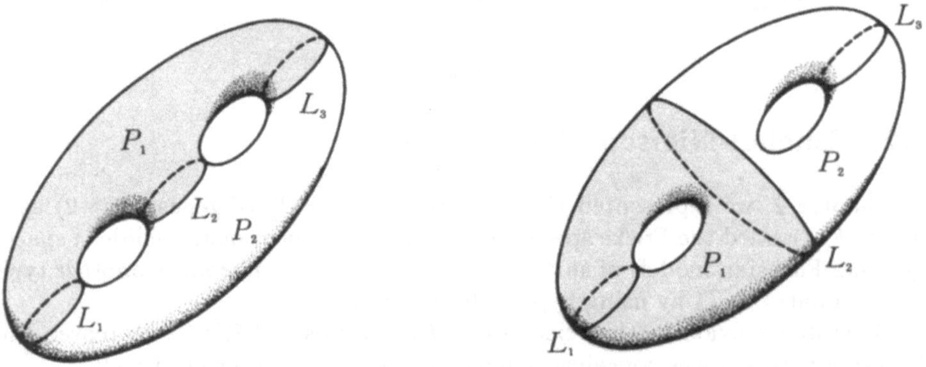

Fig. 3.5.

Proposition 3.6. *Let $\mathcal{L} = \{L_j\}_{j=1}^N$ be an arbitrary system of decomposing curves on a closed Riemann surface R of genus g (≥ 2), and let $\mathcal{P} = \{P_k\}_{k=1}^M$ be the pants decomposition of R corresponding to $\mathcal{L}$. Then M and N satisfy*

$$N = 3g - 3 \quad \textit{and} \quad M = 2g - 2.$$

Proof. Cut R along an element L_1 of $\mathcal{L}$. Let n_1 be the number of connected components of $R - L_1$, and g_1 be the sum of genera of all connected components of $R - L_1$. Then we have

$$g_1 - n_1 = (g - 1) - 1.$$

Clearly, the number of boundary components of $R - L_1$ is two.

Moreover, we can see inductively that, whenever we add a cut along a new element of $\mathcal{L}$, the number of boundary components increases by two, and the sum of genera of all connected components minus the number of connected components decreases by one. Hence, we have

$$3M = 2N \quad \text{and} \quad 0 - M = (g - 1) - N,$$

which imply the assertion. □

3.2.2. Geodesic Length Functions

Fix a point $[R, \Sigma]$ of T_g, and a system $\mathcal{L} = \{L_j\}_{j=1}^N$ of decomposing curves on R. For every t in the Fricke space F_g, we denote by $[R_t, \Sigma_t]$ the point in T_g corresponding to t. Then, we can determine uniquely a system $\mathcal{L}_t = \{L_j(t)\}_{j=1}^N$ of decomposing curves on R_t. Namely, take a marking-preserving homeomorphism $f_t : R \longrightarrow R_t$ (cf. Theorem 1.4). For every L_j in $\mathcal{L}$, let $L_j(t)$ be the unique closed geodesic in the free homotopy class of the closed curve $f_t(L_j)$ on R_t. It is not difficult to show that $L_j(t)$ is simple, and that $L_j(t)$ and $L_{j'}(t)$ are mutually disjoint when $j \neq j'$. Hence, $\mathcal{L}_t = \{L_j(t)\}_{j=1}^N$ is a system of decomposing curves on R_t.

Let Γ_t be the Fuchsian model of R_t represented by t. Note that $L_j(t)$ is the projection of the axis of an element of Γ_t which covers $f_t(L_j)$ for every j.

Now, for every t in F_g and every j, we denote the hyperbolic length $\ell(L_j(t))$ of $L_j(t)$ simply by $\ell_j(t)$. We consider $\ell_j(t)$ as a function on F_g (or equivalently, on T_g) and call it the *geodesic length function* for L_j. Then Proposition 3.3 implies the following lemma.

Lemma 3.7. *Every geodesic length function $\ell_j(t)$ is real-analytic on F_g.*

3.2.3. Twisting Parameters

Next, for every j, let $P_{j,1}$ and $P_{j,2}$ be two pairs of pants in $\mathcal{P}$ having L_j as a boundary component. Here we allow the case where $P_{j,1} = P_{j,2}$. Recall that $P_{j,1}$ and $P_{j,2}$ admit the reflection J_1 and J_2, respectively, by the Corollary to Theorem 3.5. Take a fixed point of J_k on L_j for each $P_{j,k}$ $(k = 1, 2)$, and denote it by $c_{j,k}$. Fix also an orientation on L_j (see Fig. 3.6).

As before, let $[R_t, \Sigma_t]$ be the point of T_g corresponding to t for every t in F_g. For every t and j, let $P_{j,1}(t)$ and $P_{j,2}(t)$ be the connected components of $R_t - \cup_{j=1}^N L_j(t)$ (which are pants of R_t) corresponding to $P_{j,1}$ and $P_{j,2}$, respectively. Recall that each $c_{j,k}$ $(k = 1, 2)$ is the end point on L_j of the geodesic $D_{j,k}$ joining L_j and another boundary component, say $L_{j,k}$, in $P_{j,k}$. Let $L_{j,k}(t)$ be the boundary component of $P_{j,k}(t)$ corresponding to $L_{j,k}$. Denote by $D_{j,k}(t)$ the geodesic joining $L_j(t)$ and $L_{j,k}(t)$ in $P_{j,k}(t)$ with minimal length, and by $c_{j,k}(t)$ the point of $D_{j,k}(t)$ on $L_j(t)$. Then each $c_{j,k}(t)$ $(k = 1, 2)$ is a fixed point of the reflection of $P_{j,k}(t)$.

Let $T_j(t)$ be the oriented arc on $L_j(t)$ from $c_{j,1}(t)$ to $c_{j,2}(t)$. Since $L_j(t)$ has the natural orientation determined from that of L_j, we can define the signed hyperbolic length $\tau_j(t)$ of $T_j(t)$ (so that $\tau_j(t)$ is positive or negative according to whether the orientation of $T_j(t)$ is compatible with that of $L_j(t)$ or not). Set

$$\theta_j(t) = 2\pi \frac{\tau_j(t)}{\ell_j(t)}.$$

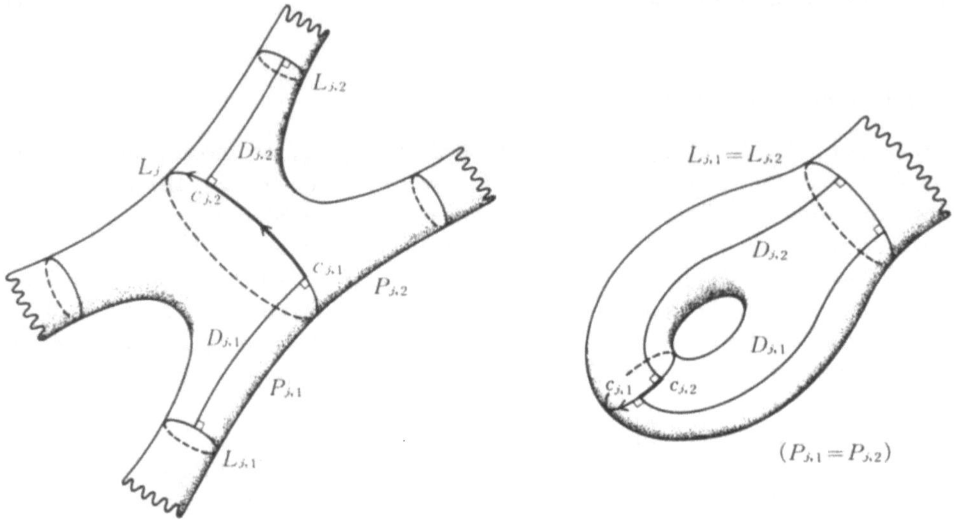

Fig. 3.6.

Then $\theta_j(t)$ is well-defined modulo 2π. We call $\theta_j(t)$ the *twisting parameter* with respect to L_j.

Lemma 3.8. *For every j, $\exp(i\theta_j(t))$ is well-defined and real-analytic on F_g.*

Proof. Fix j. For every t in F_g, let Γ_t be the Fuchsian group represented by t. Take an element of Γ_t which covers $L_j(t)$, and denote it by $\gamma_j(t)$. Next, for each k $(= 1, 2)$, let $\gamma_{j,k}(t)$ be the element of Γ_t which covers $L_{j,k}(t)$ and satisfies that the geodesic $\tilde{D}_{j,k}(t)$, connecting $A_j(t)$ and $A_{j,k}(t)$ with the minimal length, is projected onto $D_{j,k}(t)$, where $A_j(t)$ and $A_{j,k}(t)$ are the axes of $\gamma_j(t)$ and $\gamma_{j,k}(t)$, respectively (see Fig. 3.7).

Here, we may assume that the fixed points of $\gamma_j(t)$, $\gamma_{j,1}(t)$, and $\gamma_{j,2}(t)$ move real-analytically on F_g. Hence, when we take a conjugation of Γ_t by an element of $Aut(H)$ so that $\gamma_j(t)$ goes to $\tilde{\gamma}_j(t)(z) = \lambda_j(t) \cdot z$ $(\lambda_j(t) > 1)$, the fixed points of $\tilde{\gamma}_{j,k}(t)$ corresponding to $\gamma_{j,k}(t)$ move also real-analytically on F_g for each k.

Now, $c_{j,k}(t)$ is the projection of the end point $\tilde{c}_{j,k}(t)$ of $\tilde{D}_{j,k}(t)$ to $A_j(t)$. Hence, if we show that $\tilde{c}_{j,k}(t)$ moves real-analytically on F_g, the assertion follows by the definition of $\tau_j(t)$ and Lemma 3.7.

To show this, fix k, and let p_1 and p_2 be the fixed points of $\tilde{\gamma}_{j,k}(t)$. Set $\tilde{c}_{j,k}(t) = iy_k$ $(y_k > 0)$. Since

$$y_k^2 + \left(\frac{p_1 - p_2}{2}\right)^2 = \left(\frac{p_1 + p_2}{2}\right)^2,$$

we see real-analyticity of $\tilde{c}_{j,k}(t)$. □

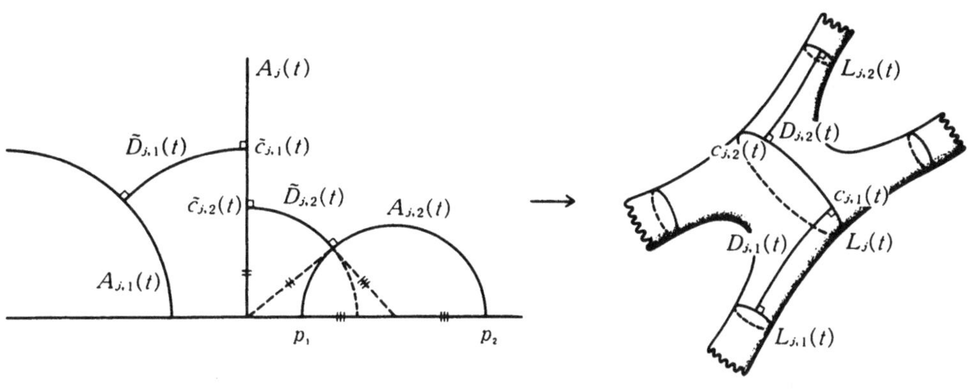

Fig. 3.7.

3.2.4. Fenchel-Nielsen Coordinates

So far, we have defined a real-analytic mapping $\Psi : F_g \longrightarrow (\mathbf{R}^+)^{3g-3} \times (S^1)^{3g-3}$;

$$\Psi(t) = (\ell_1(t), \cdots \ell_{3g-3}(t), \exp(i\theta_1(t)), \cdots, \exp(i\theta_{3g-3}(t))) .$$

Here we set $\mathbf{R}^+ = \{x \in \mathbf{R} \mid x > 0\}$ and $S^1 = \{z \in \mathbf{C} \mid |z| = 1\}$.

On the other hand, Teichmüller's theorem (Theorem 5.15) states that F_g is a simply connected domain. Accordingly, every $\theta_j(t)$ has a single-valued continuous branch on F_g (the monodromy theorem). Thus we have the following:

Lemma 3.9. *Fix a single-valued continuous branch of the twisting parameter $\theta_j(t)$ on F_g for every j. Then*

$$\tilde{\Psi}(t) = (\ell_1(t), \cdots \ell_{3g-3}(t), \theta_1(t), \cdots, \theta_{3g-3}(t))$$

is real-analytic on F_g.

Moreover, we shall now prove the following theorem.

Theorem 3.10. *This mapping $\tilde{\Psi}$ is a homeomorphism of F_g onto $(\mathbf{R}^+)^{3g-3} \times \mathbf{R}^{3g-3}$. In particular, $\tilde{\Psi}$ gives a system of global coordinates on F_g, and hence on T_g.*

We call these coordinates $\tilde{\Psi}$ *Fenchel-Nielsen coordinates* of T_g associated with the pants decomposition $\mathcal{P}$ (or, with the system $\mathcal{L}$ of decomposing curves).

Remark. Actually, $\tilde{\Psi} : F_g \longrightarrow (\mathbf{R}^+)^{3g-3} \times \mathbf{R}^{3g-3}$ is a diffeomorphism (cf. Abikoff [A-1], and Wolpert [251]).

Proof. First, we show that $\tilde{\Psi}$ is injective. Suppose that $\tilde{\Psi}(t_1) = \tilde{\Psi}(t_2)$ for some t_1 and t_2 in F_g. Let R_{t_i} be the Riemann surface represented by t_i (with the natural marking), and $\mathcal{P}_i = \{P_k(i)\}_{k=1}^{3g-3}$ be the pants decomposition of R_{t_i} corresponding to $\mathcal{P}$, for each i $(= 1, 2)$. By Theorem 3.5 and the assumption, there is a conformal mapping, say g_k, of $P_k(1)$ onto $P_k(2)$ which respects the boundary correspondence for every k. Moreover, the proof of Theorem 3.5 implies that

$$ds_1^2 = (g_k)^*(ds_2^2).$$

Here, ds_i^2 $(i = 1, 2)$ is the hyperbolic metric on R_{t_i}. In particular, every g_k is a hyperbolic isometry of the closure of $P_k(1)$ onto the closure of $P_k(2)$. Since

$$\theta_j(t_1) = \theta_j(t_2), \quad j = 1, \cdots, 3g-3,$$

all g_k can be glued together into a marking-preserving homeomorphism, say h, of R_{t_1} onto R_{t_2}. Since h is holomorphic on R_{t_1} except for a finite number of analytic curves, so is h on the entire R_{t_1} by Painlevé's theorem. Hence, h is a biholomorphic mapping of R_{t_1}, which implies that $t_1 = t_2$. Thus we have proved that $\tilde{\Psi}$ is injective.

Next, we show that $\tilde{\Psi}$ is surjective. For this purpose, we begin with fixing a point $(a_1, \cdots, a_{3g-3}, \alpha_1, \cdots, \alpha_{3g-3})$ of $(\mathbf{R}^+)^{3g-3} \times \mathbf{R}^{3g-3}$ arbitrarily. For every P_k in the pants decomposition $\mathcal{P}$ of R, denote by $\{L_{k,j}\}_{j=1}^3$ $(\subset \mathcal{L})$ the boundary components of P_k. From §1.5, there is a unique pair of pants, say P'_k, such that the triple of the hyperbolic lengths of the boundary components of P'_k is equal to the given triple $\{a_{k,j}\}_{j=1}^3$. Set $\mathcal{P}' = \{P'_k\}_{k=1}^{3g-3}$. As before, let $P_{j,1}$ and $P_{j,2}$ be the elements of $\mathcal{P}$ neighboring each other along L_j for every j, and let $P'_{j,\ell}$ be the element of $\mathcal{P}'$ corresponding to $P_{j,\ell}$ $(\ell = 1, 2)$. Let $c'_{j,\ell}$ be the point on the boundary of $P'_{j,\ell}$ corresponding to $c_{j,\ell}$ for every j and ℓ.

Now, by gluing $P'_{j,1}$ and $P'_{j,2}$ suitably along curves corresponding to L_j for every j, we obtain a Riemann surface, say R'. We need to choose a suitable gluing (and a suitable marking of R') so that R' corresponds to a point t' of F_g such that $\tilde{\Psi}(t')$ is equal to the given $(a_1, \cdots, a_{3g-3}, \alpha_1, \cdots, \alpha_{3g-3})$. This can be achieved by gluing $P'_{j,1}$ and $P'_{j,2}$ so that the twisting parameter becomes the given α_j for every j.

We shall explain this procedure more rigorously by using Fuchsian models. In the rest of this proof, we consider only the case where $P_{j,1} \neq P_{j,2}$, for the other case can be considered similarly.

Fix j, and let $\Gamma_{j,k}$ be a Fuchsian model of the Nielsen extension $\hat{P}'_{j,k}$ of $P'_{j,k}$ for each k. Here we assume that every $\Gamma_{j,k}$ acts on the upper half-plane H, and that the transformation

$$\gamma(z) = \lambda z, \quad \lambda = \exp a_j > 1$$

belongs to both $\Gamma_{j,1}$ and $\Gamma_{j,2}$, and γ covers the boundary component, say $L'_{j,k}$, of $P'_{j,k}$ corresponding to L_j for each k $(k = 1, 2)$. We also assume that the natural orientation of the axis $A = \{z \in H \mid z = iy, y > 0\}$ of γ corresponds to the prescribed orientation of L_j, and that the point $i \in H$ lies over $c'_{j,k}$ with respect to $\Gamma_{j,k}$ for each k.

Set $d_j = \exp(a_j \alpha_j / 2\pi)$, and consider the element

$$\delta(z) = d_j z, \quad d_j > 0$$

of $Aut(H)$. Identify every z on the axis A of γ considered as an element of $\Gamma_{j,2}$ with $\delta(z)$ on the axis A of γ considered as an element of $\Gamma_{j,1}$. (See Fig. 3.8.)

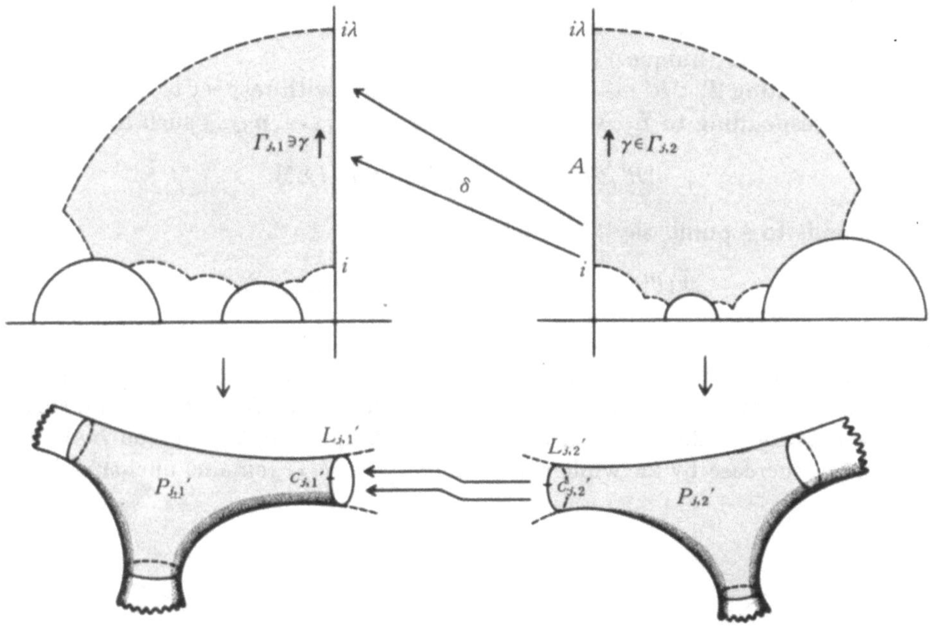

Fig. 3.8.

Then we have a gluing of $P'_{j,1}$ and $P'_{j,2}$ (by the identification of $L'_{j,1}$ and $L'_{j,2}$) such that the translation length from $c'_{j,1}$ to $c'_{j,2}$ is equal to $\log d_j$ modulo a_j. Since $\delta \in Aut(H)$, the hyperbolic metric on the Nielsen extension of the resulting domain W_j (i.e., the Riemann surface obtained from W_j by attaching a suitable ring domain, which was used in the construction of the Nielsen extension of $P'_{j,k}$ in §1.4, along every remaining boundary curve) is coincident with $ds^2_{\hat{P}'_{j,k}}$ on $P'_{j,k}$ for each k. In other words, the Fuchsian group generated by $\Gamma_{j,1}$ and $\delta \Gamma_{j,2} \delta^{-1}$ is a Fuchsian model of the Nielsen extension of W_j. (This is classically known as *Klein's combination theorem* (for connecting two surfaces). For combination theorems, see for instance Maskit [A-71], Chap. 7.)

For the proof of surjectivity of $\tilde{\Psi}$, let R' be the Riemann surface obtained by gluing $P'_{j,1}$ and $P'_{j,2}$ as before. Fix a homeomorphism h of R onto R' such that

$$h(P_k) = P'_k, \quad k = 1, \cdots, 2g-2.$$

Then $[R', h_*(\Sigma)]$ determines a point of T_g. Let t' be the corresponding point of F_g. From the preceding construction, it is clear that

$$\ell_j(t') = a_j, \quad j = 1, \cdots, 3g-3,$$

and that

$$\theta_j(t') \equiv \frac{2\pi}{a_j} \log d_j (= \alpha_j) \pmod{2\pi}, \quad j = 1, \cdots, 3g-3.$$

(Note that, at present, we cannot say that $\theta_j(t') = \alpha_j$, because the choice of branch of θ_j is not unique.)

Hence, letting $T_j : R' \longrightarrow R'$ be the *Dehn twist* with respect to L'_j, the curve on R' corresponding to L_j, we can find integers $n_1, \cdots, n_{3g-3}$ such that

$$[R', (T_1^{n_1} \circ \cdots \circ T_{3g-3}^{n_{3g-3}} \circ h)_*(\Sigma)]$$

corresponds to a point, say t'', which satisfies

$$\tilde{\Psi}(t'') = (a_1, \cdots, a_{3g-3}, \alpha_1, \cdots, \alpha_{3g-3}).$$

Thus we have shown that $\tilde{\Psi}$ is surjective.

Here, the Dehn twist T_j with respect to L'_j is, by definition, a homeomorphism of R' onto R' corresponding to the following surgery: cut R' along L'_j and reglue after a rotation of 2π (see Fig. 3.9). Note that applying T_j, we can make the value of θ_j increase by 2π while every other θ'_j $(j' \neq j)$ remains unchanged.

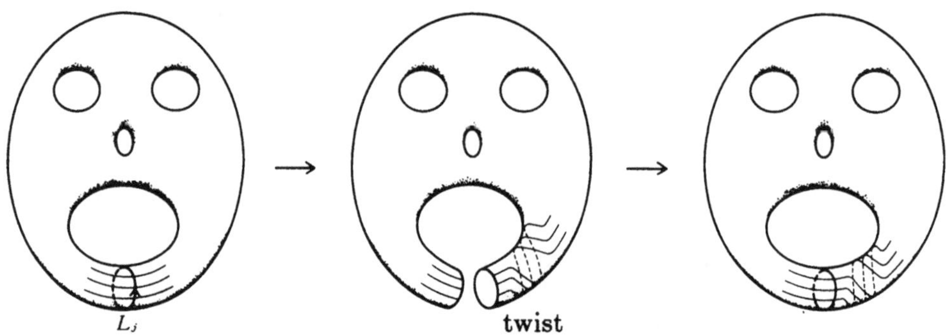

Fig. 3.9.

Now, we have proved that $\tilde{\Psi} : F_g \longrightarrow (\mathbf{R}^+)^{3g-3} \times \mathbf{R}^{3g-3}$ is bijective. By Lemma 3.9, $\tilde{\Psi}$ is also continuous. On the other hand, Teichmüller's theorem (Theorem 5.15) states that F_g is homeomorphic to $\mathbf{R}^{6g-6}$. Hence the following theorem, Brouwer's theorem on invariance of domains, implies that $\tilde{\Psi}$ is actually a homeomorphism. □

Theorem 3.11. (Brouwer's theorem on invariance of domains) *Let n be an integer not less than two. Let $\varphi : \mathbf{R}^n \longrightarrow \mathbf{R}^n$ be a continuous injection of $\mathbf{R}^n$ into $\mathbf{R}^n$. Then $D = \varphi(\mathbf{R}^n)$ is a domain, and φ is a homeomorphism of $\mathbf{R}^n$ onto D.*

Proof (A sketch). (Cf. Bers [A-12], and Newman [A-85].) Fix a point y in D arbitrarily, and set $x = \varphi^{-1}(y)$. Fix arbitrarily a closed ball B with center x. Since B is compact, the restriction of φ on B is a homeomorphism of B onto $B' = \varphi(B)$. In particular, B' and $S' = \varphi(\partial B)$ are a topological closed ball and a topological sphere, respectively, in $\mathbf{R}^n$.

Hence, by the n-dimensional Jordan's theorem, $\mathbf{R}^n - S'$ has two connected components. Denote by D_1 and D_2, respectively, the interior domain and the exterior domain of S'.

On the other hand, it can be shown that $\mathbf{R}^n - B'$ is connected. Hence, the image $\varphi(\operatorname{Int} B)$ of the interior $\operatorname{Int} B$ of B by φ should be coincident with D_1. In particular, y is an interior point of D, and φ^{-1} is continuous on D_1. Since y is arbitrary, we have the assertion. □

Fix a point $(a_1, \cdots, a_{3g-3}, \alpha_1, \cdots, \alpha_{3g-3})$ of $(\mathbf{R}^+)^{3g-3} \times \mathbf{R}^{3g-3}$, and take one θ_j of the twisting parameters. By Theorem 3.10, the preimage

$$\tilde{\Psi}^{-1}(\{(a_1, \cdots, a_{3g-3}, \alpha_1, \cdots, \alpha_{j-1}, \xi, \alpha_{j+1}, \cdots, \alpha_{3g-3}) \mid \xi \in \mathbf{R}\})$$

corresponds to a continuous curve in T_g. The variation of a Riemann surface represented by such a curve is called a *Fenchel-Nielsen deformation.* We shall investigate the deformation of this type in Chapter 8 by using quasiconformal mappings. We shall also give a direct proof of continuity of $\tilde{\Psi}^{-1}$ with respect to ξ.

3.3. Fricke-Klein Embedding

In this section, we consider the problem of finding a set of simple closed geodesics on a closed Riemann surface of genus g (≥ 2) whose hyperbolic lengths determine the surface. It would be most desirable if there was a set of $6g - 6$ simple closed geodesics whose length functions give global coordinates on T_g. However, this is impossible (cf. Seppälä and Sorvali [194]).

Here, we construct a set of $9g - 9$ simple closed geodesics whose length functions separate points of T_g, by replacing each twisting parameter in fixed Fenchel-Nielsen coordinates with two geodesic length functions. Later, in Proposition 6.17, we shall give another kind of parametrization.

As in the previous section, fix a Riemann surface R of genus g (≥ 2) and a system $\mathcal{L}$ of decomposing curves on R. Let $\mathcal{P}$ be the pants decomposition of R corresponding to $\mathcal{L}$. For every $L_j \in \mathcal{L}$, denote by $P_{j,1}$ and $P_{j,2}$ the elements of $\mathcal{P}$ having L_j as a boundary component. (Again, we allow the case that $P_{j,1} = P_{j,2}$.)

For every j, set $W_j = P_{j,1} \cup L_j \cup P_{j,2}$. Fix a simple closed geodesic in W_j which intersects L_j, and denote it by Δ_j^0. Further, let Δ_j^1 be the (unique) simple closed geodesic which is freely homotopic to the simple curve obtained from Δ_j^0 by applying the Dehn twist with respect to L_j (see Fig. 3.10).

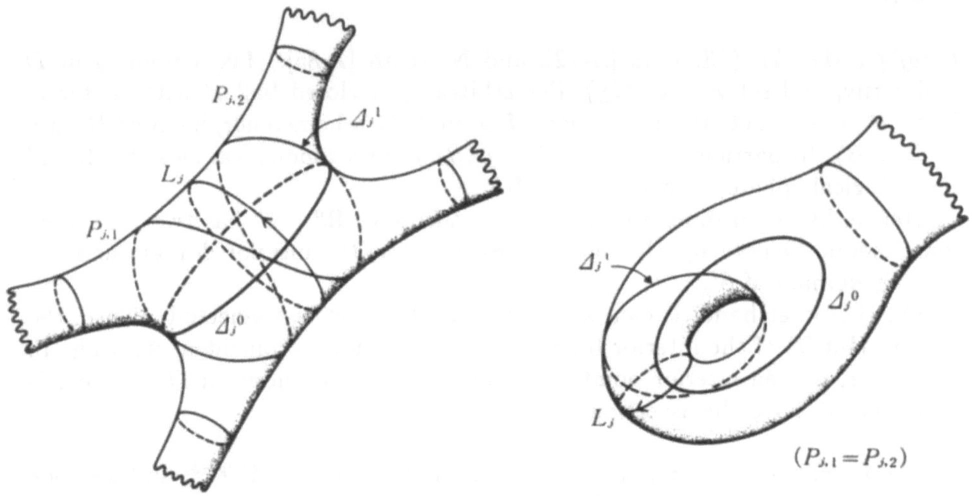

Fig. 3.10.

For every $t \in F_g$, let $[R_t, \Sigma_t]$ be the corresponding point of T_g. For every closed geodesic L on R, we express as $L(t)$ the corresponding closed geodesic on R_t, and denote by $\ell(L(t))$ the hyperbolic length of $L(t)$. Set

$$\ell_j(t) = \ell(L_j(t)), \quad \ell_{3g-3+j}(t) = \ell(\Delta_j^0(t)), \quad \ell_{6g-6+j}(t) = \ell(\Delta_j^1(t))$$

for every j, and set

$$\tilde{\ell}(t) = (\ell_1(t), \cdots, \ell_{9g-9}(t)).$$

We have the following:

Theorem 3.12. *The mapping $\tilde{\ell}$ is a proper embedding of F_g into $(\mathbf{R}^+)^{9g-9}$. (That is, $\tilde{\ell}$ is a homeomorphism onto the image $\tilde{\ell}(F_g)$, and the preimage of any compact set in $(\mathbf{R}^+)^{9g-9}$ under $\tilde{\ell}$ is compact.)*

To prove this theorem, first we fix a point t_0 of F_g arbitrarily, and write

$$\tilde{\Psi}(t_0) = (a_1, \cdots, a_{3g-3}, \alpha_1, \cdots, \alpha_{3g-3}) \in (\mathbf{R}^+)^{3g-3} \times \mathbf{R}^{3g-3}.$$

Fix j, and for every $s \in \mathbf{R}$, define a point $t(s)$ of F_g by

$$t(s) = \tilde{\Psi}^{-1}(a_1, \cdots, a_{3g-3}, \alpha_1, \cdots, \alpha_{j-1}, \alpha_j + s, \alpha_{j+1}, \cdots, \alpha_{3g-3}).$$

Then we have the following Proposition.

Proposition 3.13. *The positive function $f(s) = \ell(\Delta_j^0(t(s)))$ is strictly convex and proper on* $\mathbf{R}$ *(as a mapping to* $\mathbf{R}$*). In particular, there is a unique value of s at which f attains its minimum.*

This proposition shall be derived by using convexity of the Poincaré distance ρ, which in turn is stated in the following form.

Lemma 3.14. *Let two mutually disjoint geodesics L_1 and L_2 in H be given. Then $\rho(z,w)$ is strictly convex on $\{(z,w) \in \mathbf{C}^2 \mid z \in L_1, w \in L_2\}$. Equivalently, for every $z, z' \in L_1$ and every $w, w' \in L_2$, the inequality*

$$\rho(z_0, w_0) \leq \frac{\rho(z', w') + \rho(z, w)}{2}$$

holds, where z_0 and w_0 are the middle-points of z and z' and of w and w' (with respect to the hyperbolic distance), respectively.

Moreover, the equality holds if and only if $z = z'$ and $w = w'$.

Proof. It suffices to show the inequality when $z \neq z'$. Let L be the geodesic on H passing through z_0 and w_0. Let $\hat{z}'$, $\hat{z}$, and $\hat{z}_0$ be points symmetric to z, z', and z_0, respectively, with respect to w_0 (in the sense of the hyperbolic geometry). (See Fig. 3.11, where we replace H by the unit disk and assume that $w_0 = 0$.)

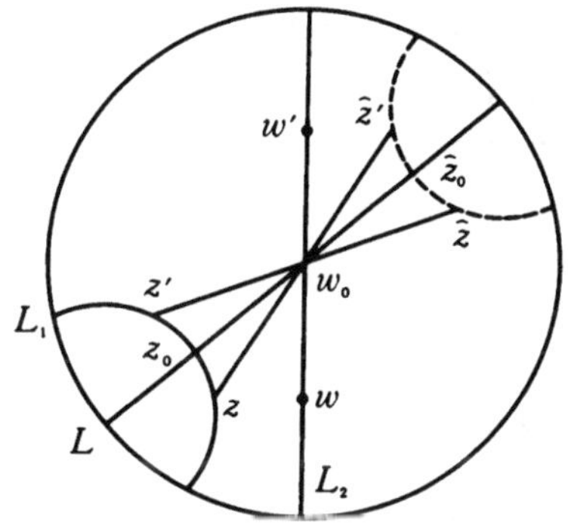

Fig. 3.11.

By the definitions, $\rho(z,w) = \rho(\hat{z}', w')$, and it is clear that

$$\rho(z', \hat{z}') \leq \rho(z', w') + \rho(w', \hat{z}').$$

On the other hand, let γ_1 and γ_2 be the elliptic elements of $Aut(H)$ of order two which have z_0 and w_0, respectively, as fixed points. Then the composition $\gamma_2 \circ \gamma_1$ is a hyperbolic element with the axis L, and sends z_0 and z' to $\hat{z}_0$ and $\hat{z}'$, respectively. Hence we have

$$2\rho(z_0, w_0) = \rho(z_0, \hat{z}_0) < \rho(z', \hat{z}').$$

Thus we obtain

$$2\rho(z_0, w_0) < \rho(z', w') + \rho(z, w).$$

□

Proof of Proposition 3.13. We consider the case where $P_{j,1} \neq P_{j,2}$ and Δ_j^0 intersects L_j at two points. The other cases can be treated similarly.

Let Γ_0 be the Fuchsian group represented by t_0, and γ_j be an element of Γ_0 which covers $L_j(t_0)$. On the axis A_j of γ_j, fix a point z_0 which is projected to an intersection point, say p, of $L_j(t_0)$ and $\Delta_j^0(t_0)$. Let δ_j^0 be the element of Γ_0 which covers $\Delta_j^0(t_0)$ and whose axis passes through z_0, and B_j be the axis of δ_j^0. By the assumption, the projection of the geodesic I contained in B_j and connecting z_0 to $z_0' = \delta_j^0(z_0)$ should intersect $L_j(t_0)$ at some point q other than p. Let z_1 be the lift of q on I, and A_j' be the lift of $L_j(t_0)$ passing through z_1 (see Fig. 3.12).

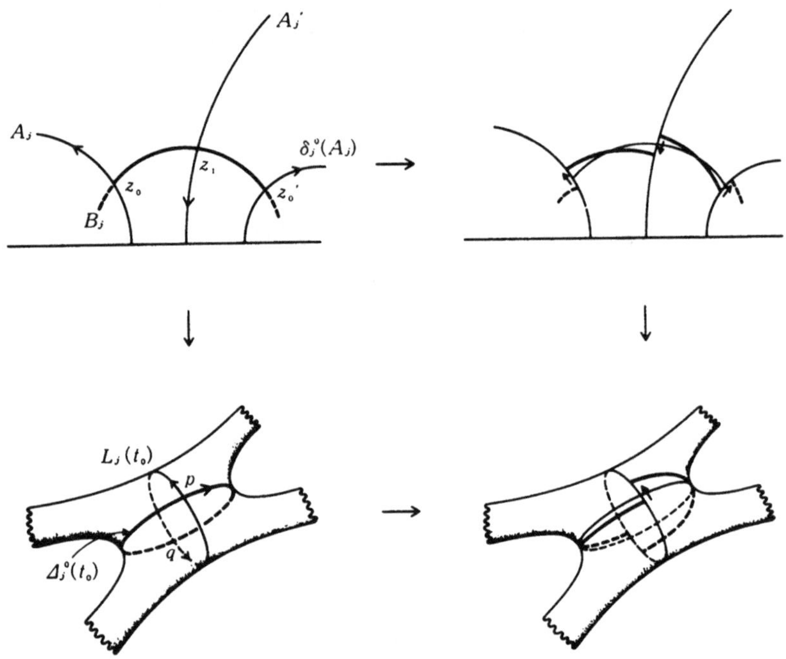

Fig. 3.12.

For any point z on an oriented geodesic L on H and any $\alpha \in \mathbf{R}$, let $z(\alpha)_L$ be the point on L obtained by translating z in the positive direction along L by hyperbolic length α. Since $t(s)$ represents the marked surface obtained from the surface represented by t_0 by twisting along $L_j(t_0)$ by hyperbolic length

$$\frac{1}{2\pi}\ell(L_j(t_0))s = \hat{s},$$

we can see that

$$f(s) = \inf_{z\in A_j, w\in A'_j}\left\{\rho(z,w) + \rho\left(w(\hat{s})_{A'_j}, \delta_j^0(z)(\hat{s})_{\delta_j^0(A_j)}\right)\right\}.$$

Hence, we define a function $F(z,w,\hat{s})$ on $A_j \times A'_j \times \mathbf{R}$ by setting

$$F(z,w,\hat{s}) = \rho(z, w(-\hat{s})_{A'_j}) + \rho\left(w, \delta_j^0(z)(\hat{s})_{\delta_j^0(A_j)}\right).$$

By Lemma 3.14, we can see that F is strictly convex. Also, F is proper, as can be easily seen from the fact that $\rho(z,w) \longrightarrow +\infty$ as either z or w tends to the boundary of H.

Thus for any fixed $\hat{s}$, $F(z,w,\hat{s})$ is strictly convex and proper on $A_j \times A'_j$. In particular, it attains the minimum $m(\hat{s})$ at exactly one point. Furthermore, it is easy to show that $m(\hat{s})$ is strictly convex and proper on $\mathbf{R}$. Since $f(s) = m(\hat{s})$ as mensioned before, we conclude the assertion. □

Proof of Theorem 3.12. Since $\Delta_j^1(t)$ is freely homotopic to the curve obtained from $\Delta_j^0(t)$ by applying the Dehn twist along $L_j(t)$ for every j, we have

$$\ell(\Delta_j^1(t(s))) = \ell(\Delta_j^0(t(s+2\pi))) = f(s+2\pi).$$

Since $f(s)$ is strictly convex and proper on $\mathbf{R}$ by Proposition 3.13, the mapping

$$s \mapsto (f(s), f(s+2\pi))$$

of $\mathbf{R}$ into $\mathbf{R}^2$ is proper, injective, and continuous. Hence it is a proper embedding of $\mathbf{R}$ into $(\mathbf{R}^+)^2$.

Since j is arbitrary, the assertion follows from Theorem 3.10. □

3.4. Thurston's Compactification

In this section, we shall give an outline of the construction of a useful (ideal) boundary for the Teichmüller space T_g of genus g (≥ 2), which is due to W. Thurston. We omit almost all proofs, which can be found, for instance, in Fathi, Laudenbach and Poénaru [A-29].

First, an embedding of F_g such one as in Theorem 3.12 is modified as follows. Denote by $\mathcal{S}$ the set consisting of all free homotopy classes of simple closed curves on a closed surface R of genus g (≥ 2). For every $t \in F_g$, let R_t be the Riemann surface represented by t. For every $C \in \mathcal{S}$, let $\ell_*(t)(C)$ be the hyperbolic length $\ell(C_t)$ of the unique geodesic C_t in the free homotopy class on R_t corresponding to C. Then we have a mapping

$$\ell_*(t) : \mathcal{S} \longrightarrow \mathbf{R}^+$$

for every $t \in F_g$, or equivalently, we have a mapping

$$\ell_* : F_g \longrightarrow (\mathbf{R}^+)^{\mathcal{S}}$$

of F_g into $(\mathbf{R}^+)^{\mathcal{S}}$. Hence Theorem 3.12 implies the following:

Corollary. *The mapping $\ell_* : F_g \longrightarrow (\mathbf{R}^+)^{\mathcal{S}}$ is a proper embedding.*

Moreover, this mapping remains an embedding, even when we take the quotient space $P(\mathbf{R}^+)^{\mathcal{S}} = (\mathbf{R}^+)^{\mathcal{S}}/\mathbf{R}^+$ as the target space. In fact, letting $P\ell_*$ be the composed mapping of ℓ_* with the projection π of $(\mathbf{R}^+)^{\mathcal{S}}$ into the projective space $P(\mathbf{R}^+)^{\mathcal{S}}$, we have the following theorem.

Theorem 3.15. (cf. [A-29], Exposé 7) *The mapping $P\ell_* : F_g \longrightarrow P(\mathbf{R}^+)^{\mathcal{S}}$ is an embedding.*

Next, there is another natural mapping of $\mathcal{S}$ into $(\overline{\mathbf{R}^+})^{\mathcal{S}}$, which is defined by using the geometric intersection numbers of curves. Here and in the remainder of this section, we set $\overline{\mathbf{R}^+} = \{x \in \mathbf{R} \mid x \geq 0\}$.

For any two α_1 and $\alpha_2 \in \mathcal{S}$, the *geometric intersection number* $i(\alpha_1, \alpha_2)$ of α_1 and α_2 is, by definition, the infimum of the number of intersection points of L_1 and L_2, where L_j moves in the free homotopy class of α_j for each j. In particular, $i(\alpha_1, \alpha_2) = i(\alpha_2, \alpha_1)$. Also, note that $i(\alpha, \alpha) = 0$ for every $\alpha \in \mathcal{S}$.

Define a mapping $i_* : \mathcal{S} \longrightarrow (\overline{\mathbf{R}^+})^{\mathcal{S}} - \{0\}$ by setting

$$i_*(\alpha)(\cdot) = i(\alpha, \cdot), \quad \alpha \in \mathcal{S},$$

where we denote $\{0\}^{\mathcal{S}}$ simply by 0. Then we can show (cf. [A-29], Exposé 3) that

$$Pi_* = \pi \circ i_* : \mathcal{S} \longrightarrow P(\overline{\mathbf{R}^+})^{\mathcal{S}}$$

is an injection.

Here, note that i_* is extended to a mapping of $\mathbf{R}^+ \times \mathcal{S}$ into $(\overline{\mathbf{R}^+})^{\mathcal{S}} - \{0\}$ by setting

$$i_*(a, \alpha)(\cdot) = a \cdot i(\alpha, \cdot)$$

for every $(a, \alpha) \in \mathbf{R}^+ \times \mathcal{S}$. It is clear that

$$\pi\left(\overline{i_*(\mathbf{R}^+ \times \mathcal{S})} - \{0\}\right) = \overline{Pi_*(\mathcal{S})}.$$

Now, we know the following:

Theorem 3.16. (cf. [A-29], Exposé 4) *The subset $\overline{Pi_*(\mathcal{S})}$ of $P(\overline{\mathbf{R}^+})^{\mathcal{S}}$ is homeomorphic to $S^{6g-7} = \{x \in \mathbf{R}^{6g-6} \mid |x| = 1\}$.*

Remark. The set $\overline{i_*(\mathbf{R}^+ \times \mathcal{S})} - \{0\}$ can be identified with (and is written hereafter as) the set $\mathcal{MF}$ of all Whitehead equivalence classes (or more precisely, equivalence classes under isotopy and Whitehead's operations) of measured foliations on R. (See [A-29], Exposé 6.) Hence we can also express $\overline{Pi_*(\mathcal{S})}$ as $P\mathcal{MF} = \pi(\mathcal{MF})$.

Here, for the sake of convenience, we recall the definition of a measured foliation on a compact surface R (cf. [A-29], Exposé 5). Let F be a foliation on R with isolated singularities, each of which should have the same local differentiable structure as that of a foliation on $\mathbf{C} - \{0\}$, along every leaf of which $z^k dz^2$ is positive with some positive integer k, near the singularity $z = 0$. Next, a transversal measure μ for F is a measure on the set of all transversal subarcs of leafs of F which satisfies that, if a transversal arc $\alpha : [0,1] \longrightarrow R$ can be moved to another one β by an isotopy, each orbit of which is contained in a single leaf of F, then

$$\mu(\alpha) = \mu(\beta).$$

A *measured foliation* (F, μ) is, by definition, a pair of a foliation F on R with such isolated singularities, and a transversal measure μ for F.

Remark 1. When $i_*(a, \alpha)$ ($a \in \mathbf{R}^+, \alpha \in \mathcal{S}$) is identified with a measured foliation (F, μ), we have

$$\mu[\beta] = i_*(a, \alpha)(\beta), \quad \beta \in \mathcal{S},$$

where $\mu[\beta] = \inf_{L \in \beta} \mu(L)$. From here on, $\mu[\beta]$ is written as $i_*(p)(\beta)$ ($\beta \in \mathcal{S}$) even for a general $p = (F, \mu) \in \mathcal{MF}$.

Remark 2. A typical example of a measured foliation on R, when R is equipped with a complex structure, is a foliation, along every leaf of which a prescribed non-zero holomorphic quadratic differential φ on R is positive (cf. §2 of Chapter 4), equipped with the canonical transversal measure

$$\mu = |\mathrm{Im}\sqrt{\varphi}|.$$

On the other hand, Hubbard and Masur [101] showed that, for any given complex structure on R, $\mathcal{MF} \cup \{0\}$ can be identified with the space of holomorphic quadratic differentials on R (with respect to the given complex structure), which in turn is known to be homeomorphic to $\mathbf{R}^{6g-6}$.

Proposition 3.17. *The image $\ell_*(F_g)$ is disjoint from $\mathcal{MF}$ in $(\overline{\mathbf{R}^+})^{\mathcal{S}}$.*

Proof. For every $t \in F_g$, the values $\ell_*(t)(\alpha)$ ($\alpha \in \mathcal{S}$) are bounded from below by a positive constant (depending on t).

On the other hand, by using Poincaré's recurrence theorem, we can show that the set $\{\mu[\alpha] \mid \alpha \in \mathcal{S}\}$ contains a sequence converging to 0 for every $p = (F, \mu) \in \mathcal{MF}$. In fact, for every $\epsilon > 0$, take a transversal arc γ such that $\mu(\gamma) < \epsilon$. By Poincaré's recurrence theorem, there is a subarc of a leaf of F which starts with and ends at γ. Hence we have a simple closed curve L such that $\mu(L) < \epsilon$, which implies the assertion. □

Now, the crucial fact to construct Thurston's compactification is the following theorem.

Theorem 3.18. (cf. [A-29], Exposé 8) *For every system $\mathcal{L} = \{L_j\}_{j=1}^{3g-3}$ of decomposing curves of R, there is a natural homeomorphism*

$$q_{\mathcal{L}} : \ell_*(F_g) \longrightarrow U(\mathcal{L}) \subset \mathcal{MF},$$

where

$$U(\mathcal{L}) = \{p = (F, \mu) \in \mathcal{MF}(\subset (\overline{\mathbf{R}^+})^{\mathcal{S}}) \mid \mu(L_j) > 0 \quad \text{for every} \quad L_j \in \mathcal{L}\}.$$

The construction of $q_{\mathcal{L}}$ is as follows. Here and in the remainder of this discussion, we use the same notation both for a curve and for the free homotopy class of it. For every $t \in F_g$, we can construct a measured foliation $p_t = (F_t, \mu_t)$ such that F_t is transversal to L_j for every j, and that

$$\ell_*(t)(L_j) = i_*(p_t)(L_j), \quad j = 1, \cdots, 3g-3,$$

or equivalently,

$$\ell(L_j(t)) = \mu_t[L_j], \quad j = 1, \cdots, 3g-3,$$

where $L_j(t)$ is the geodesic, on the marked Riemann surface represented by t, which corresponds to L_j as before. We set

$$q_{\mathcal{L}}(\ell_*(t)) = p_t.$$

It is known that this mapping $q_{\mathcal{L}}$ is actually a homeomorphism onto $U(\mathcal{L})$. Using these natural "projections" $q_{\mathcal{L}}$, we can derive the following:

Theorem 3.19. (cf. [A-29], Exposé 8) *The subset*

$$P\ell_*(F_g) \cup P\mathcal{MF}$$

of $P(\overline{\mathbf{R}^+})^{\mathcal{S}}$ with the relative topology is a compact manifold with boundary.

Moreover, $P\ell_*(F_g) \cup P\mathcal{MF}$ is homeomorphic to the real $(6g-6)$-dimensional closed ball $\{x \in \mathbf{R}^{6g-6} \mid |x| \leq 1\}$, and the boundary is coincident with $P\mathcal{MF}$ (which is homeomorphic to S^{6g-7}).

We shall show how to construct local coordinates in a neighborhood of an arbitrary point of the boundary $P\mathcal{MF}$.

Fix $\overline{p_0} \in P\mathcal{MF}$ and $p_0 \in \pi^{-1}(\overline{p_0})$ arbitrarily. Then there is a system $\mathcal{L} = \{L_j\}_{j=1}^{3g-3}$ of decomposing curves of R such that

$$i_*(p_0)(L_j) > 0, \quad j = 1, \cdots, 3g-3.$$

(See [A-29], Exposé 6.) From this $\mathcal{L}$, we construct a family $\{L_j, \Delta_j^0, \Delta_j^1\}_{j=1}^{3g-3}$ of simple closed curves which gives an embedding of F_g into $(\mathbf{R}^+)^{9g-9}$ as in §3.3.

For every $\epsilon > 0$, set

$$U(\mathcal{L},\epsilon) = \{x \in \ell_*(F_g) \mid x(L_j) > \epsilon, \quad j = 1, \cdots, 3g-3\},$$

$PU(\mathcal{L},\epsilon) = \pi(U(\mathcal{L},\epsilon))$ and $W = \pi(U(\mathcal{L}))$.

Clearly, $PU(\mathcal{L},\epsilon)$ is open in $P\ell_*(F_g)$, and $\pi \circ q_{\mathcal{L}}(U(\mathcal{L},\epsilon)) = W$. We can also show that $W \cup PU(\mathcal{L},\epsilon)$ is open in the closure of $P\ell_*(F_g)$. Recalling that π is injective on $\ell_*(F_g)$, we define a mapping φ of $W \cup PU(\mathcal{L},\epsilon)$ into $W \times [0,1)$ as follows: We set

$$\varphi(\bar{p}) = (\bar{p}, 0), \quad \bar{p} \in W.$$

For every $x \in U(\mathcal{L},\epsilon)$, we put

$$\varphi(\pi(x)) = \left(\pi \circ q_{\mathcal{L}}(x), \exp\left(-\sum_{j=1}^{3g-3} \{i_*(q_{\mathcal{L}}(x))(L_j) + i_*(q_{\mathcal{L}}(x))(\Delta_j^0) + i_*(q_{\mathcal{L}}(x))(\Delta_j^1)\}\right)\right).$$

Then, by using a fundamental inequality, which states that, for every $\alpha \in \mathcal{S}$, there is a constant A such that

$$i_*(q_{\mathcal{L}}(x))(\alpha) \leq x(\alpha) \leq i_*(q_{\mathcal{L}}(x))(\alpha) + A, \quad x \in U(\mathcal{L},\epsilon),$$

we can show that this mapping φ is a homeomorphism onto its image, and hence it gives local coordinates in a neighborhood of $\overline{p_0}$.

This compactification of the Teichmüller space T_g is called *Thurston's compactification* of T_g, and its boundary $P\mathcal{MF}$ is called *Thurston's boundary* of T_g. The importance of this compactification may be shown by the following:

Corollary. *Every orientation-preserving homeomorphism of R onto itself induces a homeomorphism of $\overline{P\ell_*(F_g)} = P\ell_*(F_g) \cup P\mathcal{MF}$ onto itself.*

Proof. Every orientation-preserving homeomorphism φ of R onto itself induces a natural action of $\mathcal{S}$ onto itself, and hence induces a continuous self-mapping Φ of $P(\overline{\mathbf{R}^+})^{\mathcal{S}}$. Here, $P\ell_*(F_g)$ and $P\mathcal{MF}$ are clearly invariant under this self-mapping. The inverse of Φ is obtained from φ^{-1}. □

Notes

We call a Riemann surface, obtained from a closed Riemann surface of genus g by deleting mutually disjoint n points and m closed disks, a *Riemann surface of type* (g,n,m). In particular, a Riemann surface of type $(g,n,0)$ is said to be of *analytically finite type* (g,n).

If such a surface R admits the hyperbolic metric, we can define, similarly to definitions in §2 and §3, the Fenchel-Nielsen coordinates and Fricke-Klein's embeddings of the Teichmüller space $T(R)$. (See, for instance, Abikoff [A-1].) In particular, we know that the Teichmüller space of a Riemann surface of type

(g, n, m) is homeomorphic to $\mathbf{R}^{6g-6+2n+3m}$. Also, Thurston's compactification is considered for such a surface (cf. Fathi, Laudenbach and Poénaru [A-29]).

Relating to §3, we recall the inverse problem, of whether we can determine a closed Riemann surface by the length spectra, i.e., the set of the hyperbolic lengths, of all simple closed geodesics on the surface. This problem is equivalent to the *problem of M. Gel'fand*: whether a closed surface is determined by the eigenvalue spectra, i.e., the set of all eigenvalues, of the Laplace-Beltrami operator on the surface with respect to the hyperbolic metric. On this problem, see Venkov [A-110], McKean [152], Sunada [218], Vignéras [242], and Wolpert [246].

For Fricke-Klein's embeddings, we further refer to Keen [110] and Okumura [173]. The argument in §3 follows that in Fathi, Laudenbach and Poénaru [A-29], Exposé 7, which is due to A. Douady. As for more advanced investigations on convexity of geodesic length functions, see Kerckhoff [112] and Wolpert [256].

A survey on Thurston's compactification by Thurston himself is given in Thurston [234]. Fathi, Laudenbach and Poénaru [A-29] is a good introduction to measured foliations and Thurston's boundary. See also Gardiner [A-34] Chapter 11, Strebel [A-102], Hubbard and Masur [101], Marden and Strebel [137], and Masur [144].

As a generalization of Fenchel-Nielsen deformations, the earthquake deformations have been considered. See Kerckhoff [112] and [114], and Thurston [233]. We also cite Bonahon [45].

Finally, there are many proofs of Teichmüller's theorem stated in this chapter. For example, see Chapter 5 and Notes of that chapter. There are also proofs by Wolpert [256] using geodesic length functions, and by Fischer and Tromba [74] from a differential geometric viewpoint.

Chapter 4

Quasiconformal Mappings

In this chapter, we shall explain basic properties of quasiconformal mappings which are needed in the later chapters. In Chapter 1, we have defined differentiable quasiconformal mappings. Here, we weaken the differentiability condition in the definition. Such a technical improvement allows us to use quasiconformal mappings as a tool in considerably more general situations, while formal treatment of such mappings remains the same as in the differentiable case.

Note that, by the uniformization theorem, we can reduce the case of mappings between Riemann surfaces to the case of those between planar domains. Accordingly, in this chapter, we restrict ourselves to the case of quasiconformal mappings between planar domains.

In Section 1, we give several definitions of general quasiconformal mappings, show equivalence of those definitions, and state basic facts following easily from the definitions. In Section 2, we prove the existence theorem of the solution of the Beltrami differential equation. In Section 3, we give a fundamental variational formula for quasiconformal mappings, due to L. Ahlfors and L. Bers.

We use several fundamental theorems in the theory of Lebesgue integral without proofs, which will be found in standard text books.

4.1. Definitions and Elementary Properties

4.1.1. Analytic Definition *A* of Quasiconformal Mappings

We want to consider an orientation-preserving homeomorphism f which is not necessarily differentiable, but satisfies the Beltrami equation $f_{\bar{z}} = \mu f_z$ in some suitable sense. Because of this requirement, f should be, at least, partially differentiable a.e. (almost everywhere) in the considered domain. This is guaranteed, for example, by assuming that f is ACL: here, a function $f(z) = f(x, y)$ on a planar domain D is said to be *ACL* (***absolutely continuous on lines***) if the following condition holds for every rectangle $R = [a, b] \times [c, d]$ in D whose sides are parallel either to the real axis or to the imaginary axis:

As a function of x, $f(x, y)$ is absolutely continuous on $[a, b]$ for almost every $y \in [c, d]$, and as a function of y, $f(x, y)$ is absolutely continuous on $[c, d]$ for almost every $x \in [a, b]$.

Remark 1. An absolutely continuous function $F(t)$ on an interval I is differentiable at almost every $t \in I$. Hence, when a function $f(z)$ on a domain D is ACL, the partial derivatives f_z and $f_{\bar{z}}$ of f are well-defined and finite at almost every $z \in D$. It is not difficult to show that they are measurable.

Now, as a natural generalization of the notion of conformal mappings, we make the following definition:

Analytic definition *A* of quasiconformal mappings. Let $f(z)$ be an orientation-preserving homeomorphism of a planar domain D into the complex plane. We say that $f(z)$ is *quasiconformal (qc)* on D if f satisfies the following conditions:

(i) f is ACL on D.
(ii) There exists a constant k with $0 \le k < 1$ such that

$$|f_{\bar{z}}| \le k|f_z| \quad \text{a.e. on} \quad D.$$

Setting $K = (1+k)/(1-k)$, we say that f is K-*qc* on D. We call the infimum of $K(\ge 1)$ such that f is K-qc the *maximal dilatation* of f, and denote it by K_f or $K(f)$.

Example 1. A conformal mapping of a domain D is quasiconformal on D. Since we can take 0 as k in the above condition (ii), a conformal mapping f is 1-qc and $K_f = 1$.

Example 2. An affine mapping $f(z) = az + b\bar{z} + c$ $(a, b, c \in \mathbf{C}, |b| < |a|)$ is quasiconformal. We can take $|b|/|a|$ as k, and hence $K_f = (|a| + |b|)/(|a| - |b|)$.

Example 3. For a given k, we set

$$f(z) = \begin{cases} z, & z \in H \\ x + iKy, & z = x + iy \in \mathbf{C} - H,\ K > 1. \end{cases}$$

Then f is a quasiconformal mapping of $\mathbf{C}$, and $K_f = K$.

Remark 2. Set

$$f(z) = \frac{z}{1 - |z|^2}, \quad z \in \Delta.$$

Then f is an orientation-preserving diffeomorphism of the unit disk Δ onto $\mathbf{C}$, but not quasiconformal. Actually, there are no quasiconformal mappings of Δ onto $\mathbf{C}$. See Proposition 4.32 later in this chapter.

Remark 3. Let f be a K-qc mapping of a domain D onto another domain D', and g be a conformal mapping of D'. Then the composite mapping $g \circ f$ is K-qc. In fact, it is easy to see from the definition that $g \circ f$ satisfies (i) and (ii), and that $K_{g \circ f} = K_f$.

In general, existence of the partial derivatives f_z and $f_{\bar{z}}$ is not enough to guarantee good properties of f applicable to further investigation. However, in the case of homeomorphisms, the following remarkable result due to Gehring and Lehto has been obtained.

Proposition 4.1. *If a homeomorphism f of a domain D into $\mathbf{C}$ has the partial derivatives f_x and f_y a.e. on D, then f is totally differentiable a.e. on D.*

Proof. It suffices to prove the assertion when D has a finite area. Fix a positive constant ϵ arbitrarily. For every n, set

$$g_n(z) = \sup_{0<|h|<1/n, h\in\mathbf{R}} \left| \frac{f(z+h)-f(z)}{h} - f_x(z) \right| .$$

Note that $g_n \to 0$ a.e. on D as $n \to \infty$. Applying Egoroff's theorem to the sequence $\{g_n\}_{n=1}^{\infty}$, we can find a measurable subset E of D such that the area of $D-E$ is less than ϵ and that $(f(z+h)-f(z))/h$ converges to $f_x(z)$ uniformly on E as $h \to 0$. By a similar argument, we may also assume that $(f(z+ih)-f(z))/h$ converges to $f_y(z)$ uniformly on E as $h \to 0$. In the rest of this proof, we fix such an E. Note that the uniform convergence implies that f_x and f_y are continuous on E, for f is continuous.

To prove the assertion, it suffices to show that f is totally differentiable a.e. on E, for ϵ is arbitrary.

First, Lebesgue's density theorem and Fubini's theorem imply that, for almost every $x_0 + iy_0 \in E$, x_0 and y_0 are density points of the subsets

$$E_{y_0} = \{x \in \mathbf{R} \mid x + iy_0 \in E\} \quad \text{and} \quad E_{x_0} = \{y \in \mathbf{R} \mid x_0 + iy \in E\},$$

respectively. (Here we say that x_0 is a density point of E_{y_0} if

$$\lim_{h>0, h\to 0} \frac{1}{2h} \int_{x_0-h}^{x_0+h} 1_{E_{y_0}}(x)dx = 1,$$

where $1_{E_{y_0}}$ is the characteristic function of E_{y_0}. Similarly, we define density points of E_{x_0}.) Fix such a point $z_0 = x_0 + iy_0$, and we shall show that f is totally differentiable at z_0. For the sake of simplicity, we assume that $z_0 = 0$.

Fix η with $0 < \eta < 1$ arbitrarily. Set

$$I(z) = |f(z) - f(0) - xf_x(0) - yf_y(0)|.$$

Since

$$\begin{aligned} I(z) \le\ & |f(x+iy) - f(x) - yf_y(x)| \\ & + |f(x) - f(0) - xf_x(0)| + |y\,(f_y(x) - f_y(0))\,|, \end{aligned}$$

there is a positive δ such that $I(z) \le \eta|z|$ for every z with $|z| < \delta$ and $x \in E$. By a similar argument, taking a smaller δ if necessary, we may assume also that $I(z) \le \eta|z|$ for every z with $|z| < \delta$ and $iy \in E$.

Further, taking a smaller δ if necessary, we may assume that, for every $z = x + iy$ with $x, y > 0$ and $|z| < \delta/2$, there exist points x_1, x_2, iy_1, and iy_2 of E such that

$$(1-\eta)x < x_1 < x < x_2 < (1+\eta)x,$$

$$(1-\eta)y < y_1 < y < y_2 < (1+\eta)y.$$

(Note that $x_0 = 0$ and $y_0 = 0$ are density points of E_{y_0} and E_{x_0}, respectively.) Then, $I(z^*) \leq \eta|z^*|$ for every z^* on the boundary of the rectangle $R = [x_1, x_2] \times [y_1, y_2]$. On the other hand, the maximal principle holds for f, since f is a homeomorphism, and hence is an open mapping. Therefore, for some suitable z^* on the boundary of R, we have

$$\begin{aligned} I(z) &\leq |f(z^*) - f(0) - xf_x(0) - yf_y(0)| \\ &\leq I(z^*) + |z - z^*|\left(|f_x(0)| + |f_y(0)|\right). \end{aligned}$$

Since $|z - z^*| \leq \eta|z|$, we conclude that, for every z with $x, y > 0$ and $|z| < \delta/2$,

$$I(z) \leq 2\eta|z|\left(1 + |f_x(0)| + |f_y(0)|\right).$$

A similar argument shows that the last inequality still holds for every z with $|z| < \delta/2$, which implies the total differentiability of f at $z_0 = 0$. □

Proposition 4.2. *If f is a quasiconformal mapping of a domain D, then the partial derivatives f_z and $f_{\bar{z}}$ are locally square integrable on D.*

Proof. First, we have a completely additive set function A by attaching the area $A(E)$ of $f(E)$ to every Borel subset E of D. Let $J_f(z)$ be the density function of A with respect to the two-dimensional Lebesgue measure $dxdy$. By Lebesgue's decomposition theorem, we have

$$\int_E J_f(z)dxdy \leq A(E) \tag{4.1}$$

for every measurable subset E of D.

On the other hand, by Proposition 4.1 we see that f is totally differentiable at almost every $z \in D$, and at such a point z we can show that

$$J_f(z) = |f_z(z)|^2 - |f_{\bar{z}}(z)|^2.$$

Since the condition (ii) on f implies that

$$|f_{\bar{z}}|^2 \leq |f_z|^2 \leq \frac{1}{1-k^2}J_f \quad \text{a.e. on} \quad D,$$

the assertion follows by (4.1). □

Remark 4. Actually, the set function A in the above proof is absolutely continuous, and hence the equality holds in (4.1). See Lemma 4.12 in §1.3.

Proposition 4.3. *For every quasiconformal mapping f of a domain D, the partial derivatives f_z and $f_{\bar{z}}$ are coincident with those in the sense of distribution. Namely, for every element φ of $C_0^\infty(D)$, the set of all smooth functions on D with compact supports, it follows that*

$$\iint_D f_z \cdot \varphi dxdy = -\iint_D f \cdot \varphi_z dxdy, \tag{4.2}$$

and

$$\iint_D f_{\bar{z}} \cdot \varphi dxdy = -\iint_D f \cdot \varphi_{\bar{z}} dxdy. \tag{4.3}$$

Proof. By Proposition 4.2 and Fubini's theorem, the left hand sides of (4.2) and (4.3) can be rewritten as the repeated integrals. Since f is ACL, the assertions follow by integration by parts. □

4.1.2. Analytic Definition A' of Quasiconformal Mappings

Noting Propositions 4.2 and 4.3, we can consider a modification of the definition A of quasiconformal mappings, such as the one which follows:

Analytic definition A' of quasiconformal mappings. Let f be a homeomorphism of a domain D into $\mathbf{C}$ which preserves orientation. We say that f is *quasiconformal* on D if f satisfies the following two conditions:

(i)′ The distributional partial derivatives of f with respect to z and $\bar{z}$ can be represented by locally integrable functions f_z and $f_{\bar{z}}$, respectively, on D.

(ii) There exists a constant k with $0 \leq k < 1$ such that

$$|f_{\bar{z}}| \leq k|f_z| \quad \text{a.e. on} \quad D.$$

Theorem 4.4. *Definitions A and A' of quasiconformal mappings are mutually equivalent.*

Proof. We have already shown as Propositions 4.2 and 4.3 that a quasiconformal mapping f in the sense of the definition A satisfies the condition (i)′. Hence, f is quasiconformal in the sense of the definition A'.

Let f be quasiconformal on D in the sense of the definition A'. To show that f is quasiconformal in the sense of the definition A, it suffices to show that f is ACL.

For this purpose, fix a rectangle $R = [a, b] \times [c, d]$ in D arbitrarily. Fix a point x_0 on $[a, b]$, and set $R(x_0) = [a, x_0] \times [c, d]$. Denote by $[f_x]$ the distributional partial derivative of f with respect to x. As an element of $C_0^\infty(R(x_0))$, take a function with form $\varphi_1(x)\varphi_2(y)$. Then we have

$$\iint_{R(x_0)} [f_x](x,y)\varphi_1(x)\varphi_2(y)dxdy = -\iint_{R(x_0)} f(x,y)(\varphi_1)'(x)\varphi_2(y)dxdy.$$

Here, let $\varphi_2(y)$ tend monotonously to the characteristic function of (c,η) ($\eta \in [c,d]$), and we have

$$\int_c^\eta \int_a^{x_0} [f_x](x,y)\varphi_1(x)dxdy = -\int_c^\eta \int_a^{x_0} f(x,y)(\varphi_1)'(x)dxdy.$$

Since η is arbitrary, we conclude that

$$\int_a^{x_0} [f_x](x,y)\varphi_1(x)dxdy = -\int_a^{x_0} f(x,y)(\varphi_1)'(x)dxdy$$

for almost every y on $[c,d]$. Next, for every sufficiently large integer n, take as $\varphi_1 = \varphi_{1,n}$ in the above equality a suitable function which is identically 1 on $[a+1/n, x_0-1/n]$, and is monotonously increasing and decreasing, respectively, on $[a, a+1/n]$ and $[x_0-1/n, x_0]$. Letting $n \to \infty$, then by the above equality, we have

$$\int_a^{x_0} [f_x](x,y)dxdy = f(x_0,y) - f(a,y) \quad \text{almost every } y \in [c,d]. \tag{4.4}$$

Here, the exceptional set of y depends on x_0. To get rid of this dependence, consider the set, say E, of all rational numbers in $[a,b]$. Since E is countable, (4.4) holds a.e. on $[c,d]$ for every x_0 in E. Since both sides of (4.4) are continuous with respect to x_0, and since E is dense on $[a,b]$, we conclude that (4.4) holds a.e. on $[c,d]$ for every x_0 in $[a,b]$. Note that for every y where (4.4) holds for every x_0, $f(x,y)$ is absolutely continuous with respect to x, and that $[f_x]$ is coincident with the usual partial derivative f_x a.e. on $[a,b]$.

As for the partial derivative of f with respect to y, we can confirm a similar assertion. Thus we conclude that f is ACL on D and that the distributional derivatives are coincident with the usual ones. □

Corollary 1. *Let g be a conformal mapping of a domain D onto another D', and f be a K-qc mapping of D'. Then $f \circ g$ is a K-qc mapping of D.*

Proof. Let $w = u + iv$ be the variable on D'. By the assumption, there exist the distributional derivatives f_w and $f_{\bar{w}}$, and they are locally integrable on D'.

Since $\varphi \circ g^{-1}$ belongs to $C_0^\infty(D')$ for every $\varphi \in C_0^\infty(D)$, it is easily seen that there exist the distributional derivatives $(f \circ g)_z$ and $(f \circ g)_{\bar{z}}$, and that they are coincident with locally integrable functions

$$(f_w \circ g) \cdot g' \quad \text{and} \quad (f_{\bar{w}} \circ g) \cdot \overline{g'},$$

respectively, on D. Moreover, the condition (ii) for $f \circ g$ clearly holds with the same k as in that for f. Thus we have the assertion. □

Corollary 2. *A 1-qc mapping is conformal.*

Proof. A 1-qc mapping f of a domain D satisfies $f_{\bar z} = 0$ on D in the sense of distribution. Hence, f is holomorphic by the classical lemma of Weyl. □

Here, for the sake of convenience, we give a proof of *Weyl's lemma*, where we use the following approximation lemma.

Lemma 4.5. *Let $p \geq 1$ be given. Let f be a continuous function on a domain D whose distributional partial derivatives f_z and $f_{\bar z}$ are locally L^p on D, namely, satisfy that $|f_z|^p$ and $|f_{\bar z}|^p$ are locally integrable on D. Then for every compact subset F of D, there is a sequence $\{f_n\}_{n=1}^{\infty}$ in $C_0^{\infty}(D)$ such that f_n converges to f uniformly on F as $n \to \infty$, and that*

$$\begin{aligned} \lim_{n\to\infty} \iint_F |(f_n)_z - f_z|^p \, dxdy = 0, \\ \lim_{n\to\infty} \iint_F |(f_n)_{\bar z} - f_{\bar z}|^p \, dxdy = 0. \end{aligned} \tag{4.5}$$

Proof. Fix an element $\eta \in C_0^{\infty}(D)$ which is identically equal to 1 in some neighborhood of F. Then ηf has a compact support in D. Further, $(\eta f)_z$ and $(\eta f)_{\bar z}$ exist and belong to L^p on D.

We set

$$\varphi(z) = \begin{cases} C \cdot \exp\left(-\frac{1}{1-|z|^2}\right), & z \in \Delta \\ 0, & z \in \mathbf{C} - \Delta. \end{cases}$$

Here, Δ is the unit disk. Choose a constant C so that

$$\iint_{\Delta} \varphi(z) dxdy = 1.$$

Further, for every positive integer n, set

$$\varphi_n(z) = n^2 \cdot \varphi(nz), \quad z \in \mathbf{C}$$

and

$$f_n(w) = \varphi_n * (\eta f)(w) = \iint_{\mathbf{C}} \varphi_n(w - z)\eta(z) f(z) dxdy, \quad w \in \mathbf{C}.$$

Then for every n, we have

$$(f_n)_z = \varphi_n * (\eta f)_z \quad \text{and} \quad (f_n)_{\bar z} = \varphi_n * (\eta f)_{\bar z}. \tag{4.6}$$

Moreover, f_n belongs to $C_0^{\infty}(D)$ for every sufficiently large n, and f_n converges to f uniformly on F as $n \to \infty$. Since

$$(\eta f)_z = f_z \quad \text{and} \quad (\eta f)_{\bar z} = f_{\bar z}$$

on F, and since we can show by noting (4.6) that

$$\lim_{n\to\infty} \iint_D |(f_n)_z - (\eta f)_z|^p \, dxdy = 0,$$

and

$$\lim_{n\to\infty}\iint_D |(f_n)_{\bar z} - (\eta f)_{\bar z}|^p\, dxdy = 0,$$

we obtain (4.5). □

Hereafter, we call such a sequence $\{f_n\}$ as in Lemma 4.5 an *L^p-smoothing sequence* for f with respect to F.

Lemma 4.6. (Weyl's lemma) *Let f be a continuous function on D whose distributional derivative $f_{\bar z}$ is locally integrable on D. If $f_{\bar z} = 0$ in the sense of distributions on D, then f is holomorphic on D.*

Proof. Fix a relatively compact subdomain D_1 of D arbitrarily, and construct an L^1-smoothing sequence for f with respect to $\overline{D_1}$ as in the proof of Lemma 4.5. From the construction there, we see that $(\eta f)_{\bar z} = 0$ in some neighborhood of $\overline{D_1}$. By (4.6), we see that $(f_n)_{\bar z} = 0$ on D_1 for every sufficiently large n. Thus, f_n is holomorphic on D_1 for every sufficiently large n. Since f_n converges to f uniformly on $\overline{D_1}$ as $n \to \infty$, f is holomorphic on D_1. Since D_1 is arbitrary, we obtain the assertion. □

4.1.3. Geometric Definition G of Quasiconformal Mappings

We give a geometric definition of quasiconformality. For this purpose, we first introduce the notion of quadrilaterals.

The closure of a domain bounded by a Jordan curve is called a Jordan closed domain. A *quadrilateral* is, by definition, a pair $(Q; q_1, q_2, q_3, q_4)$ of a Jordan closed domain Q and four points q_1, q_2, q_3, q_4 on the boundary ∂Q of Q which are mutually distinct and located in this order with respect to the positive orientation of ∂Q. We call each q_j a *vertex* of the quadrilateral. If there is no confusion, we denote a quadrilateral $(Q; q_1, q_2, q_3, q_4)$ simply by Q.

Proposition 4.7. *For every quadrilateral $(Q; q_1, q_2, q_3, q_4)$, there is a homeomorphism h of Q onto some rectangle $R = [0, a] \times [0, b]$ $(a, b > 0)$ which is conformal in the interior $\operatorname{Int} Q$ of Q, and satisfies*

$$h(q_1) = 0, \quad h(q_2) = a, \quad h(q_3) = a + ib, \quad \textit{and} \quad h(q_4) = ib.$$

Moreover, a/b is independent of h.

We call the value a/b the *module* of the quadrilateral $(Q; q_1, q_2, q_3, q_4)$, and denote it by $M(Q; q_1, q_2, q_3, q_4)$, or simply by $M(Q)$.

Proof. First, by Riemann's mapping theorem, there exists a conformal mapping h_1 of $\operatorname{Int} Q$ onto the upper half-plane H. By Carathéodory's theorem, h_1 is extended to a homeomorphism of Q onto $H \cup \hat{\mathbb{R}}$. Here, by composing a suitable Möbius transformation, we may assume that

$$h_1(q_1) = -1, \quad h_1(q_2) = 1, \quad \text{and} \quad h_1(q_3) = -h_1(q_4) > 1.$$

Set $k = 1/h_1(q_3)$, and

$$h_2(z) = \int_0^z \frac{dz}{\sqrt{(1-z^2)(1-k^2z^2)}}, \quad z \in H.$$

Then h_2 is a conformal mapping of H onto the interior of some rectangle $[-K, K] \times [0, K']$ $(K, K' > 0)$. Hence we see that

$$h(z) = h_2 \circ h_1(z) + K, \quad z \in Q,$$

is a desired mapping (Fig. 4.1).

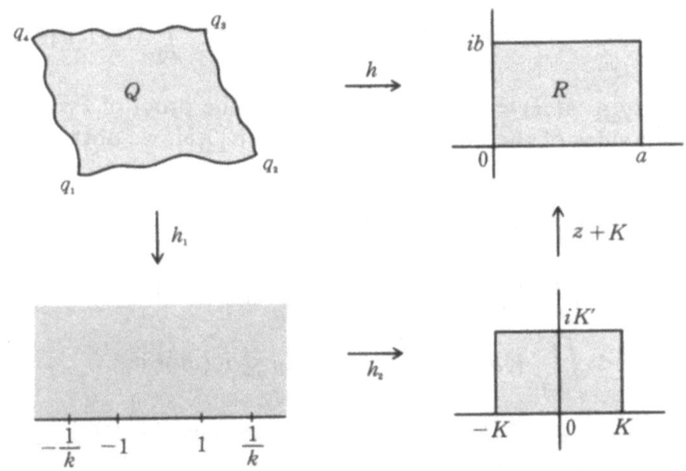

Fig. 4.1.

Next, let $\tilde{h} : Q \longrightarrow \tilde{R} = [0, \tilde{a}] \times [0, \tilde{b}]$ be another mapping which satisfies the same conditions as in Proposition 4.7. Applying Schwarz' reflection principle repeatedly, we see that $\tilde{h} \circ h^{-1}$ can be extended to an element of $Aut(\mathbf{C})$. Hence, with suitable complex numbers c and d, we have $\tilde{h}(z) = ch(z) + d$ $(z \in \text{Int}\, Q)$. Since

$$\tilde{h}(q_1) = h(q_1) = 0, \quad h(q_2) > 0, \quad \text{and} \quad \tilde{h}(q_2) > 0,$$

we conclude that $c > 0$ and $d = 0$, which implies the second assertion of Proposition 4.7. □

Hereafter, for every quadrilateral $(Q; q_1, q_2, q_3, q_4)$ and a homeomorphism f of Q into $\mathbf{C}$, we consider $f(Q)$ as a quadrilateral with vertices $f(q_1), \cdots, f(q_4)$. Then we have the following lemma.

Lemma 4.8. *Every K-qc mapping f of a domain D satisfies*

$$\frac{1}{K}M(Q) \leq M(f(Q)) \leq KM(Q)$$

for all quadrilaterals Q.

Proof. Fix mappings $h : Q \longrightarrow R = [0,a] \times [0,b]$ and $\tilde{h} : f(Q) \longrightarrow \tilde{R} = [0,\tilde{a}] \times [0,\tilde{b}]$, presented in Proposition 4.7. Then

$$F = \tilde{h} \circ f \circ h^{-1}$$

is a quasiconformal mapping of the interior of R onto the interior of $\tilde{R}$ which maps 0, a, ib, and $a+ib$ to 0, $\tilde{a}$, $i\tilde{b}$, and $\tilde{a}+i\tilde{b}$, respectively (cf. Remark 3 in §1.1 and Corollary 1 to Theorem 4.4). In particular, $F(z)$ is ACL on R, and hence for almost every $y \in [0,b]$, we have

$$\tilde{a} \leq |F(a+iy) - F(iy)| = \left|\int_0^a \frac{\partial F}{\partial x}(x+iy)dx\right| \leq \int_0^a (|F_z| + |F_{\bar{z}}|)dx.$$

Since $\iint_R J_F dxdy \leq A(R) = \tilde{a}\tilde{b}$ (as was stated in the proof of Proposition 4.2), integrating both sides of the above inequality over $[0,b]$, we obtain

$$\begin{aligned}(\tilde{a}b)^2 &\leq \left(\iint_R (|F_z| + |F_{\bar{z}}|)dxdy\right)^2 \\ &\leq \iint_{R'} \frac{|F_z| + |F_{\bar{z}}|}{|F_z| - |F_{\bar{z}}|}dxdy \cdot \iint_{R'} (|F_z|^2 - |F_{\bar{z}}|^2)dxdy \\ &\leq \iint_{R'} Kdxdy \cdot \iint_{R'} J_F dxdy \leq K(ab)(\tilde{a}\tilde{b}).\end{aligned}$$

Here, we set $R' = \{w \in R \mid |F_z(w)| \neq 0\}$. Thus we have $M(f(Q)) \leq KM(Q)$.

Next, replacing $F = \tilde{h} \circ f \circ h^{-1}$ by $(i\tilde{h}) \circ f \circ (ih)^{-1}$ (or considering $(Q; q_2, q_3, q_4, q_1)$), we can show by the same argument that

$$\frac{1}{M(f(Q))} \leq \frac{K}{M(Q)}.$$

Thus we have the assertion. □

Now, by noting Lemma 4.8, we give another definition of quasiconformal mappings without using partial derivatives.

Geometric definition G of quasiconformal mappings. Let f be a homeomorphism of a domain D into $\mathbf{C}$ which preserves orientation. We say that f is *quasiconformal* on D if f satisfies the following condition:

(iii) There is a constant $K \geq 1$ such that

$$M(f(Q)) \leq KM(Q)$$

holds for every quadrilateral Q in D.

By Lemma 4.8 and Lemma 4.15 which is proved later, we see that the infimum of such K is equal to the maximal dilatation K_f of f.

Remark. Condition (iii) in the foregoing definition is equivalent to the following one:

(iii)′ There is a constant $K \geq 1$ such that

$$\frac{1}{K}M(Q) \leq M(f(Q)) \leq KM(Q)$$

holds for every quadrilateral Q in D.

Theorem 4.9. *Two definitions A and G of quasiconformal mappings are mutually equivalent.*

We shall postpone the proof of Theorem 4.9 to §1.4. Here, we give several corollaries of Definition G and Theorem 4.9. (Recall that (ii) of the following Theorem 4.10 has been already proved as Remark 3 in §1.1 and Corollary 1 to Theorem 4.4.)

Theorem 4.10.

(i) *The inverse of a K-qc mapping is also K-qc.*
(ii) *K-quasiconformality is conformally invariant. Namely, for every conformal mapping h and $\tilde{h}$ of domains D and $\tilde{D}$, respectively, and for every K-qc mapping f of D onto $\tilde{D}$, the composed mapping $\tilde{h} \circ f \circ h^{-1}$ is also K-qc.*
(iii) *For every K_1-qc mapping f of a domain D and every K_2-qc mapping g of $f(D)$, the composed mapping $g \circ f$ is K_1K_2-qc.*

Proof. The first assertion follows from the preceding Remark. The second is seen by the conformal invariance of the module. (See the proof of Proposition 4.7). The third is clear from Definition G. □

Proposition 4.11. *If f is quasiconformal on a domain D, then $f_z \neq 0$ a.e. on D.*

Proof. Set $E = \{z \in D \mid f_z = 0\}$. Then E is measurable and $f_{\bar{z}} = 0$ a.e. on E. Hence by the next Lemma 4.12, the set $f(E)$ has area zero. Since f^{-1} is also quasiconformal by (i) of Theorem 4.10, we see again by Lemma 4.12 that the set $E = f^{-1}(f(E))$ has area zero. □

Lemma 4.12. *Every quasiconformal mapping f of a domain D satisfies*

$$\iint_E (|f_z|^2 - |f_{\bar{z}}|^2)\,dxdy = A(E) \tag{4.7}$$

for every measurable subset E of D, where $A(E) = \iint_{f(E)} dxdy$ as before.

Proof. First, we consider the case that E is a rectangle contained in D and that f is absolutely continuous on the boundary ∂E of E.

In this case, in view of Proposition 4.2, we find an L^2-smoothing sequence $\{f_n\}_{n=1}^{\infty}$ for f with respect to E (cf. Lemma 4.5). For every n, put $f_n = u_n + iv_n$. By Green's formula, we have

$$\iint_E \{(u_m)_x(v_n)_y - (u_m)_y(v_n)_x\}\, dxdy = \int_{\partial E} u_m\, dv_n$$

for every n and m. Let $m \to \infty$. Next, to the right hand side, apply the formula for integration by parts for the Stieltjes integral. Letting $n \to \infty$, we obtain

$$\iint_E (|f_z|^2 - |f_{\bar z}|^2) dxdy = \iint_E (u_x v_y - u_y v_x) dxdy = \int_{\partial E} udv.$$

Here, we write $f = u+iv$. The right hand side of the above equality is interpreted as the line integral of udv along the Jordan curve $\partial f(E)$ on the $w(= u+iv)$-plane. By the assumption, $\partial f(E)$ is rectifiable. Hence, we can show that

$$\int_{\partial E} udv = \iint_{f(E)} dudv = A(E).$$

Thus we have the assertion in the case stated at the beginning of the proof.

Since f is ACL, every rectangle contained in D can be approximated by such rectangles. Hence, (4.7) holds for every rectangle contained in D. By a routine argument, it is proved that (4.7) holds for every measurable subset E of D. □

Now, by Propiosition 4.11, for every quasiconformal mapping f of a domain D, we can consider a quantity

$$\mu_f = \frac{f_{\bar z}}{f_z}$$

a.e. on D. This μ_f is a bounded measurable function on D, and satisfies

$$\operatorname*{ess.sup}_{z \in D} |\mu_f(z)| \le \frac{K_f - 1}{K_f + 1} < 1.$$

We call μ_f the *complex dilatation* of f on D.

Proposition 4.13. *For every quasiconformal mapping f and g of a domain D, the complex dilatation $\mu_{g\circ f^{-1}}$ of the composed mapping $g \circ f^{-1}$ is given by*

$$\mu_{g\circ f^{-1}} \circ f = \frac{f_z}{\overline{f_z}} \frac{\mu_g - \mu_f}{1 - \overline{\mu_f}\mu_g}, \quad a.e.\ on\ \ D. \tag{4.8}$$

Proof. By (i) and (iii) of Theorem 4.10, $g\circ f^{-1}$ is quasiconformal on $f(D)$. Hence by Proposition 4.1, $g \circ f^{-1}$ is totally differentiable on $f(D)$ except for a subset E of measure zero. Applying Lemma 4.12 to the quasiconformal mapping f^{-1}, we see that $f^{-1}(E)$ is also of measure zero. Hence by Proposition 4.1, both f

and $g \circ f^{-1}$ are totally differentiable at z and at $f(z)$, respectively, for almost every z on D. At such a point z, the chain rule is valid. Accordingly, writing $w = f(z)$, we have

$$g_z = (g \circ f^{-1})_w \circ f \cdot f_z + (g \circ f^{-1})_{\bar{w}} \circ f \cdot \bar{f}_z$$

and

$$g_{\bar{z}} = (g \circ f^{-1})_w \circ f \cdot f_{\bar{z}} + (g \circ f^{-1})_{\bar{w}} \circ f \cdot \overline{(f_z)}.$$

Using a similar argument, we can show by Proposition 4.11 and Lemma 4.12 that

$$f_z \neq 0, \quad g_z \neq 0, \quad \text{and} \quad (g \circ f^{-1})_w \circ f \neq 0$$

for almost every $z \in D$. Thus the above equalities imply the assertion. □

4.1.4. Proof of Theorem 4.9

We have already shown in Lemma 4.8 that a quasiconformal mapping in the sense of the definition A satisfies the condition (iii). Conversely, the following Lemmas 4.14 and 4.15 show that a quasiconformal mapping in the sense of the definition G satisfies the conditions (i) and (ii).

Lemma 4.14. *Every quasiconformal mapping f of a domain D in the sense of the definition G is ACL on D.*

Proof. Fix a rectangle $R = [a, b] \times [c, d]$ in D arbitrarily. Denote by $F(y)$ the area of $f([a, b] \times [c, y])$ for every y on $[c, d]$. Since $F(y)$ is a non-decreasing function, F is differentiable at almost every $y \in [c, d]$. Fix such a $y = y_0$, and set $\eta = y - y_0$. From here on, assume that $\eta > 0$.

Take a finite family $\{I_j\}_{j=1}^m$ of mutually disjoint subintervals of $[a, b]$, and set

$$R_j = I_j \times [y_0, y] \quad \text{and} \quad Q_j = f(R_j)$$

for every j. Let ℓ_j and ℓ'_j be the lengths of I_j and $f(I_j \times \{y_0\})$, respectively. Fix j and ϵ (> 0) arbitrarily. Then there is a finite set $\{\zeta_k\}_{k=1}^n$ of points on $f(I_j \times \{y_0\})$ such that

$$\zeta_0 = f(a_j + iy_0), \quad \zeta_n = f(b_j + iy_0)$$

and

$$\sum_{k=1}^{n} |\zeta_k - \zeta_{k-1}| \geq \ell'_j - \epsilon/2.$$

Here, we set $I_j = [a_j, b_j]$, and we replace the right hand side of the inequality by $1/\epsilon$ when $\ell'_j = \infty$. (See Fig. 4.2.)

Now, since f is uniformly continuous on the compact set R, taking η sufficiently small, we may assume that

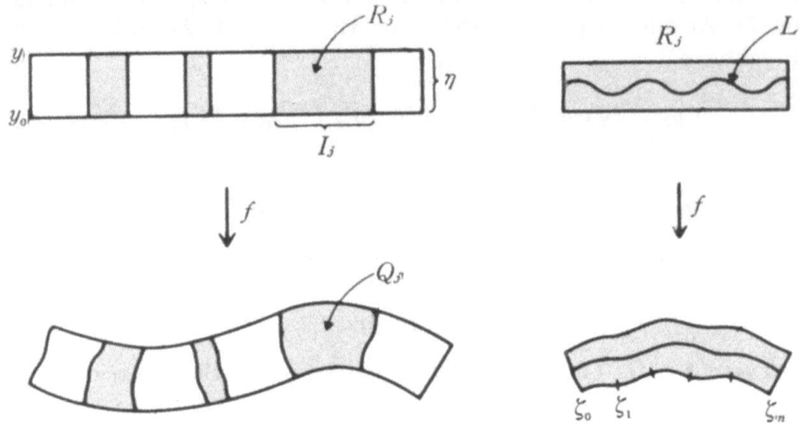

Fig. 4.2.

$$|f(x_0 + i(y_0 + \xi)) - f(x_0 + iy_0)| \leq \frac{\epsilon}{4n}$$

for every x_0 on $[a, b]$ and every ξ with $0 \leq \xi \leq \eta$. Take any curve L in R_j connecting two sides of ∂R_j which are parallel to the y-axis. Then we can see that the length of $f(L)$ is not less than

$$\tilde{\ell}_j = \sum_{k=1}^{n} |\zeta_k - \zeta_{k-1}| - \frac{\epsilon}{2}.$$

On the other hand, let $\tilde{h}$ be a homeomorphism of some rectangle $\tilde{R}_j = [\tilde{a}_j, \tilde{b}_j] \times [\tilde{c}_j, \tilde{d}_j]$ onto the quadrilateral Q_j which is conformal in the interior of $\tilde{R}_j$ and respects the vertices suitably. Then we have

$$\tilde{\ell}_j^2 \leq \left(\int_{\tilde{a}_j}^{\tilde{b}_j} |\tilde{h}'| dx \right)^2 \leq (\tilde{b}_j - \tilde{a}_j) \cdot \int_{\tilde{a}_j}^{\tilde{b}_j} |\tilde{h}'|^2 dx.$$

Integrating both sides with respect to y on $[\tilde{c}_j, \tilde{d}_j]$, we obtain

$$\tilde{\ell}_j^2 \leq M(Q_j) \cdot \tilde{A}_j.$$

Here, we denote by $\tilde{A}_j$ the area of Q_j.

Now, suppose that f is K-qc in the sense of the definition G. Since

$$M(Q_j) \leq K M(R_j) = K \frac{\ell_j}{\eta},$$

we get

$$\left(\sum_j \tilde{\ell}_j \right)^2 \leq \sum_j \frac{(\tilde{\ell}_j)^2}{\ell_j} \cdot \sum_j \ell_j \leq K \frac{F(y) - F(y_0)}{y - y_0} \cdot \sum_j \ell_j.$$

Since F is differentiable at y_0, $(F(y) - F(y_0))/(y - y_0)$ tends to a finite value as $\eta \to 0$. In particular, we can see that every ℓ_j' is finite. Since $\tilde{\ell}_j \geq \ell_j' - \epsilon$, we conclude that $\sum_{j=1}^m \ell_j' \to 0$ as $\sum_{j=1}^m \ell_j \to 0$. Consequently, $f(x, y_0)$ is absolutely continuous on $[a, b]$ as a function of x.

By the same argument, we can show that $f(x_0, y)$ is absolutely continuous on $[c, d]$ as a function of y for almost every $x_0 \in [a, b]$. Thus we conclude that f is ACL on D. □

Lemma 4.15. *If f is a quasiconformal mapping of a domain D in the sense of the definition G with the maximal dilatation K, then*

$$|f_{\bar{z}}| \leq k|f_z| \quad \text{a.e. on} \quad D,$$

where $k = (K-1)/(K+1)$.

Proof. Lemma 4.14 implies that f satisfies the assumption of Proposition 4.1. Hence, f is totally differentiable at almost every $z \in D$. Fix such a $z = z_0$. Here we may assume, without loss of generality, that $z_0 = 0$. By considering $e^{i\theta_1} \cdot f(e^{i\theta_2} \cdot z)$ with suitable real numbers θ_1 and θ_2, we may further assume that $f_z(0)$ and $f_{\bar{z}}(0)$ are non-negative. As we noted in the proof of Proposition 4.2, $f_z(0) \geq f_{\bar{z}}(0)\,(\geq 0)$. If $f_z(0) = 0$, then the assertion is clear. Hence, we assume that $f_z(0) > 0$.

In this case, note that f is expanded as

$$f(z) = f(0) + f_z(0) \cdot z + f_{\bar{z}}(0) \cdot \bar{z} + o(|z|)$$

in a neighborhood of the origin. Consider a rectangle $R_\epsilon = [0, \epsilon] \times [0, \epsilon]$ for every positive ϵ. Then $f(R_\epsilon)$ is approximately the rectangle

$$[a, a + (f_z(0) + f_{\bar{z}}(0))\,\epsilon] \times [b, b + (f_z(0) - f_{\bar{z}}(0))\,\epsilon],$$

where $f(0) = a + ib$. Hence, by Rengel's inequality (cf., for instance, Lehto and Virtanen [A-69], I 4.3), we obtain

$$K = KM(R_\epsilon) \geq M(f(R_\epsilon)) \geq \frac{f_z(0) + f_{\bar{z}}(0)}{f_z(0) - f_{\bar{z}}(0)} + o(1).$$

Thus, letting ϵ tend to 0, we conclude that

$$K \geq \frac{f_z(0) + f_{\bar{z}}(0)}{f_z(0) - f_{\bar{z}}(0)},$$

or equivalently, $f_{\bar{z}}(0) \leq kf_z(0)$. □

4.1.5. Other Fundamental Properties of Quasiconformal Mappings

We state here, without proofs, two of the fundamental and important properties on continuity of quasiconformal mappings.

Theorem 4.16. **(Mori's theorem** [157]**)** *If f is a K-qc mapping of the unit disk Δ onto itself with $f(0) = 0$, then*

$$|f(z_1) - f(z_2)| < 16|z_1 - z_2|^{1/K}, \quad z_1, z_2 \in \Delta,\ z_1 \neq z_2.$$

Theorem 4.17. *Every sequence of K-qc mappings of $\mathbf{C}$ onto itself fixing* 0 *and* 1 *contains a subsequence which converges uniformly with respect to the spherical distance.*

Moreover, the limit function of such a subsequence is again K-qc.

For proofs of these theorems and further information on quasiconformal mappings, see, for instance, Ahlfors [A-2], and Lehto and Virtanen [A-69].

4.2. Existence Theorem on Quasiconformal Mappings

We have seen that a quasiconformal mapping f of a domain D induces a bounded measurable function μ_f on D which satisfies $\mathrm{ess.sup}_{z\in D}|\mu_f(z)| < 1$. In this section, we shall show the converse. Namely, for every measurable μ with $\mathrm{ess.sup}_{z\in D}|\mu(z)| < 1$, we construct a quasiconformal mapping whose complex dilatation is equal to μ.

4.2.1. Preliminary Considerations

We denote by $L^\infty(D)$ the complex Banach space of all bounded measurable functions on a domain D. Here, the norm is given by

$$\|\mu\|_\infty = \mathrm{ess.sup}_{z\in D}|\mu(z)|, \quad \mu \in L^\infty(D).$$

Let $B(D)_1$ be the unit open ball $\{\mu \in L^\infty(D) \mid \|\mu\|_\infty < 1\}$ of $L^\infty(D)$, and call any element of $B(D)_1$ a *Beltrami coefficient* on D.

First, we note that a quasiconformal mapping with the prescribed complex dilatation is essentially unique. More precisely, we have the following:

Proposition 4.18. *Let μ be an arbitrary element of $B(D)_1$. Suppose that there exists a quasiconformal mapping f with the complex dilatation $\mu_f = \mu$. Then for every conformal mapping h of $f(D)$, the composed mapping $h \circ f$ has the same complex dilatation μ.*

Conversely, for every quasiconformal mapping g with $\mu_g = \mu$, the composed mapping $g \circ f^{-1}$ is a conformal mapping of $f(D)$.

Proof. First, as in the proof of Proposition 4.13, we have $\mu_{h\circ f} = \mu_f = \mu$. Next, since $\mu_{g\circ f^{-1}}$ is equal to 0 a.e. on $f(D)$ by Proposition 4.13, it follows that $g\circ f^{-1}$ is 1-qc, and hence is conformal by Corollary 2 to Theorem 4.4. □

Now, for any given $\mu \in B(\mathbf{C})_1$, we consider how to solve the Beltrami differential equation

$$f_{\bar{z}} = \mu f_z.$$

For this purpose, we consider first solving the $\bar{\partial}$-problem, i.e., finding a suitable f from the given $f_{\bar{z}}$. Actually, if we get a suitable representation $f = G(f_{\bar{z}})$, then we have a relation

$$f_z = G(f_{\bar{z}})_z = G(\mu f_z)_z$$

between μ and f_z. Rewriting it in the form $f_z = F(\mu)$, we obtain a solution $f = G(\mu F(\mu))$ of the given Beltrami equation.

On the other hand, a way to reconstruct f from $f_{\bar{z}}$ is classically known as the Cauchy transformation. For every p with $1 \le p < \infty$, let $L^p(\mathbf{C})$ be the complex Banach space of all measurable functions f on $\mathbf{C}$ such that $|f|^p$ are integrable on $\mathbf{C}$, where the norm is given by

$$\|h\|_p = \left(\iint_{\mathbf{C}} |h|^p dxdy\right)^{1/p}, \quad h \in L^p(\mathbf{C}).$$

The following classical *Pompeiu's formula* is essential to solve the $\bar{\partial}$-problem.

Proposition 4.19. *Fix p with $2 < p < \infty$. Let f be a continuous function on $\mathbf{C}$ whose distributional partial derivatives are represented by elements f_z and $f_{\bar{z}}$ of $L^p(\mathbf{C})$. Then f satisfies*

$$f(\zeta) = \frac{1}{2\pi i}\int_{\partial B} \frac{f(z)dz}{z-\zeta} - \frac{1}{\pi}\iint_B \frac{f_{\bar{z}}(z)}{z-\zeta}dxdy, \quad \zeta \in B$$

for every open disk B in $\mathbf{C}$.

Proof. First, we note that the integrand of the second term on the right hand side is absolutely integrable. In fact, let $q\ (< 2)$ be defined by $1/p + 1/q = 1$. Since $|1/(z-\zeta)|^q$ is integrable on B, the assertion follows by Hölder's inequality.

Next, take an L^p-smoothing sequence $\{f_n\}_{n=1}^{\infty}$ for f with respect to $\bar{B}$, and fix a point $\zeta \in B$ arbitrarily. For every n, Green's formula gives

$$f_n(\zeta) = \frac{1}{2\pi i}\int_{\partial B} \frac{f_n(z)dz}{z-\zeta} - \frac{1}{2\pi i}\iint_B \frac{(f_n)_{\bar{z}}(z)}{z-\zeta}d\bar{z}\wedge dz.$$

Since $f_n \to f$ uniformly on $\bar{B}$, and $(f_n)_{\bar{z}} \to f_{\bar{z}}$ in $L^p(B)$, respectively, as $n \to \infty$, we have the assertion. □

We define a linear operator P on $L^p(\mathbf{C})$ by setting

$$Ph(\zeta) = -\frac{1}{\pi}\iint_{\mathbf{C}} h(z)\left(\frac{1}{z-\zeta} - \frac{1}{z}\right)dxdy, \quad h \in L^p(\mathbf{C}), \zeta \in \mathbf{C}. \tag{4.9}$$

Then we have the following:

Lemma 4.20. *For every p with $2 < p < \infty$ and for every $h \in L^p(\mathbf{C})$, Ph is a uniformly Hölder continuous function on $\mathbf{C}$, with exponent $(1 - 2/p)$, and satisfies $Ph(0) = 0$.*

Moreover, Pf satisfies

$$(Ph)_{\bar{\zeta}} = h$$

on $\mathbf{C}$ in the sense of distribution.

Proof. First, as in the proof of Proposition 4.19, we shall show that the integral on the right hand side of (4.9) is well-defined. For this purpose, define q by the equation $1/p + 1/q = 1$. Since

$$\frac{1}{z-\zeta} - \frac{1}{z} = \frac{\zeta}{z(z-\zeta)}$$

belongs to $L^q(\mathbf{C})$, Hölder's inequality implies that

$$|Ph(\zeta)| \le \frac{1}{\pi}||h||_p \cdot ||\frac{\zeta}{z(z-\zeta)}||_q < \infty.$$

Further, when $\zeta \neq 0$, by changing the variable, we have

$$\iint_{\mathbf{C}} \left|\frac{1}{z(z-\zeta)}\right|^q dxdy = |\zeta|^{2-2q} \iint_{\mathbf{C}} \left|\frac{1}{z(z-1)}\right|^q dxdy.$$

Hence, there is a constant K_p depending only on p such that

$$|Ph(\zeta)| \le K_p||h||_p \cdot |\zeta|^{1-2/p}, \quad \zeta \in \mathbf{C}, \zeta \neq 0. \tag{4.10}$$

Since $Ph(0) = 0$ by the definition, (4.10) is valid even when $\zeta = 0$.

Next, set $h_1(z) = h(z + \zeta_1)$. Then we obtain

$$\begin{aligned} Ph_1(\zeta_2 - \zeta_1) &= -\frac{1}{\pi}\iint_{\mathbf{C}} h(z+\zeta_1)\left(\frac{1}{z+\zeta_1-\zeta_2} - \frac{1}{z}\right) dxdy \\ &= -\frac{1}{\pi}\iint_{\mathbf{C}} h(z)\left(\frac{1}{z-\zeta_2} - \frac{1}{z-\zeta_1}\right) dxdy \\ &= Ph(\zeta_2) - Ph(\zeta_1). \end{aligned}$$

Combining this with (4.10), we conclude that

$$|Ph(\zeta_1) - Ph(\zeta_2)| \le K_p||h||_p \cdot |\zeta_1 - \zeta_2|^{1-2/p}, \quad \zeta_1, \zeta_2 \in \mathbf{C}, \tag{4.11}$$

or equivalently, that Ph is a uniformly Hölder continuous function with exponent $1 - 2/p$.

To show the second assertion, take a sequence $\{h_n\}_{n=1}^{\infty}$ in $C_0^\infty(\mathbf{C})$ such that $||h - h_n||_p \to 0$ as $n \to \infty$. (Such a sequence is constructed, for example, as in the proof of Lemma 4.5.) Then for every h_n, we have

$$(Ph_n)_{\bar\zeta}(\zeta) = \frac{\partial}{\partial\bar\zeta}\left(-\frac{1}{\pi}\iint_{\mathbf{C}} \frac{h_n(z+\zeta)}{z}dxdy\right) \tag{4.12}$$
$$= -\frac{1}{\pi}\iint_{\mathbf{C}} \frac{(h_n)_{\bar z}(z)}{z-\zeta}dxdy.$$

Hence Green's formula gives

$$(Ph_n)_{\bar\zeta}(\zeta) = \lim_{\epsilon\to 0}\frac{1}{2\pi i}\oint_{\{|z-\zeta|=\epsilon\}} \frac{h_n(z)}{z-\zeta}dz = h_n(\zeta).$$

In particular, for every $\varphi \in C_0^\infty(\mathbf{C})$, we get

$$\iint_{\mathbf{C}} h_n\varphi dxdy = -\iint_{\mathbf{C}} Ph_n \cdot \varphi_{\bar z}dxdy.$$

Since $||h - h_n||_p \to 0$ as $n \to \infty$, Ph_n converges to Ph locally uniformly on (i.e., uniformly on any compact subsets of) $\mathbf{C}$ by (4.10). Hence, when we let $n \to \infty$, the above equality gives

$$\iint_{\mathbf{C}} h\varphi dxdy = -\iint_{\mathbf{C}} Ph \cdot \varphi_{\bar z}dxdy, \quad \varphi \in C_0^\infty(\mathbf{C}).$$

Thus we have proved the second assertion. □

Next, we need to obtain a suitable integral representation for $(Ph)_z$. For this purpose, we examine the case where h belongs to $C_0^\infty(\mathbf{C})$. In this case, Green's formula gives

$$(Ph)_\zeta(\zeta) = \frac{1}{2\pi i}\iint_{\mathbf{C}} \frac{h_z(z)}{z-\zeta}dz\wedge d\bar z \tag{4.13}$$
$$= \lim_{\epsilon\to 0}\left\{\frac{-1}{2\pi i}\oint_{\{|z-\zeta|=\epsilon\}} \frac{h(z)}{z-\zeta}d\bar z + \frac{1}{2\pi i}\iint_{\{|z-\zeta|>\epsilon\}} \frac{h(z)}{(z-\zeta)^2}dz\wedge d\bar z\right\}.$$

Since the first term on the right hand side converges to 0 as $\epsilon \to 0$, the second term is essential. Hence, we pay our attention to the linear operator T defined by the singular integral

$$Th(\zeta) = \lim_{\epsilon\to 0}\left\{-\frac{1}{\pi}\iint_{\{|z-\zeta|>\epsilon\}} \frac{h(z)}{(z-\zeta)^2}dxdy\right\}, \quad h \in C_0^\infty(\mathbf{C}).$$

Lemma 4.21. *Every $h \in C_0^\infty(\mathbf{C})$ satisfies*

$$(Ph)_z = Th \quad on \quad \mathbf{C}, \tag{4.14}$$

and

$$||Th||_2 = ||h||_2. \tag{4.15}$$

Proof. We have already verified (4.14). Next, we can see by (4.12) and (4.13) that Ph belongs to $C^\infty(\mathbf{C})$ for every $h \in C_0^\infty(\mathbf{C})$, and that

$$\begin{aligned}
\|Th\|_2^2 &= -\frac{1}{2i}\iint_{\mathbf{C}} (Ph)_z(\overline{Ph})_{\bar{z}}dz \wedge d\bar{z} \\
&= \frac{1}{2i}\iint_{\mathbf{C}} Ph \cdot (\overline{Ph})_{\bar{z}z}dz \wedge d\bar{z} \\
&= \frac{1}{2i}\iint_{\mathbf{C}} Ph \cdot (\bar{h})_{\bar{z}}dz \wedge d\bar{z} \\
&= -\frac{1}{2i}\iint_{\mathbf{C}} \bar{h}(Ph)_{\bar{z}}dz \wedge d\bar{z} = \|h\|_2^2.
\end{aligned}$$

Thus we have proved (4.15). □

Lemma 4.21 implies, in particular, that the operator T is extended to a bounded linear operator on $L^2(\mathbf{C})$ into itself with norm 1. Since we have considered the operator P as that on $L^p(\mathbf{C})$ with $p > 2$, we consider T also as such an operator on $L^p(\mathbf{C})$. Then we see by the following classical *Calderón-Zygmund's theorem* that T gives a bounded linear operator on $L^p(\mathbf{C})$ ($p > 2$) into itself.

Proposition 4.22. (Calderón and Zygmund) *For every p with $2 \le p < \infty$,*

$$C_p = \sup_{h \in C_0^\infty(\mathbf{C}), \|h\|_p = 1} \|Th\|_p$$

is finite. Hence, the operator T is extended to a bounded linear operator of $L^p(\mathbf{C})$ into itself with norm C_p.

Moreover, C_p is continuous with respect to p. In particular, C_p satisfies

$$\lim_{p \to 2} C_p = 1. \tag{4.16}$$

In §4, we shall include a proof of this basic result for the sake of convenience. Here, assuming this proposition, we solve the Beltrami equation. Note that Proposition 4.22 gives the following:

Proposition 4.23. *For an arbitrarily given p (> 2) and every $h \in L^p(\mathbf{C})$,*

$$(Ph)_z = Th$$

on $\mathbf{C}$ in the sense of distribution.

Proof. Take a sequence $\{h_n\}_{n=1}^\infty$ in $C_0^\infty(\mathbf{C})$ approximating h in $L^p(\mathbf{C})$ (cf. the proof of Lemma 4.20). For every n, (4.14) implies that

$$\iint_{\mathbf{C}} Th_n \cdot \varphi dxdy = -\iint_{\mathbf{C}} Ph_n \cdot \varphi_z dxdy, \quad \varphi \in C_0^\infty(\mathbf{C}).$$

Here, $Ph_n \to Ph$ locally uniformly on $\mathbf{C}$ by (4.10) and $Th_n \to Th$ in $L^p(\mathbf{C})$ by Proposition 4.22, respectively, as $n \to \infty$. Hence, we obtain

$$\iint_{\mathbf{C}} Th \cdot \varphi dxdy = -\iint_{\mathbf{C}} Ph \cdot \varphi_z dxdy, \quad \varphi \in C_0^\infty(\mathbf{C}).$$

□

4.2.2. Existence of the Normal Solutions

Now, we are ready to prove the following fundamental theorem.

Theorem 4.24. *Fix k such that $0 \leq k < 1$ arbitrarily. Take a positive p (> 2) with $kC_p < 1$. Then for every $\mu \in B(\mathbf{C})_1$ with $\|\mu\|_\infty \leq k$ and with compact support, there exists a continuous function f such that $f(0) = 0$, $f_z - 1$ belongs to $L^p(\mathbf{C})$, and f satisfies*

$$f_{\bar{z}} = \mu f_z$$

on $\mathbf{C}$ in the sense of distribution.

Moreover, such an f is determined uniquely by these conditions.

We call this f in Theorem 4.24 the *normal solution* of the Beltrami differential equation for μ.

Proof. First, we derive a condition which the partial derivative f_z of the normal solution f for μ should satisfy. Since $f_{\bar{z}} = \mu f_z$ has a compact support, and since $f_z - 1$ belongs to $L^p(\mathbf{C})$, $f_{\bar{z}}$ also belongs to $L^p(\mathbf{C})$. Thus we can consider $P(f_{\bar{z}})$. Set

$$F(z) = f(z) - P(f_{\bar{z}})(z), \quad z \in \mathbf{C}.$$

Then by Lemma 4.20, $F(z)$ is continuous and $F(0) = 0$. Moreover, $F_{\bar{z}} = 0$ in the sense of the distribution. Hence, Weyl's lemma (Lemma 4.6) implies that $F(z)$ is holomorphic on the whole $\mathbf{C}$. On the other hand, since $f_z - 1$ and $P(f_{\bar{z}})_z = T(f_{\bar{z}})$ belong to $L^p(\mathbf{C})$, so does $F' - 1$. Thus we can conclude that $F'(z) = 1$, i.e., $F(z) = z + a$ $(a \in \mathbf{C})$. Since $f(0) = 0$, we have $a = 0$, and hence

$$f(z) = P(f_{\bar{z}})(z) + z, \quad z \in \mathbf{C}. \tag{4.17}$$

Taking the derivatives with respect to z and noting Proposition 4.23, we finally obtain the equation

$$f_z = T(\mu f_z) + 1. \tag{4.18}$$

Next, using equation (4.18), we shall show the uniqueness of the normal solution. Suppose that there is another normal solution g. Then, as above, we have

$$g_z = T(\mu g_z) + 1.$$

By Proposition 4.22, we obtain

$$\|f_z - g_z\|_p = \|T(\mu f_z) - T(\mu g_z)\|_p \leq kC_p \|f_z - g_z\|_p.$$

Since $kC_p < 1$ by the assumption, we get $f_z = g_z$ a.e. on $\mathbf{C}$. Then the Beltrami equation also gives $f_{\bar{z}} = g_{\bar{z}}$ a.e. on $\mathbf{C}$. Hence, again by Weyl's lemma we conclude that $f - g$ and $\bar{f} - \bar{g}$ are holomorphic on $\mathbf{C}$, which in turn implies that $f - g$ should be a constant. Since $f(0) = g(0) = 0$, we conclude that $f = g$, which implies the uniqueness of the normal solution.

Finally, the existence of the normal solution follows also from (4.18). In fact, repeat substituting the whole right hand side for f_z on the right hand side. Then, we have the following formal series for $f_z - 1$:

$$f_z - 1 = T\mu + T(\mu T\mu) + T(\mu T(\mu T\mu)) + \cdots .$$

This series actually converges in $L^p(\mathbf{C})$, since the linear operator which sends $h \in L^p(\mathbf{C})$ to $T(\mu h) \in L^p(\mathbf{C})$ has the operator norm not greater than kC_p (< 1). We set

$$h = T\mu + T(\mu T\mu) + \cdots . \tag{4.19}$$

Then h belongs to $L^p(\mathbf{C})$. We shall show that

$$f(z) = P(\mu(h+1))(z) + z$$

is a desired solution.

In fact, $\mu(h+1)$ belongs to $L^p(\mathbf{C})$, for μ has a compact support. Hence, Lemma 4.20 implies that f is continuous, $f(0) = 0$, and $f_{\bar{z}} = \mu(h+1)$. Moreover, Proposition 4.23 implies that

$$f_z = T(\mu(h+1)) + 1 = h + 1.$$

Hence, f satisfies the Beltrami equation $f_{\bar{z}} = \mu f_z$, and $f_z - 1$ belongs to $L^p(\mathbf{C})$. □

4.2.3. Basic Properties of Normal Solutions

From the construction of the normal solution in the proof of Theorem 4.24, we have immediately the following:

Corollary 1. *Under the same circumstances as in Theorem 4.24, the following inequalities hold:*

$$\|f_{\bar{z}}\|_p \leq \frac{1}{1 - kC_p}\|\mu\|_p, \tag{4.20}$$

and

$$|f(\zeta_1) - f(\zeta_2)| \leq \frac{K_p}{1 - kC_p}\|\mu\|_p|\zeta_1 - \zeta_2|^{1-2/p} + |\zeta_1 - \zeta_2| \tag{4.21}$$

for every $\zeta_1, \zeta_2 \in \mathbf{C}$. Here, K_p is the constant given in the proof of Lemma 4.20.

Proof. Let h be as in the proof of Theorem 4.24. Since $h = T(\mu h) + T\mu$ by (4.19), we have

$$\|h\|_p \leq kC_p\|h\|_p + C_p\|\mu\|_p.$$

Since $f_{\bar{z}} = \mu(h+1)$, we obtain (4.20).

Next, by (4.17) we have

$$|f(\zeta_1) - f(\zeta_2)| \le |P(f_{\bar{z}})(\zeta_1) - P(f_{\bar{z}})(\zeta_2)| + |\zeta_1 - \zeta_2|.$$

Hence (4.21) follows by (4.11) and (4.20). □

Furthermore, we have dependence of the normal solutions on the Beltrami coefficients as follows.

Corollary 2. *Let k and p be as in Theorem 4.24. Let $\{\mu_n\}_{n=1}^{\infty}$ be a sequence in $B(\mathbf{C})_1$ satisfying the following conditions:*

(i) $\|\mu_n\|_\infty < k$ *for every n,*
(ii) *every μ_n has a support contained in $\{z \in \mathbf{C} \mid |z| < M\}$ with a suitable constant M independent of n, and*
(iii) *μ_n converges to some $\mu \in B(\mathbf{C})_1$ a.e. on $\mathbf{C}$ as $n \to \infty$.*

Let f_n be the normal solution for μ_n, and f be the normal solution for μ. Then $f_n \to f$ uniformly on $\mathbf{C}$ as $n \to \infty$, and

$$\lim_{n\to\infty} \|(f_n)_z - f_z\|_p = 0. \tag{4.22}$$

Proof. First, for every n, (4.18) gives

$$\begin{aligned}\|f_z - (f_n)_z\|_p &\le \|T(\mu_n(f_z - (f_n)_z))\|_p + \|T((\mu - \mu_n)f_z)\|_p \\ &\le kC_p\|f_z - (f_n)_z\|_p + C_p\|(\mu - \mu_n)f_z\|_p.\end{aligned}$$

Hence, we have

$$\|f_z - (f_n)_z\|_p \le \frac{C_p}{1 - kC_p}\|(\mu - \mu_n)f_z\|_p.$$

Since the supports of all μ_n are uniformly bounded, and since μ_n converges to μ a.e. on $\mathbf{C}$ as $n \to \infty$, we get (4.22).

Next, by (4.17) and (4.10), we obtain

$$\begin{aligned}|f(\zeta) - f_n(\zeta)| &= |P(f_{\bar{z}} - (f_n)_{\bar{z}})(\zeta)| \\ &\le K_p\left\{\|(\mu - \mu_n)f_z\|_p + k\|f_z - (f_n)_z\|_p\right\}|\zeta|^{1-\frac{2}{p}}\end{aligned}$$

for every $\zeta \in \mathbf{C}$. Thus we see that $f_n \to f$ locally uniformly on $\mathbf{C}$ as $n \to \infty$. Since $f_n - f$ is holomorphic in a fixed neighborhood of ∞ for every n, we conclude that f_n converges to f uniformly on $\mathbf{C}$. □

Now, we shall show that the normal solution is, in fact, a homeomorphism of $\mathbf{C}$ onto itself, and hence is a quasiconformal mapping of $\mathbf{C}$.

Theorem 4.25. *Under the same circumstances as in Theorem 4.24, the normal solution f for μ is a quasiconformal mapping of $\mathbf{C}$, and satisfies $\mu_f = \mu$ (a.e. on $\mathbf{C}$).*

To prove this theorem, we need the following generalization of Weyl's lemma (Lemma 4.6).

Lemma 4.26. *Let u and v be continuous functions on a simply connected domain D whose distributional partial derivatives can be represented by locally integrable functions. Further, suppose that $u_{\bar{z}} = v_z$. Then there exists a function f which is continuously differentiable (i.e., of class C^1) and satisfies $f_z = u$ and $f_{\bar{z}} = v$.*

Proof. Fix a rectangle R in D arbitrarily. Take an L^1-smoothing sequence $\{u_n\}_{n=1}^{\infty}$ and $\{v_n\}_{n=1}^{\infty}$ on R for u and v, respectively, with respect to R such that $(u_n)_{\bar{z}} = (v_n)_z$ for every n. (This is possible. See the proof of Lemma 4.5.) Then for every n, Green's formula gives

$$\int_{\partial R} (u_n dz + v_n d\bar{z}) = \iint_R \{(u_n)_{\bar{z}} - (v_n)_z\} \, d\bar{z} \wedge dz = 0.$$

Hence, letting $n \to \infty$, we have

$$\int_{\partial R} (udz + vd\bar{z}) = 0.$$

Since R is arbitrary, we conclude that the indefinite integral of $udz + vd\bar{z}$ is well-defined, and gives a desired function. □

Returning to the proof of Theorem 4.25, we take a sufficiently large M so that $\{z \in \mathbf{C} \mid |z| < M\}$ contains the support of μ, and fix it in the rest of this proof. Fix also a sequence $\{\mu_n\}_{n=1}^{\infty}$ in $C_0^{\infty}(\mathbf{C})$ with $\|\mu_n\|_{\infty} \leq k$ such that the support of μ_n is contained in $\{z \in \mathbf{C} \mid |z| < M\}$ for every n, and that $\mu_n \to \mu$ a.e. on $\mathbf{C}$ as $n \to \infty$. We denote by f_n the normal solution for μ_n for every n.

Lemma 4.27. *In the foregoing situation, $f_n \colon \mathbf{C} \longrightarrow \mathbf{C}$ is a homeomorphism belonging to $C^1(\mathbf{C})$ for every n. In particular, every f_n is quasiconformal.*

Proof. Consider a function g with $g_{\bar{z}} = \mu_n g_z$. Set

$$u = g_z \quad \text{and} \quad v = g_{\bar{z}} = \mu_n u.$$

To show that g belongs to $C^1(\mathbf{C})$, it suffices by Lemma 4.26 to see that u is a continuous function which satisfies $u_{\bar{z}} = (\mu_n u)_z$. If we set $\sigma = \log u$, this is equivalent to proving that σ is a continuous function which satisfies

$$\sigma_{\bar{z}} = \mu_n \sigma_z + (\mu_n)_z. \tag{4.23}$$

Now, differential equation (4.23) is solved in a similar way to the case of the Beltrami equation. In fact, first as in the case of (4.18), we can construct a solution $\tilde{h}$ in $L^p(\mathbf{C})$ of the equation

$$\tilde{h} = T(\mu_n \tilde{h}) + T((\mu_n)_z).$$

Next, set

$$\sigma = P(\mu_n \tilde{h} + (\mu_n)_z) + C.$$

Here we choose C so that $\lim_{z\to\infty} \sigma(z) = 0$. (Note that σ is holomorphic in a neighborhood of $z = \infty$.) Since

$$\sigma_z = T(\mu_n \tilde{h} + (\mu_n)_z) = \tilde{h} \quad \text{and} \quad \sigma_{\bar{z}} = \mu_n \tilde{h} + (\mu_n)_z,$$

this σ is actually a continuous solution of (4.23).

Now, set $\tau = e^{\sigma}$. Then τ is continuous and $\tau_{\bar{z}} = (\mu_n \tau)_z$. Thus Lemma 4.26 implies that there exists a function $g \in C^1(\mathbf{C})$ satisfying $g_{\bar{z}} = \mu_n g_z (= \mu_n \tau)$.

Here, we may assume that $g(0) = 0$. Since $g_z = \tau$ is holomorphic in a neighborhood of $z = \infty$ and $\lim_{z\to\infty} \tau(z) = 1$ by the choice of C, we see that $g_z - 1$ belongs to $L^p(\mathbf{C})$. Hence, the uniqueness of the normal solution implies that $g = f_n$. Thus we conclude that f_n is of class C^1.

Finally, since the Jacobian

$$|(f_n)_z|^2 - |(f_n)_{\bar{z}}|^2 = (1 - |\mu_n|^2) \cdot |e^{2\sigma}|$$

of f_n is positive everywhere, f_n is locally homeomorphic on $\mathbf{C}$. Since f_n has a simple pole at $z = \infty$, f_n is also locally homeomorphic on the whole $\hat{\mathbf{C}}$. Therefore, the next lemma implies that f_n is a homeomorphism of $\hat{\mathbf{C}}$ onto itself. □

Lemma 4.28. *If a function $f\colon \hat{\mathbf{C}} \longrightarrow \hat{\mathbf{C}}$ is locally homeomorphic, then f is actually a homeomorphism of $\hat{\mathbf{C}}$ onto $\hat{\mathbf{C}}$.*

Proof. Denote by $\hat{\mathbf{C}}_z$ and by $\hat{\mathbf{C}}_w$ the Riemann spheres which are the domain and the target of f, respectively. Since f is an open mapping, $f(\hat{\mathbf{C}}_z)$ is open. Since $\hat{\mathbf{C}}_z$ is compact, and hence so is $f(\hat{\mathbf{C}}_z)$, we conclude that $f(\hat{\mathbf{C}}_z) = \hat{\mathbf{C}}_w$.

Next, one way to show that f is a homeomorphism is to consider f as a holomorphic function. This can be done by introducing a new complex structure on $\hat{\mathbf{C}}_z$ by pulling back the structure on $\hat{\mathbf{C}}_w$ as in §1.4.1 of Chapter 1. (Note that the topology on $\hat{\mathbf{C}}_z$ is unchanged, for f is locally homeomorphic.)

Now by assumption, f^{-1} has holomorphic branches in a neighborhood of any point of $\hat{\mathbf{C}}_w$, and any branch of f^{-1} can be continued analytically along every curve on $\hat{\mathbf{C}}_w$. Since $\hat{\mathbf{C}}_w$ is simply connected, the classical monodromy theorem implies that f^{-1} has a single valued branch on the whole $\hat{\mathbf{C}}_w$. Thus, f is a homeomorphism. □

Furthermore, we need the following:

Lemma 4.29. *Every f_n satisfies*

$$|z_1 - z_2| \leq \frac{K_p}{(1 - kC_p)^{1+\frac{2}{p}}} \|\mu_n\|_p |f_n(z_1) - f_n(z_2)|^{1-2/p} + |f_n(z_1) - f_n(z_2)| \tag{4.24}$$

for every $z_1, z_2 \in \mathbf{C}$.

Proof. For a fixed n, we can see from the proof of Lemma 4.27 that $(f_n)^{-1}$ is of class C^1. Also, $(f_n)^{-1}$ is quasiconformal by (i) in Theorem 4.10. By Proposition 4.13, we obtain

$$\mu_{(f_n)^{-1}} \circ f_n = -\frac{(f_n)_z}{\overline{(f_n)_z}}\mu_n.$$

In particular, putting $\nu_n = \mu_{(f_n)^{-1}}$, we have $|\nu_n \circ f_n| = |\mu_n|$ (a.e.). Thus we have

$$\begin{aligned}\iint_{\mathbf{C}} |\nu_n|^p dxdy &= \iint_{\mathbf{C}} |\mu_n|^p(|(f_n)_z|^2 - |(f_n)_{\bar z}|^2)dxdy \\ &\leq \iint_{\mathbf{C}} |\mu_n|^p|(f_n)_z|^2 dxdy = \iint_{\mathbf{C}} |\mu_n|^{p-2}|(f_n)_{\bar z}|^2 dxdy \\ &\leq \left(\iint_{\mathbf{C}} |\mu_n|^p dxdy\right)^{\frac{p-2}{p}} \cdot \left(\iint_{\mathbf{C}} |(f_n)_{\bar z}|^p dxdy\right)^{\frac{2}{p}}.\end{aligned}$$

Then (4.20) in Corollary 1 gives

$$\|\nu_n\|_p \leq (1 - kC_p)^{-\frac{2}{p}} \cdot \|\mu_n\|_p.$$

Thus, applying (4.21) with $f = (f_n)^{-1}$ and $\zeta_j = f_n(z_j)$ $(j = 1, 2)$, we conclude the assertion. □

Proof of Theorem 4.25. Let $\{\mu_n\}_{n=1}^{\infty}$ and $\{f_n\}_{n=1}^{\infty}$ be as defined before Lemma 4.27. Then f_n converges to f uniformly on $\mathbf{C}$ by Corollary 2 to Theorem 4.24. Since $\|\mu_n\|_p \to \|\mu\|_p$ $(n \to \infty)$, (4.24) in Lemma 4.29 is still valid when we replace f_n and μ_n by f and μ, respectively. Hence we conclude that $f : \hat{\mathbf{C}} \longrightarrow \hat{\mathbf{C}}$ is a continuous bijection, and therefore is a homeomorphism. Next, since $f_z - 1$ belongs to $L^p(\mathbf{C})$, so does $f_{\bar z} = \mu f_z$. Thus, f satisfies the assumptions in Definition A', and hence is quasiconformal. □

4.2.4. Existence Theorem

We have shown the existence of a quasiconformal mapping with complex dilatation μ when $\mu \in B(\mathbf{C})_1$ has a compact support. This conclusion is valid for a general $\mu \in B(\mathbf{C})_1$ as well.

Theorem 4.30. *For every Beltrami coefficient $\mu \in B(\mathbf{C})_1$, there exists a homeomorphism f of $\hat{\mathbf{C}}$ onto $\hat{\mathbf{C}}$ which is a quasiconformal mapping of $\mathbf{C}$ with complex dilatation μ.*

Moreover, f is uniquely determined by the following normalization conditions:

$$f(0) = 0, \quad f(1) = 1, \quad \textit{and} \quad f(\infty) = \infty.$$

We call this f, uniquely determined by the normalization conditions, the *canonical μ-qc mapping of $\hat{\mathbf{C}}$*, or the *canonical quasiconformal mapping of $\hat{\mathbf{C}}$ with complex dilatation μ*, and denote it by f^{μ}.

Proof. The uniqueness follows by Proposition 4.18 and the normalization conditions. Hence, we shall show the existence of f^μ. First, suppose that μ has a compact support. In this case, let F^μ be the normal solution for μ. Then Theorem 4.25 implies that $F^\mu(z)/F^\mu(1)$ is the desired one.

Next, suppose that $\mu = 0$ (a.e.) in some neighborhood of the origin. In this case, pulling back μ by the Möbius transformation $\gamma(z) = 1/z$, we set

$$\tilde{\mu}(z) = \mu\left(\frac{1}{z}\right)\frac{z^2}{\bar{z}^2}, \quad z \in \mathbf{C}. \tag{4.25}$$

Then $\tilde{\mu}$ belongs to $B(\mathbf{C})_1$ and has a compact support. Hence as before, there exists the canonical $\tilde{\mu}$-qc mapping $f^{\tilde{\mu}}$ of $\hat{\mathbf{C}}$. By Proposition 4.1, $f^{\tilde{\mu}}$ is totally differentiable a.e. on $\mathbf{C}$. At every such point $1/z$, the quasiconformal mapping

$$f(z) = \frac{1}{f^{\tilde{\mu}}(1/z)}$$

is also totally differentiable so that we can apply the usual chain rule. Thus we have

$$\mu_f(z) = \frac{z^2}{\bar{z}^2}\tilde{\mu}\left(\frac{1}{z}\right) = \mu(z) \quad \text{a.e. on} \quad \mathbf{C}.$$

Clearly, f satisfies the normalization conditions. Hence we conclude that f is the desired one.

Finally, suppose that μ is a general Beltrami coefficient. In this case, we set

$$\mu_1(z) = \begin{cases} \mu(z), & z \in \mathbf{C} - \bar{\Delta} \\ 0, & z \in \bar{\Delta}, \end{cases} \tag{4.26}$$

where Δ is the unit disk. Then we have shown that f^{μ_1} exists. We also set

$$\mu_2 = \left(\frac{\mu - \mu_1}{1 - \overline{\mu_1}\mu}\frac{(f^{\mu_1})_z}{\overline{(f^{\mu_1})_z}}\right) \circ (f^{\mu_1})^{-1}. \tag{4.27}$$

Since μ_2 has a compact support, f^{μ_2} also exists. Moreover, $g = f^{\mu_2} \circ f^{\mu_1}$ is quasiconformal by (iii) in Theorem 4.10, and we can see that $\mu_g = \mu$ (a.e.). (In fact, by noting Proposition 4.13, we have defined μ_2 so that $\mu_g = \mu$.) Clearly, g satisfies the normalization conditions. Hence, g is the desired one. □

At the end of this section, we state several applications of the existence theorem (Theorem 4.30).

Proposition 4.31. *Every quasiconformal mapping of the unit disk Δ onto a Jordan domain D is extended to a homeomorphism of $\bar{\Delta}$ onto $\bar{D}$.*

Proof. Fix such a quasiconformal mapping $f : \Delta \longrightarrow D$, and set $\mu = \mu_f$. By setting $\mu = 0$ on $\mathbf{C} - \Delta$, we can regard μ as an element of $B(\mathbf{C})_1$. Hence by Theorem 4.30, there exists the canonical μ-qc mapping f^μ of $\hat{\mathbf{C}}$.

Set $g = f^{\mu} \circ f^{-1}$. Then g is a quasiconformal mapping of D. By Proposition 4.13, we see that $\mu_g = 0$ a.e. on D. Hence, Corollary 2 to Theorem 4.4 implies that g is a conformal mapping of D. Since $f^{\mu}(\Delta)$ is a Jordan domain, Carathéodory's theorem gives the extension of g to a homeomorphism of $\bar{D}$ onto $\overline{f^{\mu}(\Delta)}$. Since $f = g^{-1} \circ f^{\mu}$, we obtain the assertion. □

Proposition 4.32. *There exist no quasiconformal mappings of Δ onto* $\mathbf{C}$.

Proof. Suppose that there exists such a quasiconformal mapping $f : \Delta \longrightarrow \mathbf{C}$. Then f^{-1} is also quasiconformal. We set $\mu = \mu_{f^{-1}}$; then there exists the canonical μ-qc mapping f^{μ} of $\hat{\mathbf{C}}$. If we set $g = f^{\mu} \circ f$, then $\mu_g = 0$ a.e. on Δ, and hence g is a conformal mapping of Δ.

On the other hand, since $g^{-1}(\mathbf{C}) = \Delta$, Liouville's theorem implies that g^{-1} should be a constant, a contradiction. □

Proposition 4.33. *Let μ be an arbitrary element of $B(H)_1$. Then there exists a quasiconformal mapping w of H onto H with complex dilatation μ.*

Moreover, such a mapping w (which can be extended to a homeomorphism of $\bar{H} = H \cup \hat{\mathbf{R}}$ onto itself by Proposition 4.31) is uniquely determined by the following normalization conditions:

$$w(0) = 0, \quad w(1) = 1, \quad \textit{and} \quad w(\infty) = \infty.$$

We call this unique w satisfying the normalization conditions the *canonical μ-qc mapping of H*, and denote it by w^{μ}.

Proof. The uniqueness follows by Proposition 4.18 and the normalization conditions as before.

To show the existence, set

$$\hat{\mu}(z) = \begin{cases} \mu(z), & z \in H \\ 0, & z \in \mathbf{R} \\ \overline{\mu(\bar{z})}, & z \in H^* = \mathbf{C} - \bar{H}. \end{cases}$$

By the uniqueness theorem, the canonical $\hat{\mu}$-qc mapping $f^{\hat{\mu}}$ of $\hat{\mathbf{C}}$ satisfies

$$f^{\hat{\mu}}(z) = \overline{f^{\hat{\mu}}(\bar{z})}.$$

In particular, we see that $f^{\hat{\mu}}(\hat{\mathbf{R}}) = \hat{\mathbf{R}}$. Since $f^{\hat{\mu}}$ preserves orientation, $f^{\hat{\mu}}(H) = H$. Hence, the restriction of $f^{\hat{\mu}}$ onto H is the desired one. □

4.3. Dependence on Beltrami Coefficients

Some of the most important and useful facts on the canonical quasiconformal mapping f^μ of $\hat{\mathbf{C}}$ concern dependence of f^μ on the Beltrami coefficient μ. We have already shown continuity of the normal solutions F^μ with respect to μ in Corollary 2 to Theorem 4.24. For the canonical f^μ, we first prove the following lemma.

Lemma 4.34. *For every p (> 2),*

$$\|(f^\mu)_z - 1\|_{p,\Delta} \to 0 \quad \textit{as} \quad \|\mu\|_\infty \to 0,$$

where $\mu \in B(\mathbf{C})_1$ and

$$\|h\|_{p,\Delta} = \left(\iint_\Delta |h|^p dxdy\right)^{1/p}.$$

Proof. First, assume that the supports of all μ are uniformly bounded. Fix p (> 2) arbitrarily. Since $\|\mu\|_\infty \cdot C_p < 1$ for every μ with sufficiently small $\|\mu\|_\infty$, there exists the normal solution F^μ for every such μ. As was stated in the proof of Theorem 4.30, we have $f^\mu(z) = F^\mu(z)/F^\mu(1)$. Since $F^\mu(1) \to 1$ as $\|\mu\|_\infty \to 0$ by Corollary 2 to Theorem 4.24, (4.22) implies that

$$\|(f^\mu)_z - 1\|_{p,\Delta} \to 0 \quad \text{as} \quad \|\mu\|_\infty \to 0.$$

Next, set

$$\check{f}^\mu(z) = 1/f^\mu(1/z) \quad \text{and} \quad \tilde{\mu}(z) = \mu_{\check{f}^\mu}(z) = \mu(1/z) \cdot (z^2/\bar{z}^2).$$

Then $\check{f}^\mu$ is the canonical $\tilde{\mu}$-qc mapping $f^{\tilde{\mu}}$ of $\hat{\mathbf{C}}$ (cf. the proof of Theorem 4.30). Note that the supports of all μ are uniformly bounded if and only if all $\tilde{\mu}$ vanish on some fixed neighborhood of 0.

Suppose that all $\tilde{\mu}$ vanish in some neighborhood of 0. We shall show that

$$\iint_{\{|z|<2\}} |(\check{f}^\mu)_z - 1|^p dxdy \to 0 \quad \text{as} \quad \|\tilde{\mu}\|_\infty = \|\mu\|_\infty \to 0. \tag{4.28}$$

Since $\check{f}^\mu(z) = F^\mu(1)/F^\mu(1/z)$ and $F^\mu(1) \to 1$, it suffices to estimate the quantity

$$I(\mu) = \iint_{\{|z|<2\}} |(\check{F}^\mu)_z - 1|^p dxdy,$$

where $\check{F}^\mu(z) = 1/F^\mu(1/z)$. For this purpose, we divide the domain of integration as follows:

$$\begin{aligned} I(\mu) &= \left\{\iint_{\{1/2<|z|<R\}} + \iint_{\{|z|\geq R\}}\right\} \left|\frac{z^2(F^\mu)_z(z)}{(F^\mu(z))^2} - 1\right|^p \frac{dxdy}{|z|^4} \\ &= I_1 + I_2. \end{aligned}$$

Note that, for a sufficiently large R, every F^μ is holomorphic on $\{z \in \mathbf{C} \mid |z| > R\}$ and F^μ converges to $F^0(z) = z$ uniformly on $\{z \in \mathbf{C} \mid |z| > R\}$. Hence, we can see that $I_2 \to 0$ as $\|\mu\|_\infty \to 0$. On the other hand, since

$$I_1 = \iint_{\{1/2<|z|<R\}} \left| \frac{z^2((F^\mu)_z(z) - 1)}{(F^\mu(z))^2} + \frac{z^2}{(F^\mu(z))^2} - 1 \right|^p \frac{dxdy}{|z|^4},$$

Corollary 2 to Theorem 4.24 shows that $I_1 \to 0$ as $\|\mu\|_\infty \to 0$. Thus we obtain (4.28).

Finally, for a general μ, set

$$\nu(z) = \begin{cases} \mu(z), & z \in \Delta \\ 0, & z \in \mathbf{C} - \Delta. \end{cases}$$

Letting f^ν be the canonical ν-qc mapping of $\hat{\mathbf{C}}$, we set $g^\mu = f^\mu \circ (f^\nu)^{-1}$. Again, by Corollary 2 to Theorem 4.24, $f^\nu \to id$ uniformly on $\mathbf{C}$ as $\|\mu\|_\infty \to 0$. Hence we may assume that every $f^\nu(\Delta)$ is contained in $\{z \in \mathbf{C} \mid |z| < 2\}$ and contains $\{z \in \mathbf{C} \mid |z| \leq 1/2\}$. Then by Proposition 4.13, every μ_{g^μ} vanishes on some fixed neighborhood of 0. Further, since

$$(f^\mu)_z = (g^\mu)_z \circ f^\nu \cdot (f^\nu)_z + (g^\mu)_{\bar{z}} \circ f^\nu \cdot (\overline{f^\nu})_z,$$

and since either $(f^\nu)_{\bar{z}} = 0$ or $(g^\mu)_{\bar{z}} \circ f^\nu = 0$ a.e. on $\mathbf{C}$, we have

$$\|(f^\mu)_z - 1\|_{p,\Delta} \leq \|((g^\mu)_z - 1) \circ f^\nu \cdot (f^\nu)_z\|_{p,\Delta} + \|(f^\nu)_z - 1\|_{p,\Delta}.$$

Express the right hand side of this inequality as $I_3 + I_4$. We have already shown that $I_4 \to 0$ as $\|\mu\|_\infty \to 0$. As for I_3, we obtain

$$\begin{aligned}(I_3)^p &\leq \frac{1}{1-k^2} \iint_{f^\nu(\Delta)} |(g^\mu)_z - 1|^p |(f^\nu)_z \circ (f^\nu)^{-1}|^{p-2} dxdy \\ &\leq \frac{1}{1-k^2} \left\{ \iint_{\{|z|<2\}} |(g^\mu)_z - 1|^{2p} dxdy \cdot \iint_\Delta |(f^\nu)_z|^{2p-2} dxdy \right\}^{1/2}\end{aligned}$$

with $k = \|\mu\|_\infty$. Hence, by using (4.28), where we replace $\check{f}^\mu$ and p by g^μ and $2p$, respectively, we can show that $I_3 \to 0$ as $\|\mu\|_\infty \to 0$. Thus we have the assertion. □

To investigate dependence of f^μ on μ, first we shall derive the following integral formula for f^μ.

Lemma 4.35. *Fix p with $p > 8$ arbitrarily. Let μ be an element of $B(\mathbf{C})_1$ satisfying $\|\mu\|_\infty \cdot C_p < 1$. Then the canonical μ-qc mapping f^μ of $\hat{\mathbf{C}}$ satisfies the following integral formula:*

$$\begin{aligned}f^\mu(\zeta) = \zeta &- \frac{1}{\pi}\iint_\Delta (f^\mu)_{\bar z}(z)\left(\frac{1}{z-\zeta} - \frac{\zeta}{z-1} + \frac{\zeta-1}{z}\right)dxdy\\ &-\frac{1}{\pi}\iint_\Delta \frac{(\check f^\mu)_{\bar z}(z)}{(\check f^\mu(z))^2}\left(\frac{\zeta^2 z}{1-z\zeta} - \frac{\zeta z}{1-z}\right)dxdy\end{aligned} \tag{4.29}$$

for every $\zeta \in \Delta$, where $\check f^\mu(z) = 1/f^\mu(1/z)$.

Proof. First, Pompeiu's formula (Proposition 4.19) gives

$$f^\mu(\zeta) = \frac{1}{2\pi i}\int_{\partial\Delta}\frac{f^\mu(z)dz}{z-\zeta} - \frac{1}{\pi}\iint_\Delta \frac{(f^\mu)_{\bar z}(z)}{z-\zeta}dxdy, \quad \zeta \in \Delta.$$

To prove the assertion, we should replace the first integral on the right hand side by a suitable area integral on $\mathbf{C} - \bar\Delta$. For this purpose, change the variable z to $w = 1/z$. Then $f^\mu(w)/w$ is equal to $z/\check f^\mu(z)$. Hence we have

$$\begin{aligned}\int_{\partial\Delta}\frac{f^\mu(z)}{z-\zeta}dz = \int_{\partial\Delta}\frac{f^\mu(1/z)dz}{z(1-z\zeta)} &= \int_{\partial\Delta} f^\mu\left(\frac{1}{z}\right)\left(\frac{1}{z} + \zeta + \frac{\zeta^2 z}{1-z\zeta}\right)dz\\ &= A + B\zeta + \zeta^2\int_{\partial\Delta}\frac{z}{\check f^\mu(z)}\frac{dz}{1-z\zeta}.\end{aligned}$$

Here, A and B are constants.

Now, to replace the integral on the right hand side by an area integral on Δ, we apply Green's formula. To do this, we should check that $z/\check f^\mu(z)$ behaves not so badly at the origin. Actually, we can show that there exist a neighborhood V of the origin, and a positive constant m such that

$$|\check f^\mu(z)| \geq m|z|^{p/(p-2)}, \quad z \in V. \tag{4.30}$$

In fact, decompose $\check f^\mu(z)$ as $f_2 \circ f_1$, where f_j are defined similarly as in the proof of Theorem 4.30. Namely, μ_{f_2} has a compact support and μ_{f_1} vanishes in some neighborhood of the origin. Then we know that $(f_1)^{-1}$ is conformal, and hence uniformly Lipschitz continuous in some neighborhood of the origin. On the other hand, $(f_2)^{-1}$ is uniformly $(1-2/p)$-Hölder continuous in some neighborhood of the origin, as was shown in the proof of Theorem 4.25. Thus we conclude that $(\check f^\mu)^{-1}$ is $(1-2/p)$-Hölder continuous in some neighborhood of the origin, which gives (4.30). Further, note that m is independent of μ as long as $\|\mu\|_\infty$ is sufficiently small.

Now, for every sufficiently small positive δ, set $\Delta_\delta = \{z \in \mathbf{C} \mid \delta < |z| < 1\}$. By applying Green's formula, we can see as in the proof of Proposition 4.19 that

$$\int_{\partial\Delta_\delta}\frac{z}{\check f^\mu(z)}\frac{dz}{1-z\zeta} = -2i\iint_{\Delta_\delta}\frac{(\check f^\mu)_{\bar z}(z)\cdot z}{(\check f^\mu(z))^2(1-z\zeta)}dxdy.$$

Here, (4.30) implies that

$$\int_{\{|z|=\delta\}}\frac{|z|}{|\check f^\mu(z)|}\frac{|dz|}{|1-z\zeta|} = O(\delta^{\frac{p-4}{p-2}})$$

as $\delta \to 0$. Since $p > 8$, we can show by Hölder's inequality and (4.30) that the area integral on the right hand side converges absolutely as $\delta \to 0$. Hence we have

$$f^\mu(\zeta) = \frac{A + B\zeta}{2\pi i} - \frac{1}{\pi}\iint_\Delta \frac{(f^\mu)_{\bar z}(z)}{z - \zeta} dxdy - \frac{\zeta^2}{\pi}\iint_\Delta \frac{(\check f^\mu)_{\bar z}(z)}{(\check f^\mu(z))^2}\frac{z}{1 - z\zeta} dxdy$$

for every $\zeta \in \Delta$. Moreover, since both sides of the above equality, considered as functions of ζ, are continuous on $\bar\Delta$, the equality still holds for every $\zeta \in \bar\Delta$. Hence, by using the normalization conditions that $f^\mu(0) = 0$ and $f^\mu(1) = 1$, we obtain the desired formula. □

Using integral formula (4.29), we prove the following:

Proposition 4.36. *If μ converges to 0 in $B(\mathbf{C})_1$, then the canonical μ-qc mapping f^μ converges to the identity mapping locally uniformly on $\mathbf{C}$.*

Proof. Fix $p > 8$. Since $\|\mu\|_\infty \to 0$, we may assume that, for every μ considered, $\|\mu\|_\infty \cdot C_p < 1$, and hence (4.29) in Lemma 4.35 is valid. Writing the right hand side of (4.29) as $\zeta + I(\zeta) + \check I(\zeta)$, we shall show that both $I(\zeta)$ and $\check I(\zeta)$ converge to 0.

First, since

$$(f^\mu)_{\bar z} = \mu((f^\mu)_z - 1) + \mu,$$

we can find, by using Hölder's inequality, a constant M such that

$$|I(\zeta)| \le M(\|(f^\mu)_z - 1\|_{p,\Delta} + 1)\|\mu\|_\infty, \quad \zeta \in \Delta$$

for every μ. Hence by Lemma 4.34, we see that $I(\zeta)$ converges to 0 uniformly on Δ as $\|\mu\|_\infty \to 0$.

Next, recall that V and m in (4.30) can be chosen uniformly with respect to μ. Thus we can find a constant M such that

$$|\check I(\zeta)| \le M(\|(\check f^\mu)_z - 1\|_{p,\Delta} + 1)\|\tilde\mu\|_\infty, \quad \zeta \in \Delta$$

for every μ, where $\tilde\mu$ is defined by (4.25) in the proof of Theorem 4.30. Since $\check f^\mu$ is the canonical $\tilde\mu$-qc mapping of $\hat{\mathbf{C}}$, and since $\|\mu\|_\infty = \|\tilde\mu\|_\infty$, we see by Lemma 4.34 that $\check I(\zeta) \to 0$ uniformly on Δ as $\|\mu\|_\infty \to 0$.

Hence, we have proved that $f^\mu \to id$ uniformly on Δ as $\|\mu\|_\infty \to 0$.

Finally, for every positive r (< 1), set $\mu_r(z) = \mu(z/r)$ ($z \in \mathbf{C}$), and consider the canonical μ_r-qc mapping f^{μ_r} of $\hat{\mathbf{C}}$. Set

$$f_r(z) = f^{\mu_r}(rz)/f^{\mu_r}(r).$$

Since $\mu_{f_r}(z) = \mu_r(rz) = \mu(z)$ (a.e.), and since f_r satisfies the normalization conditions, the uniqueness theorem gives $f_r = f^\mu$. On the other hand, since $f^{\mu_r} \to id$ uniformly on Δ as shown before, so does f_r on $\{z \in \mathbf{C} \mid |z| < 1/r\}$. Since r is arbitrary, we conclude that $f^\mu \to id$ locally uniformly on $\mathbf{C}$. □

Theorem 4.37. *Let $\{\mu(t)\}$ be a family of Beltrami coefficients depending on a real or a complex parameter t. Suppose that $\|\mu(t)\|_\infty \to 0$ as $t \to 0$, and that $\mu(t)$ is differentiable at $t = 0$, i.e., $\mu(t)$ is written in the form*

$$\mu(t)(z) = t\nu(z) + t\epsilon(t)(z), \quad z \in \mathbf{C}$$

with suitable $\nu \in L^\infty(\mathbf{C})$ and $\epsilon(t) \in L^\infty(\mathbf{C})$ such that $\|\epsilon(t)\|_\infty \to 0$ as $t \to 0$. Then

$$\dot{f}[\nu](\zeta) = \lim_{t\to 0} \frac{f^{\mu(t)}(\zeta) - \zeta}{t} \tag{4.31}$$

exists for every $\zeta \in \mathbf{C}$, and the convergence is locally uniform on $\mathbf{C}$. Moreover, $\dot{f}[\nu]$ has the integral representation:

$$\dot{f}[\nu](\zeta) = -\frac{1}{\pi}\iint_{\mathbf{C}} \nu(z)\frac{\zeta(\zeta-1)}{z(z-1)(z-\zeta)}dxdy, \quad \zeta \in \mathbf{C}. \tag{4.32}$$

Proof. As in the proof of Proposition 4.36, we may assume that (4.29) is valid with $\mu = \mu(t)$ for every t. Express the right hand side of formula (4.29) for $\mu(t)$ as $\zeta + I_t(\zeta) + \check{I}_t(\zeta)$.

First, since

$$(f^{\mu(t)})_{\bar{z}} = \mu(t)((f^{\mu(t)})_z - 1) + \mu(t)$$

and since $\|\mu(t)/t - \nu\|_\infty \to 0$ as $t \to 0$, we can show by an argument similar to that in the proof of Proposition 4.36 that $I_t(\zeta)/t$ converges to

$$-\frac{1}{\pi}\iint_{\Delta} \nu(z)\left(\frac{1}{z-\zeta} - \frac{\zeta}{z-1} + \frac{\zeta-1}{z}\right)dxdy \tag{4.33}$$

uniformly on Δ as $t \to 0$.

Next, set

$$\tilde{\mu}(t)(z) = \mu_{\check{f}^\mu}(z) = \mu(t)(1/z)\cdot(z^2/\bar{z}^2)$$

and

$$\tilde{\nu}(z) = \nu(1/z)\cdot(z^2/\bar{z}^2).$$

Since $\|\tilde{\mu}(t)/t - \tilde{\nu}\|_\infty \to 0$ as $t \to 0$, by an argument similar to that in the proof of Proposition 4.36, we can show that

$$\frac{\check{I}_t(\zeta)}{t} - \left\{-\frac{1}{\pi}\iint_{\Delta}\frac{\tilde{\nu}(z)}{(\check{f}^{\mu(t)}(z))^2}\left(\frac{\zeta^2 z}{1-z\zeta} - \frac{\zeta z}{1-z}\right)dxdy\right\} \to 0$$

uniformly on Δ as $t \to 0$. Further, the uniform estimate (4.30) implies that the integrand of the above integral has a majorant

$$M|z|^{-2p/(p-2)}\frac{|z|}{(1-|\zeta|)|1-z|}, \quad \zeta \in \Delta$$

for every t with a suitable sufficiently large M. Since this majorant is absolutely integrable on Δ for every $\zeta \in \Delta$, we can see that $\check{I}_t(\zeta)/t$ converges to

$$-\frac{1}{\pi}\iint_{\Delta}\nu\left(\frac{1}{z}\right)\cdot\frac{1}{\bar{z}^2}\left(\frac{\zeta^2 z}{1-z\zeta}-\frac{\zeta z}{1-z}\right)dxdy \tag{4.34}$$

locally uniformly on Δ as $t \to 0$.

Thus, changing the variable z to $1/z$ in (4.34) and adding it to (4.33), we have (4.32) for every $\zeta \in \Delta$.

Finally, the same argument as in the last part of the proof of Proposition 4.36 shows that $(f^{\mu(t)}(\zeta)-\zeta)/t$ converges to the right hand side of (4.32) locally uniformly on $\mathbf{C}$ as $t \to 0$. □

Corollary. *Let $\{\mu(t)\}$ be a family of Beltrami coefficients depending on a real or a complex parameter t. Suppose that $\mu(t)$ is differentiable at $t = 0$, i.e., $\mu(t)$ is written in the form*

$$\mu(t)(z) = \mu(z) + t\nu(z) + t\epsilon(t)(z), \quad z \in \mathbf{C}$$

with suitable $\mu \in B(\mathbf{C})_1$, $\nu \in L^\infty(\mathbf{C})$, and $\epsilon(t) \in L^\infty(\mathbf{C})$ such that $\|\epsilon(t)\|_\infty \to 0$ as $t \to 0$. Then

$$f^{\mu(t)}(\zeta) = f^\mu(\zeta) + t\dot{f}^\mu[\nu](\zeta) + o(|t|), \quad \zeta \in \mathbf{C}$$

locally uniformly on $\mathbf{C}$ as $t \to 0$, where

$$\dot{f}^\mu[\nu](\zeta) = -\frac{1}{\pi}\iint_{\mathbf{C}}\nu(z)\frac{f^\mu(\zeta)(f^\mu(\zeta)-1)((f^\mu)_z(z))^2}{f^\mu(z)(f^\mu(z)-1)(f^\mu(z)-f^\mu(\zeta))}dxdy.$$

Proof. Set $f_t = f^{\mu(t)}\circ(f^\mu)^{-1}$. Then the complex dilatation $\lambda(t)$ of f_t is given by

$$\lambda(t) = \mu_{f_t} = \left(\frac{\mu(t)-\mu}{1-\bar{\mu}\cdot\mu(t)}\frac{(f^\mu)_z}{\overline{(f^\mu)_z}}\right)\circ(f^\mu)^{-1}.$$

Hence, $\lambda(t)$ is written as

$$\lambda(t) = t\dot{\lambda} + o(|t|) \quad \text{as} \quad t \to 0$$

in $L^\infty(\mathbf{C})$, where

$$\dot{\lambda} = \left(\frac{\nu}{1-|\mu|^2}\frac{(f^\mu)_z}{\overline{((f^\mu)_z)}}\right)\circ(f^\mu)^{-1}.$$

Apply Theorem 4.37 to this family $\{f_t\}$. Then we conclude that $(f_t(\zeta)-\zeta)/t$ converges to

$$\dot{f}[\dot{\lambda}](\zeta) = -\frac{1}{\pi}\iint_{\mathbf{C}}\dot{\lambda}(z)\frac{\zeta(\zeta-1)}{z(z-1)(z-\zeta)}dxdy$$

locally uniformly on $\mathbf{C}$ as $t \to 0$. Hence, changing the variable z in this integral to $(f^\mu)^{-1}(z)$ and noting that $(f^{\mu(t)}-f^\mu)/t = \{(f_t-f_0)/t\}\circ f^\mu$, we have the assertion. □

It is to be noted that, when $\mu(t)$ depends on a complex parameter t holomorphically, i.e., $\mu(t)$ is differentiable everywhere with respect to the complex parameter t, then $f^{\mu(t)}(\zeta)$ is holomorphic with respect to t for every fixed $\zeta \in \mathbf{C}$.

Remark. Some deeper investigations give the following results (cf. Ahlfors and Bers [13]): Fix a positive constant R. Set

$$\|f\|_{B_{R,p}} = \sup_{|z_1|,|z_2|\le R} \frac{|f(z_1) - f(z_2)|}{|z_1 - z_2|^{1-2/p}} + \left(\iint_{|z|\le R} |f_z|^p \, dxdy\right)^{1/p} + \left(\iint_{|z|\le R} |f_{\bar z}|^p \, dxdy\right)^{1/p}$$

for every canonical quasiconformal mapping f of $\hat{\mathbf{C}}$ and $p > 2$. Fix k with $0 \le k < 1$, and suppose that $kC_p < 1$. Then $\{\|f^\mu\|_{B_{R,p}} \mid \|\mu\|_\infty \le k\}$ is a bounded set.

Moreover, let $\{\mu_n\}_{n=1}^\infty$ be a sequence of Beltrami coefficients such that $\|\mu_n\|_\infty \le k$ for every n and that $\mu_n \to \mu$ a.e. on $\mathbf{C}$ as $n \to \infty$. Then $\|f^{\mu_n} - f^\mu\|_{B_{R,p}} \to 0$ as $n \to \infty$. In particular, f^{μ_n} converges to f^μ locally uniformly on $\mathbf{C}$ as $t \to 0$.

Next, when $\mu(t)$ is differentiable at $t = 0$ with respect to a real or a complex parameter t, then

$$\lim_{t\to 0} \left\| \frac{f^{\mu(t)} - f^{\mu(0)}}{t} - (\dot{f}^{\mu(0)}) \left[\frac{d\mu}{dt}(0)\right] \right\|_{B_{R,p}} = 0.$$

Furthermore, when $\mu(t)$ depends on a real or a complex parameter t real-analytically or holomorphically, respectively, then $f^{\mu(t)}$ depends on t real-analytically or holomorphically with respect to the norm $\|\cdot\|_{B_{R,p}}$.

4.4. Proof of Calderón-Zygmund Theorem

The operator T in Proposition 4.22 was defined as a singular integral on $\mathbf{C} = \mathbf{R}^2$. In the one-dimensional case, the operator corresponding to this T is the classical Hilbert transformation.

To prove Calderón-Zygmund's theorem (Proposition 4.22), therefore we shall first discuss properties of the Hilbert transformation, which are classically well-known.

Lemma 4.38. (Riesz) *For every p (≥ 2), there is a constant A_p such that*

(i) *A_p is continuous with respect to p, and $A_2 = 1$, and*
(ii) *for every $\varphi \in C_0^\infty(\mathbf{R})$, the Hilbert transformation $H\varphi$ of φ defined by*

$$H\varphi(\xi) = \lim_{\epsilon\to 0} \frac{1}{\pi} \int_{\{|x-\xi|>\epsilon\}} \frac{\varphi(x)}{x-\xi} dx, \quad \xi \in \mathbf{R} \tag{4.35}$$

satisfies

$$\|H\varphi\|_{p,\mathbf{R}} \le A_p \|\varphi\|_{p,\mathbf{R}},$$

where $\|\cdot\|_{p,\mathbf{R}}$ *means the* L^p*-norm on* $\mathbf{R}$.

Proof. Since the assertion for the complex-valued functions follows from that for real ones, we may assume that φ is real-valued.

First, set

$$F(\zeta) = u(\zeta) + iv(\zeta) = \frac{1}{\pi}\int_{-\infty}^{\infty} \frac{\varphi(x)}{x-\zeta}dx, \quad \zeta = \xi + i\eta, \eta > 0.$$

Since $v(\zeta)$ is the Poisson integral of φ, it is harmonic on the upper half-plane H, continuous on $H \cup \hat{\mathbf{R}}$, and coincident with φ on $\hat{\mathbf{R}}$. The real part $u(\zeta)$ is also harmonic on H. Moreover, since

$$\begin{aligned} u(\zeta) &= \frac{1}{\pi}\int_{-\infty}^{\infty} \frac{x-\xi}{(x-\xi)^2+\eta^2}\varphi(x)dx \\ &= \frac{1}{\pi}\int_{0}^{\infty} \frac{\varphi(\xi+x)-\varphi(\xi-x)}{x}\frac{x^2}{x^2+\eta^2}dx \end{aligned}$$

for every $\zeta = \xi + i\eta$ with $\eta > 0$, we see that $u(\xi+i\eta) \to H\varphi(\xi)$ as $\eta \to 0$ for every $\xi \in \mathbf{R}$.

Next, set

$$W(\zeta) = |F(\zeta)|^p - \frac{p}{p-1}|u(\zeta)|^p.$$

Then we obtain by simple computation that

$$\frac{\partial^2 W}{\partial\zeta\partial\bar{\zeta}}(\zeta) = \frac{p^2}{4}(|F(\zeta)|^{p-2} - |u(\zeta)|^{p-2})|F'(\zeta)|^2 \ge 0, \quad \zeta \in H.$$

Hence, applying Green's formula on the domain

$$D_{\epsilon,R} = \{\zeta \in H \mid \eta > \epsilon, |\zeta| < R\}, \quad 0 < \epsilon < R < \infty,$$

we have

$$\int_{\partial D_{\epsilon,R}} \left(-\frac{i}{2}\right)\frac{\partial W}{\partial \zeta}d\zeta \ge 0.$$

Now it is easy to get a rough estimate

$$\left|\frac{\partial W}{\partial \zeta}\right| = o(|\zeta|^{-2}) \quad (|\zeta| \to \infty).$$

Hence, letting $R \to \infty$, we have

$$\int_{\{\eta=\epsilon\}} \frac{\partial W}{\partial \eta}d\xi \le 0.$$

Note also that another easily obtained rough estimate

$$\left|\frac{\partial W}{\partial \eta}(\xi + i\eta)\right| \le M(|\xi|^2 + 1)^{-1}, \quad \xi \in \mathbf{R},$$

with a sufficiently large M, holds locally uniformly with respect to η, which implies that we can change the order of integration and differentiation in the last integral. Hence, integrating with respect to η on $[\epsilon, \infty)$, we conclude that

$$\int_{-\infty}^{\infty} |F(\xi + i\epsilon)|^p d\xi \ge \frac{p}{p-1} \int_{-\infty}^{\infty} |u(\xi + i\epsilon)|^p d\xi.$$

Here, by Minkowski's inequality, we see that the left hand side is bounded by

$$\left\{ \left(\int_{-\infty}^{\infty} |u(\xi + i\epsilon)|^p d\xi \right)^{2/p} + \left(\int_{-\infty}^{\infty} |v(\xi + i\epsilon)|^p d\xi \right)^{2/p} \right\}^{p/2}.$$

Thus we conclude that

$$\int_{-\infty}^{\infty} |u(\xi + i\epsilon)|^p d\xi \le \frac{1}{\left(\left(\frac{p}{p-1}\right)^{2/p} - 1\right)^{p/2}} \int_{-\infty}^{\infty} |v(\xi + i\epsilon)|^p d\xi.$$

Finally, let ϵ tend to 0. Fatou's lemma gives

$$\|H\varphi\|_{p,\mathbf{R}} \le \frac{1}{\left(\left(\frac{p}{p-1}\right)^{2/p} - 1\right)^{p/2}} \|\varphi\|_{p,\mathbf{R}}.$$

Hence,

$$A_p = \frac{1}{\left(\left(\frac{p}{p-1}\right)^{2/p} - 1\right)^{p/2}}$$

is a desired constant for every p. □

Now, take a function φ in $C_0^\infty(\mathbf{C})$ and change the variable x to $\tilde{x} = x - \xi$ in (4.35). Considering ξ as a complex variable, we regard $H\varphi$ as a function on $\mathbf{C}$. As one of the generalizations of the Hilbert transformation on $\mathbf{R}$ to the two-dimensional case, we can consider the "circular avarage" operator, i.e., the following singular integral operator:

$$\begin{aligned} S\varphi(\zeta) &= \frac{1}{2} \int_0^\pi \left(\lim_{\epsilon \to 0} \frac{1}{\pi} \int_{\{x \in \mathbf{R} \mid |x| > \epsilon\}} \frac{\varphi(xe^{i\theta} + \zeta)}{x} dx \right) e^{-i\theta} d\theta \\ &= \lim_{\epsilon \to 0} \frac{1}{2\pi} \iint_{\{|z| > \epsilon\}} \varphi(z + \zeta) \frac{dxdy}{z|z|}, \end{aligned}$$

where $\zeta \in \mathbf{C}$ and $\varphi \in C_0^\infty(\mathbf{C})$. With respect to this operator S, Lemma 4.38 gives the following lemma.

Lemma 4.39. *For a given* p (≥ 2), *the inequality*

$$\|S\varphi\|_p \leq \frac{\pi}{2} A_p \|\varphi\|_p$$

holds for every $\varphi \in C_0^\infty(\mathbf{C})$.

In particular, the operator S *is extended to a bounded linear operator on* $L^p(\mathbf{C})$ *into itself.*

Proof. For every θ, set $\varphi_\theta(z) = \varphi(ze^{i\theta})$. Then

$$\|S\varphi\|_p \leq \frac{\pi}{2} \sup_{\theta \in [0,\pi]} \|H\varphi_\theta\|_p.$$

On the other hand, for every θ, Lemma 4.38 gives

$$\begin{aligned}
\|H\varphi_\theta\|_p^p &= \int_{-\infty}^{\infty} \left(\int_{-\infty}^{\infty} |H\varphi_\theta|^p dx \right) dy \\
&\leq A_p^p \cdot \int_{-\infty}^{\infty} \left(\int_{-\infty}^{\infty} |\varphi_\theta|^p dx \right) dy \\
&= A_p^p \|\varphi_\theta\|_p^p = A_p^p \|\varphi\|_p^p.
\end{aligned}$$

Hence, the assertion follows. □

The first assertion of Calderón-Zygmund's theorem is reduced to the following lemma.

Lemma 4.40. *For every* $\varphi \in C_0^\infty(\mathbf{C})$,

$$T\varphi = -S(S\varphi).$$

Proof. First, by Green's formula we obtain

$$\begin{aligned}
S\varphi(\zeta) &= \lim_{\epsilon \to 0} \frac{1}{2\pi} \iint_{\{|z|>\epsilon\}} \varphi(z+\zeta) \frac{\partial}{\partial z}\left(\frac{-2}{|z|}\right) dxdy \\
&= \lim_{\epsilon \to 0} \frac{1}{\pi} \left\{ \iint_{\{|z|>\epsilon\}} \varphi_z(z+\zeta) \frac{1}{|z|} dxdy + \int_{\{|z|=\epsilon\}} \frac{i}{2} \frac{\varphi(z+\zeta)}{|z|} d\bar{z} \right\} \\
&= \frac{1}{\pi} \iint_{\mathbf{C}} \varphi_z(z+\zeta) \frac{1}{|z|} dxdy = \frac{1}{\pi} \iint_{\mathbf{C}} \varphi_\zeta(z+\zeta) \frac{1}{|z|} dxdy \\
&= \frac{1}{\pi} \frac{\partial}{\partial \zeta} \left(\iint_{\mathbf{C}} \frac{\varphi(z)}{|z-\zeta|} dxdy \right).
\end{aligned}$$

As in the case of the operator P, we replace the kernel $1/|z-\zeta|$ of the integral on the right hand side by

$$k(z,\zeta) = \frac{1}{|z-\zeta|} - \frac{1}{|z|}, \quad z, \zeta \in \mathbf{C}.$$

Then we have

$$S\varphi(\zeta) = \frac{1}{\pi}\frac{\partial}{\partial\zeta}\left(\iint_{\mathbf{C}} \varphi(z)k(z,\zeta)dxdy\right). \tag{4.36}$$

(This modification is to guarantee the convergence of the integral for every $\varphi \in L^p(\mathbf{C})$.) By a standard argument, we can show that (4.36) is still valid for every $\varphi \in L^p(\mathbf{C})$ $(p \geq 2)$ in the sense of distribution.

On the other hand, Lemma 4.39 implies that $S\varphi$ belongs to $L^p(\mathbf{C})$ for every $\varphi \in C_0^\infty(\mathbf{C})$. Hence, we see by using Fubini's theorem that

$$\begin{aligned} S(S\varphi)(w) &= \frac{1}{\pi}\frac{\partial}{\partial w}\left\{\iint_{\mathbf{C}} S\varphi(\zeta)k(\zeta,w)d\xi d\eta\right\} \\ &= \frac{1}{\pi^2}\frac{\partial}{\partial w}\left\{\iint_{\mathbf{C}} k(\zeta,w)\left(\iint_{\mathbf{C}} \varphi_z(z)\frac{dxdy}{|z-\zeta|}\right)d\xi d\eta\right\} \\ &= \frac{1}{\pi^2}\frac{\partial}{\partial w}\left\{\iint_{\mathbf{C}} \varphi_z(z)\left(\iint_{\mathbf{C}} \frac{k(\zeta,w)}{|z-\zeta|}d\xi d\eta\right)dxdy\right\} \end{aligned}$$

for every $w \in \mathbf{C}$. By using Green's formula, we obtain

$$S(S\varphi)(w) = -\frac{1}{\pi^2}\frac{\partial}{\partial w}\left\{\iint_{\mathbf{C}} \varphi(z)\cdot\frac{\partial}{\partial z}\left(\iint_{\mathbf{C}} \frac{k(\zeta,w)}{|z-\zeta|}d\xi d\eta\right)dxdy\right\}. \tag{4.37}$$

Here, we need to check the partial differentiability of the integral

$$\iint_{\mathbf{C}} \frac{k(\zeta,w)}{|z-\zeta|}d\xi d\eta.$$

First, we can see that

$$\frac{\partial}{\partial z}\left(\iint_{\{|\zeta-w|>R\}} \frac{k(\zeta,w)}{|z-\zeta|}d\xi d\eta\right)$$

converges to 0 locally uniformly on $\mathbf{C}$ as $R \to \infty$. Hence, we get

$$\frac{\partial}{\partial z}\left(\iint_{\mathbf{C}} \frac{k(\zeta,w)}{|z-\zeta|}d\xi d\eta\right) = \lim_{R\to\infty}\left\{\frac{\partial}{\partial z}\iint_{\{|\zeta-w|\leq R\}} \frac{k(\zeta,w)}{|z-\zeta|}d\xi d\eta\right\}.$$

Replace the right hand side by

$$\lim_{R\to\infty}\left\{\frac{\partial}{\partial z}\iint_{\{|\zeta-w|\leq R\}} \frac{d\xi d\eta}{|z-\zeta||\zeta-w|} - \frac{\partial}{\partial z}\iint_{\{|\zeta|\leq R\}} \frac{d\xi d\eta}{|\zeta||z-\zeta|}\right\}.$$

As for the first term, we see by change of variable that

$$\begin{aligned} &\frac{\partial}{\partial z}\left(\iint_{\{|\zeta-w|\leq R\}} \frac{d\xi d\eta}{|z-\zeta||\zeta-w|}\right) = \frac{\partial}{\partial z}\left(\iint_{\{|\zeta|\leq \frac{R}{|z-w|}\}} \frac{d\xi d\eta}{|\zeta||1-\zeta|}\right) \\ &= -\frac{1}{2}\frac{R}{(z-w)|z-w|}\int_0^{2\pi} \frac{d\theta}{\left|1-\frac{Re^{i\theta}}{|z-w|}\right|} \to -\frac{\pi}{z-w} \quad (R\to\infty). \end{aligned}$$

Similarly, we obtain

$$\frac{\partial}{\partial z}\left(\iint_{\{|\zeta|\leq R\}} \frac{d\xi d\eta}{|\zeta||z-\zeta|}\right) \to -\frac{\pi}{z} \quad (R\to\infty).$$

Hence, we conclude that

$$\frac{\partial}{\partial z}\left(\iint_{\mathbf{C}} \frac{k(\zeta,w)}{|z-\zeta|} d\xi d\eta\right)$$

exists and is equal to

$$-\frac{\pi}{z-w}+\frac{\pi}{z}.$$

Now, going back to (4.37), we have shown that

$$\begin{aligned} S(S\varphi)(w) =& \frac{\partial}{\partial w}\left\{\frac{1}{\pi}\iint_{\mathbf{C}} \varphi(z)\left(\frac{1}{z-w}-\frac{1}{z}\right) dxdy\right\} \\ =& -\frac{\partial}{\partial w} P\varphi(w) = -T\varphi(w). \end{aligned}$$

Thus the assertion follows. □

Finally, the last assertion of Proposition 4.22 follows from the continuity of C_p with respect to p. Noting that $C_p \geq 1$, which is verified easily, the continuity, in turn, follows from the following Riesz-Thorin's convexity theorem.

Lemma 4.41. *The function* $\log C_p$ *is convex with respect to* $1/p$ *on* $(0,1/2]$.

Proof. Fix p_1 and p_2 with $p_j \geq 2$ for each j. Set $\alpha_j = 1/p_j$ and $C_j = C_{p_j}$ $(j=1,2)$. It suffices to show that

$$\|Tf\|_{1/\alpha} \leq C_1^{1-t}\cdot C_2^t \cdot \|f\|_{1/\alpha}$$

holds for every $\alpha = (1-t)\alpha_1 + t\alpha_2$ $(0\leq t\leq 1)$ and every $f\in L^{1/\alpha}(\mathbf{C})$. Since

$$\|Tf\|_{1/\alpha} = \sup_{g\in L^{1/(1-\alpha)}(\mathbf{C}), \|g\|_{1/(1-\alpha)}\leq 1} \left|\iint_{\mathbf{C}} Tf\cdot g dxdy\right|$$

by duality, we shall estimate $\iint_{\mathbf{C}} Tf\cdot g dxdy$.

First, we assume that f and g are step functions with compact supports. For every complex value ζ, we set

$$F(\zeta) = |f|^{\alpha(\zeta)/\alpha}\frac{f}{|f|}$$

and

$$G(\zeta) = |g|^{(1-\alpha(\zeta))/(1-\alpha)}\frac{g}{|g|},$$

where $\alpha(\zeta) = (1-\zeta)\alpha_1 + \zeta\alpha_2$. Clearly, $F(\zeta)$ and $G(\zeta)$ are also step functions for every ζ. With suitable real constants λ_k, we can write

$$\Phi(\zeta) = \iint_{\mathbf{C}} TF(\zeta) \cdot G(\zeta) dxdy = \sum_k a_k e^{\lambda_k \zeta} \quad \text{(a finite sum).}$$

Consequently, $\Phi(\zeta)$ is bounded and holomorphic on $\{\zeta = \xi + i\eta \in \mathbf{C} \mid |\xi| < M\}$ for every M $(< \infty)$. It is obvious that

$$\Phi(t) = \iint_{\mathbf{C}} Tf \cdot g dxdy.$$

In particular, $\Phi(\zeta)$ is bounded and holomorphic on $\{\zeta \in \mathbf{C} \mid 0 < \xi < 1\}$. On $\{\zeta \in \mathbf{C} \mid \xi = 0\}$, we have

$$|\Phi(\zeta)| \leq \|TF(\zeta)\|_{1/\alpha_1} \|G(\zeta)\|_{1/(1-\alpha_1)} \leq C_1(\|f\|_{1/\alpha})^{\alpha_1/\alpha}.$$

Similarly, on $\{\zeta \in \mathbf{C} \mid \xi = 1\}$, we obtain

$$|\Phi(\zeta)| \leq C_2(\|f\|_{1/\alpha})^{\alpha_2/\alpha}.$$

Hence, by the classical three-lines theorem, or by applying the maximal principle directly to the subharmonic function

$$\log|\Phi(\zeta)| - (1-\xi)\log C_1 - \xi \log C_2 - \frac{(1-\xi)\alpha_1 + \xi\alpha_2}{\alpha} \log \|f\|_{1/\alpha},$$

we can conclude that

$$|\Phi(t)| \leq C_1^{1-t} \cdot C_2^t \cdot \|f\|_{1/\alpha}.$$

at $\zeta = t$.

Since f and g are arbitrary step functions with compact supports, a routine approximation argument derives the assertion. □

Notes

This chapter is based on Chapters II and V of Ahlfors' celebrated lecture note [A-2]. Some of other equivalent definitions of quasiconformal mappings, one by the circular dilatation (cf. Chapter 1, §4) and one by the extremal length, are also frequently used. For these definitions and other basic properties, see standard texts on quasiconformal mappings such as Lehto and Virtanen [A-69].

For spatial quasiconformal mappings, see, for instance, Väisälä [A-109]. We also draw attention to Agard [3], Mostow [159], and Tukia [238].

Quasiconformal mappings are, nowadays, recognized as a useful, important, and fundamental tool, and applied not only in the theory of Teichmüller spaces, but also in various fields of complex analysis of one variable such as the theories of Riemann surfaces, of Kleinian groups, of univalent functions, and of value distributions. For example, see Bers, John and Schechter [A-16], Gehring [A-35], Ławrynowicż and Krzyz [A-65], Reimann and Rychener [A-91], Sario and Nakai [A-93], Schober [A-96], Ahlfors [12], Bers [27], [37], Drasin [56], Gehring [85], Jones [109], Segawa [191], and Sullivan [213]. Also, see Notes of Chapter 5, and of Appendix A.

Recently, quasiconformal mappings have played a crucial role in new investigations in the complex dynamics. A new notable result, called the *improved λ-lemma*, has been proved, the statement of which, for the sake of convenience, we include here.

The improved λ-lemma. *Let E be a subset of $\mathbf{C}$, and let $f(\lambda, z)\colon \Delta \times E \longrightarrow \hat{\mathbf{C}}$ be admissible, i.e., let f satisfy the following conditions:*

(i) $f(0, z) = z, \quad z \in E,$
(ii) *for every fixed $\lambda \in \Delta$, the map $f(\lambda, \cdot)\colon E \longrightarrow \hat{\mathbf{C}}$ is an injection, and*
(iii) *for every fixed $z \in E$, the map $f(\cdot, z) : \Delta \longrightarrow \hat{\mathbf{C}}$ is holomorphic.*

Then there is an admissible map $\hat{f}(\lambda, z) : \Delta_{1/3} \times \hat{\mathbf{C}} \longrightarrow \hat{\mathbf{C}}$ such that $f = \hat{f}$ on $\Delta_{1/3} \times E$, where $\Delta_{1/3} = \{\lambda \in \mathbf{C} \mid |\lambda| < 1/3\}$.

Moreover, for every fixed $\lambda \in \Delta_{1/3}$, $\hat{f}(\lambda, \cdot)$ is a quasiconformal homeomorphism of $\hat{\mathbf{C}}$ onto itself.

The second assertion is the contents of the so-called *λ-lemma*. For the proof, see Bers and Royden [43], and Sullivan and Thurston [217]. See also Slodkowski [209].

As for related papers in this field, we further cite Blanchard [44], Douady [52], Douady and Hubbard [55], Mañé, Sad and Sullivan [135], Shishikura [206] and [207], and Sullivan [214], [215], and [216].

Chapter 5

Teichmüller Spaces

In this chapter, we shall construct Teichmüller spaces alternatively by using quasiconformal mappings. First, in Section 1, we give a new definition of the Teichmüller space of an arbitrary Riemann surface by using quasiconformal mappings. In Sections 2 and 3, we investigate the case of closed Riemann surfaces of genus $g\,(\geq 2)$, and prove Teichmüller's theorem, which states that the Teichmüller space of a closed Riemann surface of genus $g\,(\geq 2)$ is homeomorphic to the open unit ball in the real $(6g-6)$-dimensional Euclidean space. The key of the proof is the existence and uniqueness of the extremal quasiconformal mappings, called Teichmüller mappings.

5.1. Analytic Construction of Teichmüller Spaces

For a closed Riemann surface R, we have considered, as the Teichmüller space of R, the set of all (marked) Riemann surfaces diffeomorphic to R equipped with isomorphisms between the fundamental groups as markings (see §3 of Chapter 1). However, in the case of general Riemann surfaces, it is too rough to place all mutually diffeomorphic surfaces in a single family, at least from the viewpoint of complex analysis. For instance, the complex plane $\mathbf{C}$ and the unit disk Δ are mutually diffeomorphic, but have various different complex analytic properties. Actually, basic functions such as Green functions and nonconstant bounded holomorphic functions exist on Δ, but do not on $\mathbf{C}$.

As a consequence, to make the investigations fruitful enough, we should consider more restrictive and analytic conditions, rather than topological ones, in the definition of the Teichmüller spaces. One way to do this is to impose the uniform boundedness condition to the distortions of the complex structures, when we consider the Teichmüller space as the set of the complex structures on the base surface. Quasiconformal mappings neatly fit this purpose.

5.1.1. Teichmüller Space of an Arbitrary Riemann Surface

Fix an arbitrary, not necessarily closed, Riemann surface R. For every quasiconformal mapping f of R onto another Riemann surface S, consider a pair (S, f). We say that two pairs (S_1, f_1) and (S_2, f_2) are *equivalent* if $f_2 \circ f_1^{-1}$ is homotopic to a conformal mapping of S_1 onto S_2. Denote by $[S, f]$ the equivalence class of (S, f).

We call the set of all such equivalence classes the *Teichmüller space* of R, and denote it by $T(R)$. This $T(R)$ can be identified with the so-called reduced Teichmüller space $T^{\#}(\Gamma)$ of a Fuchsian model Γ of R, as shall be seen in §1.2. Letting id be the identity mapping of R, we call $[R, id]$ the *base point* of $T(R)$. The topology of $T(R)$ shall be introduced in §1.3 by means of the Teichmüller distance.

In this definition, we have used the notion of quasiconformal mappings between Riemann surfaces. Since quasiconformality is a local property and conformally invariant, we can naturally define quasiconformal mappings between Riemann surfaces. However, there is another way: namely, by the uniformization theorem in §1 of Chapter 2, we define them from quasiconformal mappings between planar domains as follows.

For every Riemann surface R, take a universal covering surface $\tilde{R}$ of R. By the uniformization theorem, we may assume that $\tilde{R}$ is one of $\hat{\mathbf{C}}$, $\mathbf{C}$, or the upper half-plane H. For every homeomorphism f of R onto another Riemann surface S, by Theorem 2.4, there is a homeomorphism $\tilde{f}$, a lift of f, of $\tilde{R}$ onto a universal covering surface $\tilde{S}$ of S (which is also assumed to be one of $\hat{\mathbf{C}}$, $\mathbf{C}$, or H). We say that f is *quasiconformal* or *K-qc* if the lift $\tilde{f}$ is quasiconformal or K-qc. (Here a quasiconformal mapping of $\hat{\mathbf{C}}$ means a canonical quasiconformal mapping of $\hat{\mathbf{C}}$ composed with a Möbius transformation.) Note that by conformal invariance of quasiconformal mappings, this definition is independent of the choice of a lift $\tilde{f}$.

First, we discuss briefly the Teichmüller spaces of the exceptional Riemann surfaces (cf. §4.1 of Chapter 2).

Example 1. Suppose that $\tilde{R} = \hat{\mathbf{C}}$. Then $\tilde{R} = R = \hat{\mathbf{C}}$, and the only Riemann surface homeomorphic to R is $\hat{\mathbf{C}}$. Moreover, every quasiconformal mapping of $\hat{\mathbf{C}}$ is homotopic to id. Hence $T(\hat{\mathbf{C}})$ consists of a single point.

Example 2. Suppose that $\tilde{R} = \mathbf{C}$. By Theorem 2.13, R is conformally equivalent to one of $\mathbf{C}$, $\mathbf{C} - \{0\}$, or tori. The image of $\mathbf{C}$ or $\mathbf{C} - \{0\}$ by a quasiconformal mapping is conformally equivalent to $\mathbf{C}$ or $\mathbf{C} - \{0\}$, respectively. Moreover, every quasiconformal self-mapping of $\mathbf{C}$ is homotopic to id. Hence $T(\mathbf{C})$ consists of a single point. Next, every quasiconformal self-mapping of $\mathbf{C} - \{0\}$ is homotopic to id or to the conformal mapping $z \mapsto 1/z$. Hence, $T(\mathbf{C} - \{0\})$ also consists of a single point.

Finally, in the case of a torus R, we can show by the same argument as in §2 of Chapter 1 that $T(R)$ can be identified with the upper half-plane H.

Thus, from here on, we always assume that $\tilde{R} = H$, the upper half-plane. In this case, we have noted in Remark 1 in §4.2 of Chapter 2 that the fundamental group of R is abelian if and only if R is conformally equivalent to one of Δ (the unit disk), $\Delta - \{0\}$, or ring domains $\{z \in \mathbf{C} \mid 1 < |z| < r (< \infty)\}$.

Example 3. Suppose that $R = \Delta$ or $R = \Delta - \{0\}$. It is not difficult to see that the image of R by a quasiconformal mapping is conformally equivalent to R, and that every quasiconformal self-mapping of R is homotopic to id. Hence each of $T(\Delta)$ and $T(\Delta - \{0\})$ consists of a single point.

Next, suppose that $R = \{z \in \mathbf{C} \mid 1 < |z| < r\}$. We can show that the image of R by a quasiconformal mapping is conformally equivalent to another ring domain $S = \{z \in \mathbf{C} \mid 1 < |z| < s\}$, and that every quasiconformal self-mapping of S is homotopic to id or to the conformal mapping $z \mapsto s/z$. Moreover, by the reflection principle, we conclude that ring domains corresponding to different s are not mutually conformally equivalent. Hence, $T(R)$ is identified with the open interval $(1, +\infty)$.

5.1.2. Teichmüller Space of a Fuchsian Group

Fix a Fuchsian model Γ of R acting on the upper half-plane H. Following the discussion in §1.1, we always assume that Γ is non-commutative. In this case, note that the set of all fixed points of elements of $\Gamma - \{id\}$ contains at least three points.

In fact, Γ contains two elements γ_1 and γ_2 with $\gamma_1 \circ \gamma_2 \neq \gamma_2 \circ \gamma_1$. Moreover, set $\gamma_3 = \gamma_1 \circ \gamma_2$, then any two of $\{\gamma_j\}_{j=1}^3$ are non-commutative. Also, we know that every γ_j is either hyperbolic or parabolic (cf. Lemmas 2.6 and 2.10). If all γ_j are parabolic, then letting p_j be the unique fixed point of γ_j, we can see that $\{p_j\}_{j=1}^3$ are mutually distinct by non-commutativity of $\{\gamma_j\}_{j=1}^3$. If some element, say γ_1, is hyperbolic, then any fixed point of γ_2 is distinct from each of the fixed points of γ_1 by Lemma 2.20.

Now, we may assume that each of 0, 1, and ∞ is a fixed point of some element of $\Gamma - \{id\}$. If we do not state otherwise, we always consider the lift $\tilde{f} : H \longrightarrow H$ of a quasiconformal mapping $f : R \longrightarrow S$ which fixes each of 0, 1, and ∞. (By Proposition 4.33, such an $\tilde{f}$ is determined uniquely, and is the restriction of a quasiconformal mapping of $\hat{\mathbf{C}}$ to H.) We call this $\tilde{f}$ the *canonical lift of f with respect to Γ*.

Using the canonical lift $\tilde{f}$, we have an injective homomorphism

$$\theta_{\tilde{f}} \colon \Gamma \to PSL(2, \mathbf{R}),$$

which is defined by

$$\theta_{\tilde{f}}(\gamma) = \tilde{f} \circ \gamma \circ \tilde{f}^{-1}, \quad \gamma \in \Gamma.$$

Then we have an isomorphism $\theta_{\tilde{f}}$ of Γ onto another Fuchsian group Γ_1, and $S = H/\Gamma_1$. We also express Γ_1 as $\tilde{f}\Gamma\tilde{f}^{-1}$.

The following lemma corresponds to Theorem 1.4.

Lemma 5.1. *Two points* $[S_1, f_1], [S_2, f_2] \in T(R)$ *satisfy* $[S_1, f_1] = [S_2, f_2]$ *in* $T(R)$ *if and only if* $\theta_{\tilde{f}_1} = \theta_{\tilde{f}_2}$, *where* $\tilde{f}_j$ *is the canonical lift of* f_j *for each* $j\,(= 1, 2)$.

Proof. First, suppose that $[S_1, f_1] = [S_2, f_2]$. By composing a suitable conformal mapping of S_1 onto S_2, we may assume that $S_1 = S_2$, and f_1 is homotopic to f_2. A homotopy between f_1 and f_2 is written as a 1-parameter family $\{f_t\}_{1 \le t \le 2}$ of mappings of R to S_1. Let $\tilde{f}_1$ be the canonical lift of f_1 with respect to Γ. Then the homotopy $\{f_t\}$ has a unique continuous lift, say $\{\tilde{F}_t\}$, under the condition that $\tilde{F}_1 = \tilde{f}_1$, and $\{\tilde{F}_t\}$ gives a homotopy between $\tilde{f}_1$ and a lift $\tilde{F}_2$ of f_2.

Fix an element $\gamma \in \Gamma$ and $z \in H$ arbitrarily. Then both of the paths $\{\tilde{F}_t \circ \gamma(z) \mid 1 \le t \le 2\}$ and $\{\tilde{f}_1 \circ \gamma \circ \tilde{f}_1^{-1}(\tilde{F}_t(z)) \mid 1 \le t \le 2\}$ have the same initial point $\tilde{f}_1 \circ \gamma(z)$, and have the same projection $\{f_t \mid 1 \le t \le 2\}$ on S_1. Hence, both paths actually coincide with each other. In particular, the terminal point $\tilde{F}_2 \circ \gamma(z)$ of the former is coincident with $\tilde{f}_1 \circ \gamma \circ \tilde{f}_1^{-1}(\tilde{F}_2(z))$. Since z is arbitrary, we conclude that

$$\tilde{F}_2 \circ \gamma \circ \tilde{F}_2^{-1} = \theta_{\tilde{f}_1}(\gamma).$$

Since γ is also arbitrary, and since each of $0, 1$, and ∞ is fixed by some element of $\Gamma - \{id\}$, we see that $\tilde{F}_2$ fixes $0, 1$, and ∞. In fact, assume, for instance, that 0 is the attractive fixed point of a hyperbolic element γ_0. Then $\tilde{F}_2 \circ \gamma_0 \circ \tilde{F}_2^{-1} = \theta_{\tilde{f}_1}(\gamma_0)$ is also hyperbolic, and has 0 as the attractive fixed point. Hence, we see that $\tilde{F}_2(0) = 0$.

Thus we have shown that $\tilde{F}_2$ is coincident with the canonical lift $\tilde{f}_2$ of f_2 with respect to Γ, and hence $\theta_{\tilde{f}_1} = \theta_{\tilde{f}_2}$.

Conversely, assume that $\theta_{\tilde{f}_1} = \theta_{\tilde{f}_2} = \theta$. Then, for every $\gamma \in \Gamma$ we obtain

$$\tilde{f}_j \circ \gamma = \theta(\gamma) \circ \tilde{f}_j, \quad j = 1, 2.$$

For every t in the interval $[0, 1]$ and every $z \in H$, letting g_z be the geodesic (with respect to the Poincaré metric) connecting $\tilde{f}_1(z)$ and $\tilde{f}_2(z)$, we denote by $\tilde{f}(z, t)$ the point which divides g_z in the ratio $t : (1 - t)$. Then $\{\tilde{f}_t = \tilde{f}(z, t - 1) \mid 1 \le t \le 2\}$ is a homotopy between $\tilde{f}_1$ and $\tilde{f}_2$. From the above, we have

$$\tilde{f}_t \circ \gamma = \theta(\gamma) \circ \tilde{f}_t, \quad \gamma \in \Gamma, \, t \in [1, 2].$$

Hence, every $\tilde{f}_t$ is projected to a continuous mapping f_t of R into $S_1 = S_2$, and we have a homotopy between f_1 and f_2. □

Noting this lemma, we set

$$T^{\#}(\Gamma) = \{\theta_{\tilde{f}} \mid \tilde{f} \text{ is a canonical quasiconformal mapping of } \hat{\mathbf{C}} \text{ such that } \theta_{\tilde{f}}(\Gamma) = \tilde{f}\Gamma\tilde{f}^{-1} \text{ is a Fuchsian group}\}.$$

We call this $T^{\#}(\Gamma)$ the *reduced Teichmüller space* of Γ. It can be also regarded as the set $\{\theta_{\tilde{f}}(\Gamma) \mid \theta_{\tilde{f}} \in T^{\#}(\Gamma)\}$ of Fuchsian groups equipped with isomorphisms to Γ, or equivalently, the set of "marked" Fuchsian groups obtained as deformations of Γ by canonical quasiconformal mappings of $\hat{\mathbf{C}}$.

Remark. We can show, as in the proof of Lemma 5.2 below, that two quasiconformal mappings $f_j : R \longrightarrow S_j$ $(j = 1, 2)$ satisfy $\theta_{\tilde{f}_1} = \theta_{\tilde{f}_2}$ if and only if $\tilde{f}_1 = \tilde{f}_2$ on the limit set $L(\Gamma)$ $(\subset \hat{\mathbf{R}})$ of Γ (whose definition was given in Remark 2 of §4.3 in Chapter 2).

On the other hand, we define the Teichmüller space $T(\Gamma)$ of the Fuchsian model Γ as follows.

Let $QC(\Gamma)$ be the set of all canonical quasiconformal mappings w of $\hat{\mathbf{C}}$ such that $w\Gamma w^{-1}$ are also Fuchsian groups. We say that two elements $w_1, w_2 \in QC(\Gamma)$ are *equivalent* to each other if $w_1 = w_2$ on $\mathbf{R}$. Denote by $[w]$ the equivalence class of w. Set

$$T(\Gamma) = \{[w] \mid w \in QC(\Gamma)\}.$$

We call $T(\Gamma)$ the *Teichmüller space* of Γ. (Note that the Teichmüller space $T(\Gamma)$ can be defined for an arbitrary Fuchsian group Γ in the same way.)

From the above Remark, it is clear that two mappings determining the same point of $T(\Gamma)$ give also the same point of $T^{\#}(\Gamma)$. Hence, we can say that $T^{\#}(\Gamma)$ is a kind of reduced space of $T(\Gamma)$.

Here, we note that $T(\Gamma)$ is coincident with $T^{\#}(\Gamma)$ when $R = H/\Gamma$ is a closed Riemann surface. Actually, we have the following:

Lemma 5.2. *Suppose that R is compact. Two quasiconformal mappings f_j : $R \longrightarrow S_j$ $(j = 1, 2)$ satisfy $\theta_{\tilde{f}_1} = \theta_{\tilde{f}_2}$ if and only if $\tilde{f}_1 = \tilde{f}_2$ on* $\mathbf{R}$.

Proof. First, suppose that $\tilde{f}_1 = \tilde{f}_2$ on $\mathbf{R}$. Then for every $\gamma \in \Gamma$, Möbius transformations $\theta_{\tilde{f}_1}(\gamma)$ and $\theta_{\tilde{f}_2}(\gamma)$ are coincident with each other on $\mathbf{R}$. Hence, $\theta_{\tilde{f}_1} = \theta_{\tilde{f}_2}$ for every $\gamma \in \Gamma$, which means $\theta_{\tilde{f}_1} = \theta_{\tilde{f}_2}$.

Conversely, suppose that $\theta_{\tilde{f}_1} = \theta_{\tilde{f}_2} = \theta$. Fix a point $z_0 \in H$. By Proposition 2.19, for every $\zeta \in \hat{\mathbf{R}}$, there exists a sequence $\{\gamma_n\}_{n=1}^{\infty}$ in Γ such that $\lim_{n\to\infty} \gamma_n(z_0) = \zeta$. Moreover, the proof of Lemma 2.18 shows that such a sequence $\{\gamma_n\}_{n=1}^{\infty}$ converges locally uniformly on H to a constant function ζ. Since

$$\tilde{f}_j \circ \gamma_n(z_0) = \theta(\gamma_n) \circ \tilde{f}_j(z_0),$$

we can conclude that $\tilde{f}_1(\zeta) = \tilde{f}_2(\zeta)$. Since ζ is arbitrary, we see that $\tilde{f}_1 = \tilde{f}_2$ on $\mathbf{R}$. □

Thus we conclude the following:

Proposition 5.3. *Let Γ be a Fuchsian model of a Riemann surface R. Then the Teichmüller space $T(R)$ of R is identified with $T^{\#}(\Gamma)$ (as sets).*

Further, if R is compact, then $T(R)$ is also identified with $T(\Gamma)$.

Proof. The mapping which sends a point $[S, f] \in T(R)$ to $\theta_{\tilde{f}} \in T^{\#}(\Gamma)$ as above is well-defined, and injective by Lemma 5.1. Next, for every such $\tilde{f}$ as in the definition of $T^{\#}(R)$, we set $\Gamma_1 = \theta_{\tilde{f}}(\Gamma)$. Then $\tilde{f}$ is projected to a quasiconformal mapping f of $R = H/\Gamma$ onto $S = H/\Gamma_1$, and hence determines a point $[S, f] \in T(R)$. Thus the original mapping is also surjective, and we have the first assertion.

The second assertion follows by Lemma 5.2. □

5.1.3. Teichmüller Distance

We shall now introduce a topology on $T(R)$. For this purpose, we define a distance on $T(R)$.

Take a point $[S, f] \in T(R)$. Let μ_f be the complex dilatation of the canonical lift $\tilde{f}$ of f with respect to Γ. Then we have

$$\theta_{\tilde{f}}(\gamma) \circ \tilde{f} = \tilde{f} \circ \gamma, \quad \gamma \in \Gamma.$$

Hence, for almost every $z \in H$, it follows that

$$(\theta_{\tilde{f}}(\gamma)' \circ \tilde{f}) \cdot \tilde{f}_z = (\tilde{f}_z \circ \gamma) \cdot \gamma'$$

and

$$(\theta_{\tilde{f}}(\gamma)' \circ \tilde{f}) \cdot \tilde{f}_{\bar{z}} = (\tilde{f}_{\bar{z}} \circ \gamma) \cdot \overline{\gamma'}.$$

Thus we obtain

$$\mu_{\tilde{f}} = (\mu_{\tilde{f}} \circ \gamma)\overline{\gamma'}/\gamma' \quad \text{a.e. on} \quad H, \quad \gamma \in \Gamma. \tag{5.1}$$

Conversely, if (5.1) holds for every $\gamma \in \Gamma$, then we can see that $\theta_{\tilde{f}}(\gamma) = \tilde{f} \circ \gamma \circ \tilde{f}^{-1}$ is a holomorphic homeomorphism of H, i.e., belongs to $Aut(H)$. Hence, we conclude that $\theta_{\tilde{f}}(\Gamma)$ is a Fuchsian group, which implies that $\tilde{f}$ is projected to a quasiconformal mapping of R onto another Riemann surface $H/\theta_{\tilde{f}}(\Gamma)$.

We call a bounded measurable function μ on H satisfying (5.1) with μ instead of μ_f a *Beltrami differential* on H with respect to Γ. We denote by $B(H, \Gamma)$ the set of all Beltrami differentials on H with respect to Γ. Further, we set

$$B(H, \Gamma)_1 = \{\mu \in B(H, \Gamma) \mid \|\mu\|_\infty < 1\}.$$

We call any element of $B(H, \Gamma)_1$ a *Beltrami coefficient* on H with respect to Γ.

Similarly, we call a measurable $(-1, 1)$-form $\mu = \mu(z)d\bar{z}/dz$ on R such that $\|\mu\|_\infty = \text{ess.sup}_{z \in R}|\mu(z)| < \infty$ a *Beltrami differential* on R. Denote by $B(R)$ the set of all Beltrami differentials on R. Further, we set

$$B(R)_1 = \{\mu \in B(R)_1 \mid \|\mu\|_\infty < 1\}.$$

We call any element of $B(R)_1$ a *Beltrami coefficient* on R.

Remark 1. By the definition, $B(R)$ and $B(H, \Gamma)$ are canonically identified togather with norms.

Also, for every quasiconformal mapping f of R onto another Riemann surface, the complex dilatation $\mu_{\tilde{f}} \in B(H, \Gamma)_1$ of the canonical lift $\tilde{f}$ of f determines naturally an element $\mu \in B(R)_1$. We call this μ the *Beltrami coefficient* of f, and denote it by μ_f.

Now, for any two points $p_1 = [S_1, f_1]$, $p_2 = [S_2, f_2] \in T(R)$, let $\mathcal{F}_{f_1,f_2}$ be the set of all quasiconformal mappings of S_1 onto S_2 which are homotopic to $f_2 \circ f_1{}^{-1}$. We set

$$d(p_1, p_2) = \inf_{g \in \mathcal{F}_{f_1,f_2}} \log K(g).$$

Here, $K(g)$ means the maximal dilatation of g (i.e., that of a lift of g). We call this quantity the *Teichmüller distance* on $T(R)$ between p_1 and p_2. It is easy to see that $d(p_1, p_2)$ is independent of the choice of the representatives of p_1 and p_2.

We need to check that this function satisfies the axioms of distance.

First, since $K(g) = K(g^{-1})$ in general, we can see by using the argument as in the proof of Lemma 5.1 that the Teichmüller distance d is symmetric.

Next, the triangle inequality follows by the fact that $K(g_1 \circ g_2) \leq K(g_1) \cdot K(g_2)$.

Finally, suppose that $d(p_1, p_2) = 0$. Note that there is a sequence $\{g_n\}_{n=1}^{\infty}$ in $\mathcal{F}_{f_1,f_2}$ such that $K(g_n) \longrightarrow 1$ as $n \longrightarrow \infty$. Let $\tilde{g}_n$ be the canonical lift of g_n with respect to $\theta_{\tilde{f}_1}(\Gamma)$ for every n. Since $\mu_{\tilde{g}_n} \longrightarrow 0$ as $n \longrightarrow \infty$, it follows from Proposition 4.36 that $\tilde{g}_n$ converges to id locally uniformly on H. On the other hand, since

$$\theta_{\widetilde{f_2 \circ f_1{}^{-1}}}(\gamma) = \theta_{\tilde{g}_n}(\gamma), \quad \gamma \in \Gamma,\ n \in \mathbf{N},$$

we conclude that

$$\theta_{\widetilde{f_2 \circ f_1{}^{-1}}}(\gamma) = \gamma, \quad \gamma \in \Gamma.$$

This implies that $p_1 = p_2$.

When R is compact, we define a topology on $T(R)$ by this Teichmüller distance. Since $T(\Gamma)$ is identified with $T(R)$ (Proposition 5.3), a topolgy on $T(\Gamma)$ is defined by that of $T(R)$ under this identification. For a topology on $T(\Gamma)$ of a general Fuchsian group Γ, see for instance Lehto [A-68], Chapter V.

Remark 2. Proposition 4.13 implies that

$$\begin{aligned} \log K(f \circ g^{-1}) &= \operatorname{ess.sup}_H \log \left\{ \left(1 + \left| \frac{\mu_f - \mu_g}{1 - \overline{\mu_g}\mu_f} \right| \right) \Big/ \left(1 - \left| \frac{\mu_f - \mu_g}{1 - \overline{\mu_g}\mu_f} \right| \right) \right\} \\ &= \operatorname{ess.sup}_H \rho(\mu_f, \mu_g) \end{aligned}$$

for every $f, g \in QC(R)$, where ρ is the Poincaré distance on the unit disk Δ.

Theorem 5.4. *The Teichmüller space $T(R)$ (and hence $T^{\#}(\Gamma)$ also) is complete with respect to the Teichmüller distance.*

Proof. Take any Cauchy sequence $\{p_n = [S_n, f_n]\}_{n=1}^{\infty}$ in $T(R)$ with respect to the Teichmüller distance d. For every $\epsilon > 0$, the definition of d implies that we can find a sufficiently large N_ϵ such that, for every $n, m \geq N_\epsilon$, there is a quasiconformal mapping, say $f_{n,m}$, homotopic to $f_m \circ f_n{}^{-1}$ and satisfying that $\|\mu_{n,m}\|_\infty < \epsilon$, where $\mu_{n,m} = \mu_{f_{n,m}}$. In particular, we can find a subsequence $\{p_{n_j}\}_{j=1}^\infty$ and a sequence $\{f_{n_j,n_{j+1}}\}_{j=1}^\infty$ of quasiconformal mappings such that

$$\|\mu_{n_j,n_{j+1}}\|_\infty < 2^{-j}, \quad j = 1, 2, 3, \cdots .$$

Next, let p_0 be the base point of $T(R)$. Since $\{d(p_0, p_n)\}_{n=1}^\infty$ is a bounded sequence, we may assume that $K(f_n) < K$ for every n with a sufficiently large K (> 1). Since

$$K(f_{n_j,n_{j+1}}) \leq \frac{1+2^{-j}}{1-2^{-j}} \leq 1 + 4 \cdot 2^{-j}$$

for every j, we see that

$$g_j = f_{n_{j-1},n_j} \circ f_{n_{j-2},n_{j-1}} \circ \cdots \circ f_{n_1,n_2} \circ f_{n_1}$$

is a quasiconformal mapping of R onto S_{n_j}, homotopic to f_{n_j}, and satisfies

$$K(g_j) \leq K \cdot \prod_{j=1}^{j-1} (1 + 4 \cdot 2^{-j}).$$

Hence, $\{K(g_j)\}_{j=1}^\infty$ is a bounded sequence. We denote by K_1 the supremum of $\{K(g_j)\}$.

Now, let $\tilde{g}_j$ be the canonical lift of g_j with respect to Γ for every j. Then $\mu_j = \mu_{\tilde{g}_j}$ belongs to $B(H, \Gamma)_1$, and $\|\mu_j\|_\infty \leq k_1 = (1 - K_1)/(1 + K_1)$ (< 1). Also, we have

$$\frac{1}{2}\|\mu_j - \mu_{j+1}\|_\infty \leq \left\| \frac{\mu_{j+1} - \mu_j}{1 - \overline{\mu_j}\mu_{j+1}} \right\|_\infty = \|\mu_{n_j,n_{j+1}}\|_\infty < 2^{-j}$$

for every j. In particular, $\{\mu_j\}_{j=1}^\infty$ is a Cauchy sequence in $B(H, \Gamma)$. Hence, $\mu = \lim_{j\to\infty} \mu_j$ exists in $B(H, \Gamma)$, and satisfies $\|\mu\|_\infty \leq k_1$.

Let $\tilde{f}$ be the canonical μ-qc mapping of H. Then we can show that $\tilde{f}$ belongs to $QC(\Gamma)$. Let $p = [S, f]$ be the point in $T(R)$ determined by $\theta_{\tilde{f}}$. Since

$$\tanh\left(\frac{d(p_{n_j}, p)}{2}\right) \leq \left\| \frac{\mu - \mu_j}{1 - \overline{\mu_j}\mu} \right\|_\infty \leq \frac{1}{1 - (k_1)^2}\|\mu_j - \mu\|_\infty,$$

we see that p_{n_j} converges to p. Since the limit of a Cauchy sequence is unique, p_n also converges to the same p. This implies the completeness. □

Now, fix a point $[R_1, f_1] \in T(R)$ arbitrarily. By setting

$$[f_1]_*([S, f]) = [S, f \circ f_1{}^{-1}], \quad [S, f] \in T(R),$$

we can define a mapping $[f_1]_* : T(R) \longrightarrow T(R_1)$ of $T(R)$ onto the Teichmüller space $T(R_1)$ with base point $[R_1, id]$. Moreover, we have the following proposition.

Proposition 5.5. *This mapping $[f_1]_* : T(R) \longrightarrow T(R_1)$ is an isometrical homeomorphism with respect to the Teichmüller distances.*

In particular, $T(R_1)$ is homeomorphic to $T(R)$.

Proof. First, since $[f_1{}^{-1}]_* : T(R_1) \longrightarrow T(R)$ gives the inverse mapping of $[f_1]_*$, clearly $[f_1]_*$ is a bijection.

Next, for any two points $p = [S, f], q = [S', g] \in T(R)$, the family $\mathcal{F}_{f,g}$ in the definition of the Teichmüller distance coincides with $\mathcal{F}_{f\circ f_1^{-1}, g\circ f_1^{-1}}$. Hence it is clear that $[f_1]_*$ is an isometry. □

We call a mapping such as this $[f_1]_* : T(R) \longrightarrow T(R_1)$ a *translation of the base point* of the Teichmüller space $T(R)$.

Remark 3. Proposition 5.5 implies that $T(R)$ are mutually homeomorphic for all closed Riemann surfaces R of the same genus g (≥ 2). Namely, the Teichmüller space for such R is independent of the base point. Hence, we may denote such a space by T_g, and call it the *Teichmüller space of genus* g (≥ 2).

We show in the next section that $T(R)$ and T_g defined in this section are coincident with those defined in Chapter 1.

5.2. Teichmüller Mappings and Teichmüller's Theorems

5.2.1. Holomorphic Quadratic Differentials

In the case of tori, closed Riemann surfaces of genus one, we have seen that any point in T_1, the Teichmüller space of genus 1, is sent to any other point in T_1 under the projection of an orientation-preserving affine mapping of $\mathbf{C}$.

We shall discuss similar "canonical" mappings for the case of genus $g \geq 2$.

Let $w = f(z) = \alpha z + \beta \bar{z}$ ($\alpha, \beta \in \mathbf{C}, |\alpha| > |\beta|$) be an orientation-preserving affine mapping of $\mathbf{C}$. Applying a suitable rotation to the z-plane, and suitable rotation and stretching (or compressing) to the w-plane, we can write $f(z)$ in the form

$$z \longmapsto Kx + iy = \frac{K+1}{2} z + \frac{K-1}{2} \bar{z}$$

for some $K \geq 1$. Hence, when we set $k = (K-1)/(K+1)$, the pull-back of the Euclidean metric $|dw|^2$ on the w-plane under f has the form $((K+1)/2)^2 \, |dz + k d\bar{z}|^2$. Upon changing the local parameter z by a conformal mapping $z = h(\zeta)$, we see that $|dz + kd\bar{z}|^2$ becomes

$$|h'|^2 \left| d\zeta + k \frac{\overline{(h')^2}}{|(h')^2|} d\bar{\zeta} \right|^2 .$$

Here, as we shall explain later, the quantity $(h')^2$ in this form can be considered as a representative of a so-called holomorphic quadratic differential.

A family $\varphi = \{\varphi_j\}$ of holomorphic functions φ_j on $z_j(U_j)$ for all coordinate neighborhoods (U_j, z_j) of a Riemann surface R is called a ***holomorphic quadratic differential*** on R if it satisfies

$$\varphi_k(z_k) = \varphi_j \circ z_{jk}(z_k) \cdot (z_{jk}'(z_k))^2 \quad \text{on } U_j \cap U_k, \tag{5.2}$$

where $z_{jk} = z_j \circ z_k^{-1}$.

We express (5.2) simply as

$$\varphi_k(z_k) = \varphi_j(z_j)(dz_j/dz_k)^2.$$

We also write

$$\varphi = \varphi(z)dz^2.$$

Denote by $A_2(R)$ the complex vector space of all holomorphic quadratic differentials on R. A holomorphic quadratic differential corresponds to a holomorphic automorphic form of weight -4 with respect to a Fuchsian model Γ of R acting on the upper half-plane H. Here, a ***holomorphic automorphic form*** $\varphi(z)$ ***of weight*** -4 with respect to Γ is, by definition, a holomorphic function $\varphi(z)$ on H such that

$$\varphi(\gamma(z))\gamma'(z)^2 = \varphi(z), \quad z \in H, \ \gamma \in \Gamma.$$

We denote by $A_2(H, \Gamma)$ the complex vector space of all holomorphic automorphic functions of weight -4 with respect to Γ.

Remark. From these definitions, $A_2(R)$ is canonically identified with $A_2(H, \Gamma)$. In fact, any element of $A_2(H, \Gamma)$ clearly determines an element of $A_2(R)$. Conversely, for every $\varphi = \{\varphi_j\} \in A_2(R)$, formula (5.2) implies that the family $\{\varphi_j(z_j \circ \pi)((z_j \circ \pi)')^2\}$, as a whole, determines exactly one single-valued holomorphic function on H, which belongs to $A_2(H, \Gamma)$. Here, $\pi\colon H \longrightarrow R = H/\Gamma$ is the projection.

5.2.2. Teichmüller Mappings

As a "locally affine" quasiconformal mapping of R, we take a mapping f such that for some constant $k\,(0 \le k < 1)$, it satisfies

$$f_{\bar{z}} = k f_z$$

for a suitable local coordinate z around almost every point of R. More precisely, we discuss a quasiconformal mapping f whose Beltrami coefficient μ_f satisfies that

$$\mu_f = k\frac{\overline{\varphi}}{|\varphi|}$$

with a suitable $\varphi \in A_2(R)$. (See Proposition 5.19 below.)

Let a positive k (< 1) and $\varphi \in A_2(R) - \{0\}$ be given. Then we call a quasiconformal mapping f a ***formal Teichmüller mapping*** of R for the pair (k, φ)

if the Beltrami coefficient μ_f of f is equal to $k\overline{\varphi}/|\varphi|$. Here, we regard conformal mappings also as formal Teichmüller mappings, which correspond to the case $k = 0$ or $\varphi = 0$. Note that $k\overline{\varphi}/|\varphi|$ is unchanged when we replace φ by $c\varphi$ with any positive constant c.

Set

$$A_2(R)_1 = \{\varphi \in A_2(R) \mid \|\varphi\|_1 < 1\}.$$

From here on, writing $\varphi = \varphi(z)dz^2$, we put

$$\|\varphi\|_1 = 2\iint_R |\varphi(z)|dxdy.$$

(This integral is often written simply as $\iint_R |\varphi|$.)

For an element $\varphi \in A_2(R)_1$, we may assume that $k = \|\varphi\|_1$. We call a formal Teichmüller mapping for this pair (k, φ) a *Teichmüller mapping* for φ.

In the following discussion, we consider only the case of closed Riemann surfaces of genus g (≥ 2). In this case, $\|\varphi\|_1$ is finite for any $\varphi \in A_2(R)$, and $A_2(R)$ is considered as a complex Banach space with this norm $\|\cdot\|_1$. Moreover, Riemann-Roch's theorem tells that $A_2(R)$ is actually a $(3g-3)$-dimensional complex vector space. (Recall that in the case of a torus R, $A_2(R)$ is one dimensional.)

Now, we have a mapping

$$\mathcal{T}: A_2(R)_1 \to T(R)$$

defined by

$$\mathcal{T}(\varphi) = [S, f], \quad \varphi \in A_2(R)_1,$$

where $f: R \longrightarrow S = f(R)$ is a Teichmüller mapping for $\varphi \neq 0$, and $f = id$ for $\varphi = 0$.

The main purpose of this section is to show that this mapping $\mathcal{T}: A_2(R)_1 \longrightarrow T(R)$ is a surjective homeomorphism. For this purpose, we use the Fricke space F_g defined in §5 of Chapter 2.

First, we shall show that $T(R)$ and T_g constructed in this chapter are identified with those defined in Chapter 1. In this section, we use the notation $T(R)^{old}$ and T_g^{old} for the latter. Namely, T_g^{old} is the set of all closed marked Riemann surfaces of genus g (≥ 2), and $T(R)^{old}$ is that defined in §3 of Chapter 1 by using orientation-preserving diffeomorphisms of R. Recall that T_g^{old} is identified with $T(R)^{old}$ as stated in Theorem 1.4, and that we introduce a topology on T_g^{old} in §5 of Chapter 2 so that $\mathcal{F}_g: T_g^{old} \longrightarrow F_g$ is a surjective homeomorphism. On the other hand, fixing a marking $\Sigma = \{[A_j], [B_j]\}_{j=1}^g$ on R, we have a mapping

$$\Phi_\Sigma: T(R) \to T_g^{old}$$

given by

$$\Phi_\Sigma([S, f]) = [S, f_*(\Sigma)], \quad [S, f] \in T(R),$$

where $f_*(\Sigma) = \{f_*([A_j]), f_*([B_j])\}_{j=1}^g$. Note that Lemma 5.1, Proposition 2.23 and Theorem 2.25 imply that Φ_Σ is well-defined and injective. Also, recall that the identification between $T(R)^{old}$ and T_g^{old} was given by the same Φ_Σ. Hence, Φ_Σ is clearly surjective. Thus we have the following lemma.

Lemma 5.6. *The mappings $\Phi_\Sigma : T(R) \longrightarrow T_g^{old}$ and $\mathcal{F}_g \circ \Phi_\Sigma : T(R) \longrightarrow F_g$ are bijective. In particular, $F_g = \mathcal{F}_g \circ \Phi_\Sigma(T(R))$.*

For the sake of simplicity, we set $\tilde{\mathcal{T}} = \mathcal{F}_g \circ \Phi_\Sigma \circ \mathcal{T}$. Then we obtain the following:

Lemma 5.7. *The mapping $\tilde{\mathcal{T}} : A_2(R)_1 \longrightarrow F_g$ is continuous.*

Proof. Let $\{\varphi_n\}_{n=1}^\infty$ be an arbitrary convergent sequence in $A_2(R)_1$, and φ_0 be its limit. For every n, let f_n be a Teichmüller mapping of R for φ_n, and $\tilde{f}_n$ be the canonical lift of f_n on H with respect to Γ, where Γ is the normalized Fuchsian model for $[R, \Sigma]$. Set

$$\Gamma_n = \tilde{f}_n \Gamma \tilde{f}_n^{-1}, \quad n = 0, 1, 2, \cdots.$$

Then $t_n = \tilde{\mathcal{T}}(\varphi_n)$ is a point of F_g representing Γ_n by definition.

We set $\tilde{h}_n = \tilde{f}_n \circ \tilde{f}_0^{-1}$ and $\mu_n = \mu_{\tilde{h}_n}$ for every n. Then we obtain

$$\mu_n = \left(\frac{\mu_{\tilde{f}_n} - \mu_{\tilde{f}_0}}{1 - \overline{\mu_{\tilde{f}_0}} \cdot \mu_{\tilde{f}_n}} \frac{(\tilde{f}_0)_z}{\overline{(\tilde{f}_0)_z}} \right) \circ \tilde{f}_0^{-1}.$$

Since $\lim_{n\to\infty} \|\varphi_n\|_1 = \|\varphi_0\|_1 < 1$, we can find a positive $k < 1$ such that

$$\|\mu_n\|_\infty \le k \quad (n = 0, 1, 2, \cdots).$$

When $\varphi_0 = 0$, then $\lim_{n\to\infty} \|\varphi_n\|_1 = 0$. Hence, by Proposition 4.36, $\tilde{h}_n$ converges to id locally uniformly on H. Even when $\varphi_0 \ne 0$, we can show the same assertion. In fact, since $\lim_{n\to\infty} \|\varphi_n - \varphi_0\|_1 = 0$, $\tilde{\varphi}_n(z)$ converges to $\tilde{\varphi}_0(z)$ locally uniformly on H, where $\tilde{\varphi}_n(z)$ is the element of $A_2(H, \Gamma)$ corresponding to φ_n. Hence, letting $H' = \{z \in H \mid \tilde{\varphi}_0(z) \ne 0\}$, we can show that μ_n converges to 0 locally uniformly on H', which is enough to show the locally uniform convergence of $\{\tilde{h}_n\}_{n=1}^\infty$ on H. However, since it needs a fairly long argument, we first finish the proof of Lemma 5.7.

Since $\tilde{h}_n$ converges to id locally uniformly on H in any case, $\tilde{f}_n \circ \gamma \circ \tilde{f}_n^{-1}$ converges to $\tilde{f}_0 \circ \gamma \circ \tilde{f}_0^{-1}$ for every $\gamma \in \Gamma$, which implies that t_n converges to t_0. Thus we have proved the assertion.

Now, we return to the proof of the locally uniform convergence of $\{\tilde{h}_n\}_{n=1}^\infty$ to id on H even when $\varphi_0 \ne 0$.

For every n, we set

$$\nu_n(z) = \begin{cases} \mu_n(z), & z \in H \\ 0, & z \in \mathbf{R} \\ \overline{\mu_n(\bar{z})}, & z \in H^*. \end{cases}$$

Then $\tilde{h}_n$ is the restriction of the canonical ν_n-qc mapping f^{ν_n} of $\hat{\mathbf{C}}$ to H (cf. the proof of Proposition 4.33). Also, we know that $\|\nu_n\|_\infty < k$ for every n and that $\nu_n \longrightarrow 0$ as $n \longrightarrow \infty$ a.e. on $\mathbf{C}$. Hence, if we use the fact in the Remark at the end of §3 of Chapter 4, we obtain the assertion immediately. However, since we have not given a proof of this fact, we take another fairly long but rather elementary approach, for we only use properties of normal solutions stated in §2 of Chapter 4.

For this purpose, we decompose f^{ν_n} as in the proof of Theorem 4.30. First, for every n, we set

$$\nu_n^1(z) = \begin{cases} 0, & z \in \bar{\Delta} \\ \nu_n(z), & z \in \mathbf{C} - \bar{\Delta}, \end{cases}$$

$$\tilde{\nu}_n(z) = \nu_n^1\left(\frac{1}{z}\right) \cdot \left(\frac{z^2}{\bar{z}^2}\right).$$

Then the support of every $\tilde{\nu}_n$ is contained in $\bar{\Delta}$, and we know that $f^{\nu_n^1}(z) = 1/f^{\tilde{\nu}^n}(1/z)$. Clearly, $\|\tilde{\nu}_n\|_\infty \leq k$ for every n and $\tilde{\nu}_n \longrightarrow 0$ as $n \longrightarrow \infty$ a.e. on $\mathbf{C}$. Hence, by Corollary 2 in §2.3 of Chapter 4, the normal solution $\tilde{F}_n$ for $\tilde{\nu}_n$ converges to id uniformly on $\mathbf{C}$ as n tends to ∞. Then it is easy to see that $f^{\nu_n^1}(z) = \tilde{F}(1)/\tilde{F}_n(1/z)$ converges to id locally uniformly on $\mathbf{C}$.

Next, for every n, decompose f^{ν_n} as $f^{\lambda_n} \circ f^{\nu_n^1}$. Then

$$\lambda_n(z) = \begin{cases} \left(\nu_n \dfrac{\overline{(f^{\nu_n^1})'}}{(f^{\nu_n^1})'}\right) \circ (f^{\nu_n^1})^{-1}, & z \in f^{\nu_n^1}(\Delta) \\ 0, & z \in \mathbf{C} - f^{\nu_n^1}(\Delta) \end{cases}$$

for every n. In particular, the supports of λ_n are uniformly bounded. We shall show that f^{λ_n} also converges to id locally uniformly on $\mathbf{C}$.

Now, since every $\tilde{F}_n$ satisfies (4.24) of Chapter 4, as was stated in the proof of Theorem 4.25, we can see that the family $\{(f^{\nu_n^1})^{-1}\}_{n=1}^\infty$ is locally uniformly bounded and locally equicontinuous, and hence is a normal family on $\mathbf{C}$. Also, we can show by a standard argument that $(f^{\nu_n^1})^{-1}$ converges to id locally uniformly on $\mathbf{C}$. Hence,

$$\left(\frac{\overline{(f^{\nu_n^1})'}}{(f^{\nu_n^1})'}\right) \circ (f^{\nu_n^1})^{-1}$$

converges to 1 locally uniformly on Δ, since every $(f^{\nu_n^1})'$ has no zeros on Δ. Also, since ν_n converges to 0 locally uniformly on $\{z \in \Delta \mid z \in H' \text{ or } \bar{z} \in H'\}$, we conclude that $\nu_n \circ (f^{\nu_n^1})^{-1}$ converges to 0 a.e. on Δ.

Thus we have shown that λ_n converges to 0 a.e. on Δ as $n \to \infty$. Since $f^{\nu_n^1}$ converges to id locally uniformly on $\mathbf{C}$, it is easy to see that λ_n also converges to 0 a.e. on $\mathbf{C} - \bar{\Delta}$. Hence, again by Corollary 2 in §2.3 of Chapter 4, the normal solution F_n for λ_n converges to id locally uniformly on $\mathbf{C}$, and hence so does $f^{\lambda_n}(z) = F_n(z)/F_n(1)$. □

Similarly, (but more easily by applying Proposition 4.36) we can show the following lemma.

Lemma 5.8. *The mapping $\mathcal{F}_g \circ \Phi_\Sigma : T(R) \longrightarrow F_g$ is continuous.*

5.2.3. Teichmüller's Theorems

The injectivity of $\tilde{T}$ follows from the following *Teichmüller's uniqueness theorem.*

Theorem 5.9. *Let f be a Teichmüller mapping for an element $\varphi \in A_2(R)_1$, and let $T(\varphi) = [S, f]$. Then every quasiconformal mapping f_1 of R to S which is homotopic to f satisfies*

$$\|\mu_{f_1}\|_\infty \geq \|\mu_f\|_\infty.$$

Moreover, the equality holds if and only if $f_1 = f$.

A proof of this theorem shall be given in §3, for it needs some preliminary discussions. Returning to the proof of the fact that T is a surjective homeomorphism, we note the following corollary to Theorem 5.9.

Corollary. *The mappings T and $\tilde{T}$ are injective.*

Proof. By Lemma 5.6, it suffices to show the injectivity of T. Assume that $T(\varphi_1) = T(\varphi_2)$ for some $\varphi_1, \varphi_2 \in A_2(R)_1$. Let f_j be a Teichmüller mapping for φ_j and $T(\varphi_j) = [S_j, f_j]$ for each j. Then the assumption implies that there is a conformal mapping h of S_1 onto S_2 such that $h \circ f_1$ is homotopic to f_2. Thus Theorem 5.9 gives

$$\|\mu_{f_1}\|_\infty = \|\mu_{h \circ f_1}\|_\infty \geq \|\mu_{f_2}\|_\infty.$$

Similarly, since $h^{-1} \circ f_2$ is homotopic to f_1, we have

$$\|\mu_{f_2}\|_\infty \geq \|\mu_{f_1}\|_\infty.$$

Hence we conclude that $\|\mu_{h \circ f_1}\|_\infty = \|\mu_{f_2}\|_\infty$, which implies that $h \circ f_1 = f_2$ again by Theorem 5.9. In particular, $\mu_{f_1} = \mu_{f_2}$.

Thus if $\varphi_1 = 0$, then $\varphi_2 = 0$. If $\varphi_1 \neq 0$, then $\|\varphi_1\|_1 = \|\varphi_2\|_1$, and $\varphi_1/|\varphi_1| = \varphi_2/|\varphi_2|$ a.e. on R. Hence we conclude that φ_2/φ_1 is positive a.e. on R. Since φ_2/φ_1 is meromorphic, it should be a constant. Namely, there is a positive constant c with $\varphi_1 = c\varphi_2$. Since $\|\varphi_1\|_1 = \|\varphi_2\|_1$, we conclude that $c = 1$, i.e., $\varphi_1 = \varphi_2$, which shows the injectivity of T. □

Lemma 5.10. *The image $\tilde{T}(A_2(R)_1)$ of $A_2(R)_1$ under $\tilde{T} : A_2(R)_1 \longrightarrow F_g$ is an open set, and $\tilde{T}$ is a homeomorphism onto its image.*

Proof. By Lemma 5.7 and the Corollary to Theorem 5.9, we see that $\tilde{T}$ is a continuous injection. Since $A_2(R)_1$ is homeomorphic to $\mathbf{R}^{6g-6}$, Brouwer's theorem on invariance of domains (Theorem 3.11) gives the assertion. □

Lemma 5.11. *The mapping $\mathcal{T}: A_2(R)_1 \longrightarrow T(R)$ is a homeomorphism onto its image.*

Proof. Since $\mathcal{T}$ is injective by the Corollary to Theorem 5.9, it follows that $\mathcal{T}^{-1}$ is well-defined on $E = \mathcal{T}(A_2(R)_1)$. Since $\mathcal{T}^{-1} = \tilde{\mathcal{T}}^{-1} \circ (\mathcal{F}_g \circ \Phi_\Sigma)$, Lemmas 5.8 and 5.10 imply that $\mathcal{T}^{-1}$ is continuous on E.

It remains to show that $\mathcal{T}$ is continuous. For this purpose, fix a point $\varphi \in A_2(R)_1$ arbitrarily, and set $p = \mathcal{T}(\varphi) = [R_1, f_1]$. Consider a translation $[f_1]_* : T(R) \longrightarrow T(R_1)$ of the base point, which sends p to the base point $[R_1, id]$ of $T(R_1)$. Then we know by Proposition 5.5 that $[f_1]_*$ is a surjective isometry.

We define a mapping

$$\tilde{\mathcal{T}}_1 = \mathcal{F}_g \circ \Phi_\Sigma \circ [f_1]_*^{-1} \circ \mathcal{T}_1 : A_2(R_1)_1 \longrightarrow F_g$$

in the same way as in the case of $\tilde{\mathcal{T}}$. Since $\tilde{\mathcal{T}}(\varphi) = \tilde{\mathcal{T}}_1(0)$, Lemma 5.10 implies that $(\tilde{\mathcal{T}}_1)^{-1} \circ \tilde{\mathcal{T}} = (\mathcal{T}_1)^{-1} \circ [f_1]_* \circ \mathcal{T} : A_2(R)_1 \longrightarrow A_2(R_1)_1$ is well-defined in some neighborhood of φ, and is a homeomorphism onto its image. Hence, $\mathcal{T}$ is continuous at φ if and only if this is also true for $\mathcal{T}_1 : A_2(R_1)_1 \longrightarrow T(R_1)$ (defined similarly to $\mathcal{T}$ by using Teichmüller mappings) at the origin.

Now, take any sequence $\{\psi_n\}_{n=1}^\infty$ in $A_2(R_1)_1$ such that $\|\psi_n\|_1 \longrightarrow 0$ as $n \longrightarrow \infty$. Since the maximal dilatation of a Teichmüller mapping for ψ_n is equal to $(1 + \|\psi_n\|_1)/(1 - \|\psi_n\|_1)$ for every n, we have

$$d(\mathcal{T}_1(0), \mathcal{T}_1(\psi_n)) \leq \log \frac{1 + \|\psi_n\|_1}{1 - \|\psi_n\|_1} \longrightarrow 0 \quad (n \longrightarrow \infty).$$

Thus $\mathcal{T}_1$ is continuous at the origin.

Since φ is arbitrary, we conclude that $\mathcal{T}$ is continuous. □

Finally, we shall show the surjectivity of $\mathcal{T}$. First, we recall the following:

Lemma 5.12. *The spaces $T(R)$ and F_g are connected.*

Proof. Fix an arbitrary point $[S, f] \in T(R)$, and set $\mu_f = \mu$. For every t with $0 \leq t \leq 1$, let f_t be a quasiconformal mapping of R whose Beltrami coefficient is $t\mu$. We set $S_t = f_t(R)$. Then we obtain a continuous curve $\{[S_t, f_t] \mid 0 \leq t \leq 1\}$ in $T(R)$ which connects between the base point $[R, id]$ and $[S, f]$. Thus $T(R)$ is arcwise connected.

Next, Lemmas 5.6 and 5.8 imply that F_g is also connected. □

Lemma 5.13. *The mappings $\mathcal{T}$ and $\tilde{\mathcal{T}}$ are surjective. Namely, $\mathcal{T}(A_2(R)_1) = T(R)$, and $\tilde{\mathcal{T}}(A_2(R)_1) = F_g$.*

Proof. By noting Lemma 5.6, it suffices to show the assertion for $\mathcal{T}$.

From Lemmas 5.8 and 5.10, we get

$$E = \mathcal{T}(A_2(R)_1) = (\mathcal{F}_g \circ \Phi_\Sigma)^{-1}(\tilde{\mathcal{T}}(A_2(R)_1)),$$

which is an open set in $T(R)$. Further, Lemma 5.12 implies that $T(R)$ is connected. Hence the assertion follows if we show that the relative boundary ∂E of E in $T(R)$ is empty.

Now, suppose that $\partial E \neq \phi$. Take any $[S, f] \in \partial E$. Then there is a sequence $\{\varphi_n\}_{n=1}^{\infty}$ in $A_2(R)_1$ such that $T(\varphi_n) \longrightarrow [S, f]$, and $\|\varphi_n\|_1 \longrightarrow 1$ as $n \longrightarrow \infty$. Let f_n be a Teichmüller mapping for φ_n. We set $T(\varphi_n) = [S_n, f_n]$. By the assumption, there is a quasiconformal mapping h_n of S_n onto S which is homotopic to $f \circ f_n^{-1}$ for every n such that $\|\mu_{h_n}\|_\infty \longrightarrow 0$ as $n \longrightarrow \infty$. In particular, for a suitable $k < 1$, we have

$$\|\mu_{g_n}\|_\infty \leq k, \quad n = 1, 2, \cdots,$$

where $g_n = h_n^{-1} \circ f$.

On the other hand, since g_n is homotopic to f_n, Theorem 5.9 implies that

$$\|\mu_{g_n}\|_\infty \geq \|\mu_{f_n}\|_\infty = \|\varphi\|_\infty \longrightarrow 1 \quad (n \longrightarrow \infty).$$

This is a contradiction. Thus we conclude that ∂E is empty. □

As a corollary to Lemma 5.13, we obtain the following *Teichmüller's existence theorem.*

Theorem 5.14. *For every quasiconformal mapping $f : R \longrightarrow S$, there exists a Teichmüller mapping homotopic to f.*

Lemma 5.13 finishes a proof of *Teichmüller's theorem.* Namely, we have proved the following theorem.

Theorem 5.15. *The mapping $T \colon A_2(R)_1 \longrightarrow T(R)$ is a surjective homeomorphism.*

In particular, $T(R)$ is homeomorphic to $A_2(R)_1$, and hence to $\mathbf{R}^{6g-6}$

In the course of this proof, we have also shown that all representations we have considered as the Teichmüller space of a closed Riemann surface of genus g (≥ 2) are mutually homeomorphic.

Corollary. *The spaces $T(R)$, $T(R)^{old}$, T_g, T_g^{old}, F_g, and $\mathbf{R}^{6g-6}$ are mutually homeomorphic to each other.*

5.3. Proof of Teichmüller's Uniqueness Theorem

Teichmüller's uniqueness theorem (Theorem 5.9) asserts that deformations of the complex structures with the best "efficiency" are given by applying "locally affine" mappings called Teichmüller mappings.

One of the decisive properties of affine mappings is that they send lines to lines. First, we discuss the meaning of "lines" with respect to a Teichmüller mapping.

Recall that lines on $\mathbf{C}$ are geodesics with respect to the Euclidean metric. Even in the case of Teichmüller mappings, it may be natural to consider geodesics with respect to some metric associated with the given Teichmüller mapping. As such a metric, we consider the "metric" $ds^2 = |\varphi(z)||dz|^2$ on R, where $\varphi = \varphi(z)dz^2$ is the element of $A_2(R)_1$ corresponding to the given Teichmüller mapping. This "metric" degenerates at every zero of φ. However, the notions such as area and length can be introduced with no difficulty. In the next subsection, we explain such a "metric" briefly.

5.3.1. Geometry Induced by a Holomorphic Quadratic Differential

Fix an element $\varphi \in A_2(R)_1 - \{0\}$. If $p_0 \in R$ is not a zero of φ, then $\varphi^{1/2} = \varphi(z)^{1/2}dz$ has a single valued holomorphic branch in some neighborhood U of p_0, and the function defined by

$$\zeta(p) = \int_{p_0}^{p} \varphi^{1/2}, \quad p \in U$$

gives a conformal mapping of U into $\mathbf{C}$. We call this mapping a *φ-coordinate* around p_0 (or, in U). (Recall that zeros of φ are independent of local coordinates.)

By using φ-coordinates, a Teichmüller mapping f for φ is represented locally as an affine mapping F given by

$$F(\zeta) = \zeta + k\bar{\zeta}, \quad k = \|\mu_f\|_\infty.$$

This follows from the assumption that

$$\mu_f = k\frac{\bar{\varphi}}{|\varphi|} = k\frac{d\bar{\zeta}}{d\zeta}.$$

(See also Proposition 5.19.)

At a zero, say p_0, of φ of order m (≥ 1), it is proved that φ is written in the form

$$\varphi = z^m dz^2$$

for a suitable local coordinate z on some neighborhood U of p_0. Hence, we see that

$$\zeta(p) = \int_{p_0}^{p} \varphi^{1/2} = \frac{2}{m+2} \cdot z^{(m+2)/2}$$

cannot be single-valued in any neighborhood of p_0. Anyway, we may consider that this ζ maps a domain $\{z \in \mathbf{C} \mid 0 < \arg z < 2\pi,\ 0 < |z| < r\}$, with a sufficiently small r, conformally onto a "domain" $\{\zeta \in \mathbf{C} \mid 0 < \arg\zeta < (m+2)\pi,\ 0 < |\zeta| < 2r^{(m+2)/2}/(m+2)\}$ spread over the ζ-plane. Hence, we may also call ζ a *φ-coordinate* around p_0.

Now, we consider the "metric" $ds^2 = |\varphi(z)||dz|^2$, which is nothing but the pull-back of the Euclidean metric on the ζ-plane by an arbitrary φ-coordinate ζ.

To make discussions clearer, we consider the lift $\tilde{\varphi} \in A_2(H, \Gamma)$ of φ on the upper half-plane H with respect to a Fuchsian model Γ of R.

For every piecewise smooth curve C on H, we put

$$|C|_{\tilde{\varphi}} = \int_C |\tilde{\varphi}(z)|^{1/2}|dz|.$$

We call this $|C|_{\tilde{\varphi}}$ the *$\tilde{\varphi}$-length* of C. For any two points $z_1, z_2 \in H$, denote by $\mathcal{L}_{z_1,z_2}$ the set of all piecewise smooth curves connecting z_1 and z_2 in H. We set

$$d_{\tilde{\varphi}}(z_1, z_2) = \inf_{C \in \mathcal{L}_{z_1,z_2}} |C|_{\tilde{\varphi}}.$$

We call it the *$\tilde{\varphi}$-distance* between z_1 and z_2. An element C_0 of $\mathcal{L}_{z_1,z_2}$ is called a *$\tilde{\varphi}$-geodesic* between z_1 and z_2 if it satisfies

$$|C_0|_{\tilde{\varphi}} = d_{\tilde{\varphi}}(z_1, z_2).$$

Now, we describe how a $\tilde{\varphi}$-geodesic looks. Assume that there exists a $\tilde{\varphi}$-geodesic C_0 between z_1 and z_2 in H. For every $p \in C_0$ which is not a zero of $\tilde{\varphi}$, the length-minimality implies that C_0 should be a segment near $\tilde{\zeta}(p)$ on the $\tilde{\zeta}$-plane, where $\tilde{\zeta}$ is a $\tilde{\varphi}$-coordinate, i.e., the composed mapping of a φ-coordinate and the projection of H onto R. At a zero $p \in C_0$ of $\tilde{\varphi}$ of order $m > 0$, C_0 may be broken. However, the angle at p should not be less than $2\pi/(m+2)$. (See Fig. 5.1.)

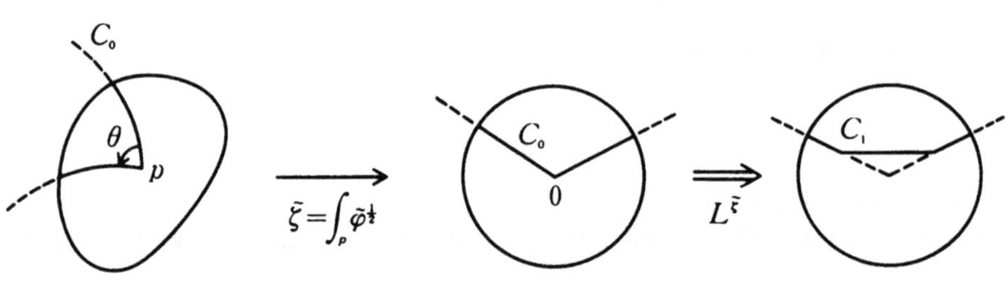

$(|C_1|_{\tilde{\varphi}} < |C_0|_{\tilde{\varphi}}$ for $\theta < 2\pi/(m+2))$

Fig. 5.1.

We call a closed arc L on H a *$\tilde{\varphi}$-segment* if, for every interior point p of L, L is mapped by a $\tilde{\varphi}$-coordinate at p to a segment. By the definition of a $\tilde{\varphi}$-segment, it is clear that $\arg\tilde{\varphi}(z(t))z'(t)^2$ (which is simply written as $\arg\tilde{\varphi}(z)dz^2$ from here on) is a constant modulo 2π along any $\tilde{\varphi}$-segment $L : z = z(t)$ $(0 \le t \le 1)$. From the foregoing, every $\tilde{\varphi}$-geodesic connecting two points of H consists of a finite number of $\tilde{\varphi}$-segments whose end points are either z_1, or z_2, or zeros of $\tilde{\varphi}$. Moreover, we have seen that two such $\tilde{\varphi}$-segments make an angle not less than $2\pi/(m+2)$ at a zero of order $m > 0$.

Here, we consider only the case of a closed Riemann surface R. Then we can easily show that H is complete with respect to this $\tilde{\varphi}$-distance. In particular, for every two points of H, there exists a $\tilde{\varphi}$-geodesic connecting them. Also, we have the following uniqueness of a $\tilde{\varphi}$-geodesic.

Proposition 5.16. *For any two distinct points $z_1, z_2 \in H$, the $\tilde{\varphi}$-geodesic connecting z_1 and z_2 is unique.*

Proof. Suppose that there are two $\tilde{\varphi}$-geodesics C_1, C_2 connecting z_1 and z_2. Then clearly, C_1 and C_2 are simple. By replacing z_1 and z_2 with a suitable pair of points if necessary, we may assume without loss of generality that $C_1 \cap C_2 = \{z_1, z_2\}$. Then $C_1 \cup C_2$ bounds a Jordan domain D in H.

As has been seen before, C_1 and C_2 consist of a finite number of $\tilde{\varphi}$-segments. We give them an order with respect to the positive orientation of the boundary ∂D of D, and denote them by $\{L_j\}_{j=1}^m$ $(m \ge 2)$. Let m_j be the order of zeros of $\tilde{\varphi}$ at $L_j \cap L_{j+1}$ for every j, where $L_{m+1} = L_1$ and we allow m_j to be zero. Also, let $\theta_j (> 0)$ be the angle between L_j and L_{j+1} in D for every j. By the argument principle, we can show that

$$\int_{\partial D} d\arg\tilde{\varphi}(z) = 2N\pi + \sum_{j=1}^{m} m_j\theta_j,$$

where N is the number of zeros of $\tilde{\varphi}$ in D including multiplicity. On the other hand, we have

$$\int_{\partial D} d(\arg dz) + \sum_{j=1}^{m}(\pi - \theta_j) = 2\pi.$$

Further, we know that

$$d\arg\tilde{\varphi}(z) + 2d(\arg dz) = 0$$

on a $\tilde{\varphi}$-segment. Hence, we conclude that

$$\sum_{j=1}^{m}(2\pi - (m_j+2)\theta_j) = 2\pi(N+2).$$

Since $N \ge 0$, it follows that θ_j should be not greater than $2\pi/(m_j+2)$ for at least three j's. However, $\theta_j \ge 2\pi/(m_j+2)$ for every j not corresponding to z_1 or z_2, which gives a contradiction. □

Corollary. *A $\tilde{\varphi}$-segment L is the unique $\tilde{\varphi}$-geodesic connecting its end points.*

Now, to prove Theorem 5.9, the following lemma due to Teichmüller plays a crucial role. To state it, we prefer to returning to R and φ. In particular, the *φ-length* $|L|_\varphi$ of a curve L on R is defined by

$$|L|_\varphi = \int_L |\varphi|^{1/2}.$$

The projection of a $\tilde{\varphi}$-segment to R is called a *φ-segment.*

Lemma 5.17. (Teichmüller) *Let $h : R \longrightarrow R$ be a quasiconformal self-mapping of R homotopic to id. Then there is a positive constant M depending only on R, h, and φ such that*

$$|h(L)|_\varphi \geq |L|_\varphi - M$$

for every φ-segment L. (Here, $h(L)$ may not be rectifiable, i.e., it may happen that $|h(L)|_\varphi = \infty$.)

Proof. Let $\tilde{h}$ be the canonical lift of h with respect to the canonical Fuchsian model Γ of R. Then it suffices to find a constant M such that

$$|\tilde{h}(\tilde{L})|_{\tilde{\varphi}} \geq |\tilde{L}|_{\tilde{\varphi}} - M$$

for every $\tilde{\varphi}$-segment $\tilde{L}$ in H.

First, by Lemma 5.1, it follows that $\tilde{h} \circ \gamma = \gamma \circ \tilde{h}$ for every $\gamma \in \Gamma$. Hence, letting C_z be the $\tilde{\varphi}$-geodesic connecting z and $\tilde{h}(z)$ for every $z \in H$, we have

$$|C_z|_{\tilde{\varphi}} = |C_{\gamma(z)}|_{\tilde{\varphi}}$$

for every $\gamma \in \Gamma$. Since R is compact, we see that

$$M = 2\sup\{|C_z|_{\tilde{\varphi}} \mid z \in H\}$$

is finite. (Note that $|C_z|_{\tilde{\varphi}}$ is continuous with respect to z.)

Now, let a $\tilde{\varphi}$-segment $\tilde{L}$ be given. Let z_1 and z_2 be the end points of $\tilde{L}$. Then the curve $C_{z_1} \cdot \tilde{h}(\tilde{L}) \cdot {C_{z_2}}^{-1}$ also connects z_1 and z_2. Since $\tilde{L}$ is a $\tilde{\varphi}$-geodesic connecting z_1 and z_2, we obtain

$$|\tilde{L}|_{\tilde{\varphi}} \leq |C_{z_1}|_{\tilde{\varphi}} + |\tilde{h}(\tilde{L})|_{\tilde{\varphi}} + |C_{z_2}|_{\tilde{\varphi}} \leq |\tilde{h}(\tilde{L})|_{\tilde{\varphi}} + M.$$

Thus M is a desired constant. □

5.3.2. Preliminary Considerations

A prototype of Teichmüller's uniqueness theorem is the following *Grötzsch's theorem*, which treats the case where R is a rectangle $\{ z = x + iy \in \mathbf{C} \mid 0 \leq x \leq r, 0 \leq y \leq 1 \}$.

Proposition 5.18. *Let f be an affine quasiconformal mapping of a closed rectangle $R = [0,r] \times [0,1]$ to another closed rectangle $S = [0,s] \times [0,1]$ defined by*

$$f(z) = \frac{z + k\,\bar{z}}{1-k} = Kx + iy,$$

which is K-qc on the interior of R. Here, it is assumed that

$$K = s/r \geq 1, \quad k = (K-1)/(K+1) < 1.$$

Then f satisfies the following extremal property: for every homeomorphism $f_1 : R \longrightarrow S$ which is $(1+k_1)/(1-k_1)$-qc on the interior of R, and maps $0, r, r+i$, and i, respectively, to $0, s, s+i$, and i, it follows that

$$k_1 \geq k.$$

Moreover, the equality holds if and only if $f_1 = f$.

Proof. First, note that

$$s = |f_1(r+iy) - f_1(iy)| \leq \int_0^r |(f_1)_x(x+iy)|\,dx$$

for almost every $y \in [0,1]$. Integrating both sides with respect to y over $[0,1]$, we get

$$s \leq \iint_R |(f_1)_x|\,dxdy.$$

Since $(f_1)_x = (f_1)_z + (f_1)_{\bar{z}}$, Schwarz' inequality gives

$$\begin{aligned} s^2 &\leq \left\{ \iint_R (|(f_1)_z| + |(f_1)_{\bar{z}}|)dxdy \right\}^2 \\ &\leq \iint_R \frac{|(f_1)_z| + |(f_1)_{\bar{z}}|}{|(f_1)_z| - |(f_1)_{\bar{z}}|}dxdy \cdot \iint_R (|(f_1)_z|^2 - |(f_1)_{\bar{z}}|^2)dxdy \\ &\leq (K_1 r) \cdot s, \end{aligned}$$

where $K_1 = (1+k_1)/(1-k_1)$. Hence we conclude that $K_1 \geq s/r = K$, or equivalently, that $k_1 \geq k$. (Compare with the proof of Lemma 4.8.)

Next, if the equality holds, then we obtain

$$|(f_1)_z + (f_1)_{\bar{z}}| = |(f_1)_z| + |(f_1)_{\bar{z}}|,$$

and

$$|(f_1)_{\bar{z}}| = k|(f_1)_z|$$

a.e. on R. Thus we see that $\mu_{f_1} = k$ a.e. on R. This implies that $g = f_1 \circ f^{-1}$ is 1-qc, and hence conformal on the interior of S. Since g fixes all vertices of S, the same argument as in the second half of the proof of Proposition 4.7 shows that $g = id$. That is, $f_1 = f$. □

Thus an affine mapping as in Proposition 5.18 is extremal. Returning to a closed Riemann surface R, we find that any Teichmüller mapping looks very similar to an affine mapping. Actually, we have the following proposition.

Proposition 5.19. *Fix $\varphi \in A_2(R)_1$ arbitrarily. Let $f : R \longrightarrow S$ be a Teichmüller mapping for φ, and set $k = \|\varphi\|_1$ (< 1). Then there exists a unique holomorphic quadratic differential ψ on S satisfying the following conditions:*

(i) *If p is a zero of φ of order m, then $f(p)$ is a zero of ψ of the same order m.*
(ii) *Let p be an arbitrary point of R which is not a zero of φ, and ζ be a φ-coordinate around p. Then there exists a ψ-coordinate ω at $f(p)$ such that*

$$\omega \circ f = \frac{\zeta + k\bar{\zeta}}{1-k}. \tag{5.3}$$

We call φ and ψ in Proposition 5.19 the *initial differential* of f and the *terminal differential* of f, respectively.

Proof. For every p which is not a zero of φ, we define a mapping $\omega = \omega_p$ in a neighborhood of $f(p)$ by (5.3). Then in some neighborhood U of p, we have

$$\mu_{\omega\circ f} = k\,\frac{d\bar{\zeta}}{d\zeta} = k\,\frac{\bar{\varphi}}{|\varphi|}.$$

Since $\mu_{\omega\circ f} = k\bar{\varphi}/|\varphi|$ for every local coordinate w on $f(U)$, we see that $\omega \circ w^{-1}$ is 1-qc, and hence conformal on $f(U)$. Thus ω is also a local coordinate around $f(p)$.

Next, for a zero p of φ of order m, we have seen that $\varphi = z^m dz^2$ with a suitable local coordinate z. Define ω as a continuous branch determined by

$$\omega \circ f = \left(\frac{z^{(m+2)/2} + k\,\bar{z}^{(m+2)/2}}{1-k}\right)^{2/(m+2)}.$$

Then we can see similarly that ω is a local coordinate in a neighborhood of $f(p)$.

Finally, consider $(d\omega_p)^2$ in a neighborhood of every point $f(p)$ such that p is not a zero of φ, where $\omega_p = \omega$ is as above. Then we can show that these $(d\omega_p)^2$ give a single holomorphic quadratic differential on S, which we denote by ψ. From the construction, ψ clearly satisfies (i) and (ii). The uniqueness of ψ follows at once from (i) and (ii). □

5.3.3. Proof of Theorem 5.9

Assume that the assumptions of Theorem 5.9 are satisfied. Let ψ be the terminal differential of f obtained in Proposition 5.19. For every $p \in R$ which is not a zero of φ, take a φ-coordinate ζ around p, and a ψ-coordinate ω around $q = f(p)$ as in Proposition 5.19. Consider the "horizontal dilatation"

$$\lambda(f_1, p) = \left|\frac{\partial(\omega \circ f_1 \circ \zeta^{-1})}{\partial \xi}\right|(0), \quad \zeta = \xi + i\eta,$$

of f_1 with respect to ζ and ω. This $\lambda(f_1, p)$ is defined a.e. on R, and is a measurable function. Recall that Proposition 5.19 implies that, for the "horizontal dilatation" $\lambda(f, p)$ of f with respect to ζ and ω, we obtain

$$\lambda(f, p) = \left|\frac{\partial(\omega \circ f \circ \zeta^{-1})}{\partial \xi}\right|(0) = \left|\frac{\partial(K\xi + i\eta)}{\partial \xi}\right|(0) = \frac{1+k}{1-k}$$

for almost every $p \in R$. In the rest of this section, we set $K = (1+k)/(1-k)$.

Now, recalling the proof of Proposition 5.18, we may regard the following inequality as one way to represent the fact that a Teichmüller mapping f has the best "efficiency" for deformation of the complex structure of R:

$$\iint_R \lambda(f_1, p) d\xi d\eta \geq \iint_R K d\xi d\eta, \tag{5.4}$$

where $d\xi d\eta = |\varphi(z)|\, dxdy$. Set $g = f_1 \circ f^{-1}$ and $\omega = \sigma + i\tau$. Then the "horizontal dilatation"

$$\lambda(g, q) = \left|\frac{\partial(\omega \circ g \circ \omega^{-1})}{\partial \sigma}\right|(0)$$

of g is defined a.e. on S. It is measurable, and satisfies

$$K\lambda(g, f(p)) = \lambda(f_1, p) \tag{5.5}$$

for almost every $p \in R$. Since $d\sigma d\tau = K d\xi d\eta$, we have

$$\begin{aligned}\iint_R \lambda(f_1, p) d\xi d\eta &= \frac{1}{K} \iint_S \lambda(f_1, f^{-1}(q)) d\sigma d\tau \\ &= \iint_S \lambda(g, q) d\sigma d\tau.\end{aligned}$$

Hence, the inequality (5.4) is equivalent to

$$\iint_S \lambda(g, q) d\sigma d\tau \geq \iint_S d\sigma d\tau. \tag{5.6}$$

Here, we assume (5.4), or equivalently (5.6), and prove Theorem 5.9 by the same argument as in the proof of Proposition 5.18.

First, we set $k_1 = \|\mu_{f_1}\|_\infty$, $K_1 = (1+k_1)/(1-k_1)$, $\hat{f}_1 = \omega \circ f_1 \circ \zeta^{-1}$, and $J(f_1)(p) = |(\hat{f}_1)_\zeta|(0)^2 - |(\hat{f}_1)_{\bar{\zeta}}|(0)^2$. Then we get

$$\lambda(f_1, p)^2 \leq \left(|(\hat{f}_1)_\zeta|(0) + |(\hat{f}_1)_{\bar{\zeta}}|(0)\right)^2 \leq K_1 J(f_1)(p) \quad \text{a.e. on} \quad R. \tag{5.7}$$

Next, applying Schwarz' inequality to (5.6), we obtain

$$\iint_S \lambda(g, q)^2 d\sigma d\tau \geq \iint_S d\sigma d\tau.$$

Hence, (5.5) and (5.7) give

$$\iint_S d\sigma d\tau \le \iint_R \left(\frac{\lambda(f_1,p)}{K}\right)^2 K d\xi d\eta$$
$$\le \frac{K_1}{K} \iint_R J(f_1)(p)\, d\xi d\eta = \frac{K_1}{K} \iint_S d\sigma d\tau.$$

Thus, $K_1 \ge K$, and hence $k_1 \ge k$.

Finally, if $k_1 = k$, then both equalities in (5.7) should hold. Namely,

$$|(\hat{f}_1)_\zeta + (\hat{f}_1)_{\bar{\zeta}}|(0) = |(\hat{f}_1)_\zeta|(0) + |(\hat{f}_1)_{\bar{\zeta}}|(0)$$

and

$$|(\hat{f}_1)_\zeta|(0) = k|(\hat{f}_1)_{\bar{\zeta}}|(0)$$

a.e. on R. This implies that $\mu_{f_1} = k\bar{\varphi}/|\varphi|$. Hence, g is 1-qc, i.e., it is conformal on R. Since g is homotopic to id, the canonical lift of g on H is coincident with id by Lemma 5.2. Thus we have $f_1 = f$.

Now, to prove Theorem 5.9, it remains to show the following lemma.

Lemma 5.20. *The inequality (5.6) holds.*

Proof. It is in this proof where we need Lemma 5.17.

For the sake of simplicity, we assume that $\psi^{1/2}$ has a single-valued global branch on S, which is a holomorphic Abelian differential on S. If not, take any local branch of $\psi^{1/2}$, and continue it analytically as far as possible. Then we can construct a two-sheeted branched covering surface $\hat{S}$ of S, with a branch point at every zero of ψ of odd order, such that $\psi^{1/2}$ becomes a single-valued (holomorphic Abelian) differential on $\hat{S}$. Applying the argument below to this differential on $\hat{S}$, we have the assertion for the general case.

When $\psi^{1/2}$ has a single-valued global branch, say θ, which is an Abelian differential, we define the geodesic flow $\{F_t \mid t \in \mathbf{R}\}$ on S with respect to the "metric" $|\psi|$. To explain the construction, we always take

$$\omega = \Psi_p(q) = \int_p^q \theta$$

as a ψ-coordinate around p which is not a zero of ψ. We continue the inverse mapping Ψ_p^{-1} along segments on $\mathbf{R}$ in both directions as far as possible. Then we get a locally biholomorphic mapping, which is denoted by the same notation Ψ_p^{-1}, of a domain containing an open interval $I_p = (u_1, u_2)$ of $\mathbf{R}$ into S, where $-\infty \le u_1 < u_2 \le \infty$. We set $H_p = \Psi_p^{-1}(I_p)$. This H_p is called the *θ-horizontal line* passing through p. It is also called a *horizontal trajectory* of ψ. (See Fig. 5.2.)

Note that, when we trace along H_p in one direction, either H_p ends at a zero of ψ, or $|\omega|$ tends to ∞.

In particular, restricting ψ-distance on H_p, we can identify H_p either with a circle or with a subinterval, say I_p of $\mathbf{R}$, preserving orientation and length. Let

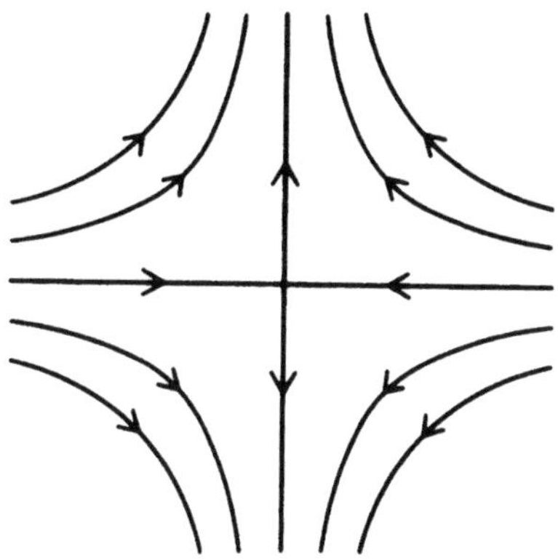

Fig. 5.2. (θ-horizontal lines near a simple zero of θ)

E be the set of $p \in S$ such that I_p is a proper subinterval of $\mathbf{R}$. Since ψ has only a finite number of zeros, we can see that E has area zero with respect to the "metric" $|\psi|$. Thus for every $p \in S - E$, setting $\ell_p = \Psi_p^{-1}|_{\mathbf{R}}$, we obtain a mapping

$$\ell_p : \mathbf{R} \to H_p, \quad \ell_p(0) = p,$$

which is locally isometric with respect to the Euclidean metric on $\mathbf{R}$ and the "metric" $|\psi|$ on S.

Now, we shall define the geodesic flow $\{F_t \mid t \in \mathbf{R}\}$ on S with respect to the "metric" $|\psi|$. Let E be as before, and set $\Omega = S - E$. We define F_t by setting

$$F_t(p) = \ell_p(t), \quad t \in \mathbf{R}, \ p \in \Omega.$$

Then we can see that every F_t is a bijective measurable self-mapping of Ω, which represents the "parallel horizontal translation" by t with respect to ψ-coordinates. In particular, it is easy to see that

$$\iint_{F_t(X)} d\sigma d\tau = \iint_X d\sigma d\tau$$

for every measurable subset X of Ω, i.e., F_t is area-preserving with respect to the "metric" $|\psi|$. From here on, we use this $d\sigma d\tau$ as the "area element" on S.

Next, we set

$$\lambda(g, q, t) = \lambda(g, F_t(q)), \quad q \in \Omega, \ t \in \mathbf{R}.$$

Then $\lambda(g, q, t)$ is a measurable function on $\Omega \times \mathbf{R}$. Since $E = S - \Omega$ has area zero with respect to $|\psi|$, we get

$$\begin{aligned} I &\equiv \int_{-L}^{L} \left(\iint_S \lambda(g,q,t)\, d\sigma d\tau \right) dt \qquad (5.8) \\ &= \int_{-L}^{L} \left(\iint_{F_t(\Omega)} \lambda(g,q)\, d\sigma d\tau \right) dt \\ &= \int_{-L}^{L} \left(\iint_S \lambda(g,q)\, d\sigma d\tau \right) dt = 2L \iint_S \lambda(g,q)\, d\sigma d\tau \end{aligned}$$

for every positive L.

On the other hand, since g is quasiconformal, and hence is ACL, we see that Fubini's theorem gives

$$\begin{aligned} I &= \iint_S \left(\int_{-L}^{L} \lambda(g,q,t)\, dt \right) d\sigma d\tau \qquad (5.9) \\ &= \iint_S |g(L_q)|_\psi \, d\sigma d\tau. \end{aligned}$$

Here, we set $L_q = \ell_q([-L,L])$, and hence $|L_q|_\psi = 2L$.

Finally, applying Lemma 5.17, we conclude from (5.8) and (5.9) that

$$2L \iint_S \lambda(g,q) d\sigma d\tau \geq (2L - M) \iint_S d\sigma d\tau.$$

Divide both sides by $2L$, and let L tend to ∞. Then we obtain the desired inequality (5.6). □

Notes

For the reduced Teichmüller spaces, see for instance Earle [57].

When $\Gamma = \{id\}$, we denote by $T(1)$ the corresponding Teichmüller space $T(\Gamma)$, and call it the ***universal Teichmüller space***. See Lehto [A-68], Chapter III. The fact that $T(1)$ is contractible was firstly shown in Earle and Eells [62]. In Douady and Earle [53], it is proved that $T(\Gamma)$ is also contractible for every Γ.

For investigations on the Teichmüller metric from the differential-geometric viewpoint, see Kravetz [128] and O'Byrne [169]. The Teichmüller metric on T_g is not smooth. See Earle and Kra [65], Royden [184], and Gardiner [A-34], §9.4. Moreover, T_g does not have non-positive "curvature" with respect to the Teichmüller metric, as is proved in Masur [142]. See also Theorem 6.21 in Chapter 6.

Teichmüller's theorem gives another compactification of the Teichmüller space T_g, which is called Teichmüller's compactification of T_g. This is different from Thurston's one defined in Chapter 3. See Kerckhoff [111] and Masur [146].

The original "proof" of Teichmüller's theorem is found in Teichmüller [A-106]. The proof in this chapter follows that in Bers [23]. We also refer to Abikoff

[A-1]. Teichmüller's uniqueness theorem states that the Beltrami coefficient of a Teichmüller mapping is extremal (minimal in some sense). Hamilton [89] proved by using functional analysis a necessary condition for a quasiconformal mapping to be extremal. We can use this Hamilton's theorem to prove Teichmüller's uniqueness theorem. Moreover, Reich and Strebel showed that this *Hamilton's condition* is also a sufficient one (cf. Strebel [211]). See Gardiner [A-34] and Lehto [A-68] for more details.

In the case of a general Riemann surface, Teichmüller mappings are still uniquely extremal quasiconformal mappings, while formal Teichmüller mappings are not necessarily extremal. Even if a formal Teichmüller mapping is extremal, it need not to be uniquely extremal. See for example Reich and Strebel [180]. Also, Hayman and Reich [97], and Sethares [196] contain results on uniqueness of extremal quasiconformal mappings. Furthermore, a lift of an extremal quasiconformal mapping to a covering surface is not necessarily extremal. See McMullen [153], Ohtake [170], and Strebel [210]. As other investigations on extremal quasiconformal mappings, we cite Krushkal' [A-60], Fehlmann and Sakan [70], Gardiner [80], and Sakan [187].

Some topics on quasiconformal deformations of general Riemann surfaces are stated in Appendix A and its Notes.

Finally, for geometrical studies of holomorphic quadratic differentials as in §3, we refer to Gardiner [A-34], Jenkins [A-47], and Strebel [A-102]. We also refer to Douady and Hubbard [54], Kerckhoff, Masur and Smillie [115], Masur [144], [145], [147], [148], [149], Taniguchi [222], and Veech [239], [240].

Chapter 6

Complex Analytic Theory of Teichmüller Spaces

We introduce a natural complex manifold structure of the Teichmüller space $T(R)$ of a closed Riemann surface R of genus $g\,(\geqq 2)$, which is realized as a bounded domain in $\mathbf{C}^{3g-3}$. Furthermore, we prove that the Teichmüller modular group $Mod(R)$ acts properly discontinuously as a group of biholomorphic automorphisms of $T(R)$.

In this chapter, unless otherwise stated, we assume that Γ is a Fuchsian model of a closed Riemann surface of genus $g\,(\geqq 2)$ and that each of 0, 1, and ∞ is fixed by an element in $\Gamma - \{\,id\,\}$ (cf. §1.2 of Chapter 5).

In Section 1, following the idea due to Bers, by using Schwarzian derivatives, we prove that the Teichmüller space $T(\Gamma)$ is realized as a bounded domain $T_B(\Gamma)$ in the space $A_2(H^*/\Gamma)$ of holomorphic quadratic differentials on the Riemann surface H^*/Γ, where H^* is the lower half-plane. The Riemann-Roch theorem shows that $A_2(H^*/\Gamma)$ is a complex $(3g-3)$-dimensional vector space. Hence $T_B(\Gamma)$ is regarded as a bounded domain in $\mathbf{C}^{3g-3}$. Identifying $T(\Gamma)$ with $T_B(\Gamma)$, we see that $T(\Gamma)$ has a complex manifold structure of dimension $3g-3$.

In Section 2, we show that this complex structure of $T(\Gamma)$ is independent of Γ, that is, $T(\Gamma)$ is biholomorphically equivalent to $T(\Gamma')$ for another Fuchsian model Γ' of a closed Riemann surface of genus g.

It is verified in Section 3 that the Teichmüller modular group $Mod(\Gamma)$ of Γ acts properly discontinuously as a group of biholomorphic automorphisms of $T(\Gamma)$. Thus we conclude that the moduli space $M_g = T(\Gamma)/Mod(\Gamma)$ has a normal complex analytic space structure of dimension $3g-3$. In Section 4, we shall explain Royden's theorem which asserts that every biholomorphic automorphism of $T(\Gamma)$ is induced by an element of $Mod(\Gamma)$.

Finally, in Section 5, we give a brief exposition of the Thurston-Bers theory on the classification of Teichmüller modular transformations.

6.1. Bers' Embedding and the Complex Structure of Teichmüller Space

6.1.1. Simultaneous Uniformization

Following Bers, we shall represent the Teichmüller space $T(\Gamma)$ by quasiconformal mappings of the Riemann sphere $\hat{\mathbf{C}}$ which are conformal on the lower half-plane H^*.

For a given element $\mu \in B(H,\Gamma)_1$, i.e., a Beltrami coefficient μ on H for Γ, we set

$$\tilde{\mu}(z) = \begin{cases} \mu(z), & z \in H \\ 0, & z \in \mathbf{C} - H. \end{cases}$$

From Theorem 4.30, there exists uniquely a canonical $\tilde{\mu}$-qc mapping of $\hat{\mathbf{C}}$, i.e., a quasiconformal mapping of $\hat{\mathbf{C}}$ which has the complex dilatation $\tilde{\mu}$, and leaves 0, 1, and ∞ fixed, respectively. In order to distinguish this quasiconformal mapping from the canonical μ-qc mapping w^μ of H (Proposition 4.33), we denote it by w_μ.

For any $\gamma \in \Gamma$, putting $\chi_\mu(\gamma) = w_\mu \circ \gamma \circ (w_\mu)^{-1}$, by similar reasoning to that in §1.3 of Chapter 5, we see that $\chi_\mu(\gamma)$ is a Möbius transformation, i.e., an element of $Aut(\hat{\mathbf{C}})$. Thus we have a subgroup $\Gamma_\mu = \{\chi_\mu(\gamma) \mid \gamma \in \Gamma\}$ of $Aut(\hat{\mathbf{C}})$, which acts properly discontinuously on both $H_\mu = w_\mu(H)$ and $H^*_\mu = w_\mu(H^*)$. Such a group Γ_μ is a typical example of quasi-Fuchsian groups. Here, a ***quasi-Fuchsian group*** is, by definition, a discrete subroup of $PSL(2,\mathbf{C})$ which leaves a directed simple closed curve in $\hat{\mathbf{C}}$ fixed. Note that every element of $\Gamma_\mu - \{id\}$ has no fixed points on both H_μ and H^*_μ.

The quasiconformal mapping w_μ induces a quasiconformal mapping of $R = H/\Gamma$ to $R_\mu = H_\mu/\Gamma_\mu$ and a biholomorphic mapping of $R^* = H^*/\Gamma$ to H^*_μ/Γ_μ, where R^* is the mirror image of $R = H/\Gamma$. Since two Riemann surfaces R_μ and R^* are represented by H_μ/Γ_μ and H^*_μ/Γ_μ, respectively, two Riemann surfaces R_μ and R^* are uniformized simultaneously by a single quasi-Fuchsian group Γ_μ. This is called ***Bers' simultaneous uniformization*** (see Bers [24]).

In particular, for any two closed Riemann surfaces R and S of genus g, we find a quasi-Fuchsian group Γ_μ which uniformizes simultaneously R and S. In fact, pick a Fuchsian model Γ of the mirror image R^* of R, and take a quasiconformal mapping f of R^* to S. Set $\mu = \mu_f$, the Beltrami coefficient of f. Then by Bers' simultaneous uniformization, we conclude that S and the mirror image R of R^* are biholomorphic to H_μ/Γ_μ and H^*_μ/Γ_μ, respectively.

Lemma 6.1. *For any two elements $\mu, \nu \in B(H,\Gamma)_1$, the following are equivalent:*

(i) $w^\mu = w^\nu$ *on* $\mathbf{R}$.
(ii) $w_\mu = w_\nu$ *on* H^*.

Proof. If $w^\mu = w^\nu$ on $\mathbf{R}$, then we have a homeomorphism $f\colon \hat{\mathbf{C}} \to \hat{\mathbf{C}}$ given by

$$f(z) = \begin{cases} (w^\mu)^{-1} \circ w^\nu(z), & z \in H \\ z, & z \in H^* \cup \hat{\mathbf{R}}. \end{cases}$$

Since $(w^\mu)^{-1} \circ w^\nu$ is quasiconformal on $\mathbf{C}$, we see that f is ACL on $\mathbf{C}$. Thus, by the analytic definition A of quasiconformal mappings (§1.1 of Chapter 4), f is quasiconformal. Hence, $g = w_\mu \circ f \circ (w_\nu)^{-1}$ is a 1-qc mapping on $\mathbf{C}$, i.e., a Möbius transformation. Since g leaves each of 0, 1, and ∞ fixed, g must be the identity. Therefore, we have $w_\mu = w_\nu$ on H^*.

Conversely, if $w_\mu = w_\nu$ on H^*, then $w_\mu = w_\nu$ on $H^* \cup \hat{\mathbf{R}}$. Thus we obtain a quasiconformal mapping $h = w^\mu \circ (w_\mu)^{-1} \circ w_\nu \circ (w^\nu)^{-1} \colon H \to H$. By the same argument as before, it follows that h must be the identity, which means that $w^\mu = w^\nu$ on $\mathbf{R}$. □

Now, for two elements $\mu, \nu \in B(H, \Gamma)_1$, w_μ and w_ν are said to be *equivalent* if $w_\mu = w_\nu$ on H^*. Denote by $[w_\mu]$ the equivalence class of w_μ for every element $\mu \in B(H, \Gamma)_1$. Let $T_\beta(\Gamma)$ be the set of these equivalence classes $[w_\mu]$. Lemma 6.1 shows that the correspondence $[w^\mu] \mapsto [w_\mu]$ is a bijection of $T(\Gamma)$ to $T_\beta(\Gamma)$. The topology of $T_\beta(\Gamma)$ is induced from that of $T(\Gamma)$ under this correspondence. In other words, this correspondence gives a homeomorphism of $T(\Gamma)$ onto $T_\beta(\Gamma)$. In this way, we can identify $T_\beta(\Gamma)$ with $T(\Gamma)$ as topological spaces. We also call $T_\beta(\Gamma)$ the *Teichmüller space* of Γ.

Let β be a mapping of $B(H, \Gamma)_1$ onto $T_\beta(\Gamma)$ given by $\beta(\mu) = [w_\mu]$. Then by the definition of topology of $T_\beta(\Gamma)$, we immediately obtain the following.

Proposition 6.2. *The mapping* $\beta \colon B(H, \Gamma)_1 \to T_\beta(\Gamma)$ *is a continuous surjection.*

One merit of the Teichmüller space $T_\beta(\Gamma)$ introduced by Bers is the applicability of the theory of univalent functions, i.e., conformal mappings of H^*.

6.1.2. Schwarzian Derivative

A quasiconformal mapping w_μ as defined in §1.1 is conformal on the lower half-plane H^*.

Now, assume that w_μ is a Möbius transformation. Since w_μ leaves each of 0, 1, and ∞ fixed, w_μ must be the identity. Thus we have $[w_\mu] = [id]$ in $T_\beta(\Gamma)$. It may be considered that the difference between $[w_\mu]$ and $[id]$ in $T_\beta(\Gamma)$ is indicated by the difference of the conformal mapping w_μ on H^* from Möbius transformations.

To measure the difference of a conformal mapping on H^* from a Möbius transformation, we shall find a differential equation which all Möbius transformations satisfy. Let $\gamma(z) = (az + b)/(cz + d)$ be a Möbius transformation, where $a, b, c, d \in \mathbf{C}$ and $ad - bc = 1$. Take derivatives of γ to eliminate a, b, c, and d. Since $\gamma'(z) = (cz + d)^{-2}$ and $\gamma''(z) = -2c(cz + d)^{-3}$, we obtain $\gamma'(z)/\gamma''(z) = -z/2 - d/2c$. Thus we have $(1/(\gamma''/\gamma'))' = -1/2$. Consequently, we get

$$\frac{\gamma'''(z)}{\gamma'(z)} - \frac{3}{2}\left(\frac{\gamma''(z)}{\gamma'(z)}\right)^2 = 0.$$

For an arbitrary conformal mapping f on a domain in $\mathbf{C}$, we define the *Schwarzian derivative* $\{f, z\}$ of f by

$$\{f, z\} = \frac{f'''(z)}{f'(z)} - \frac{3}{2}\left(\frac{f''(z)}{f'(z)}\right)^2.$$

Lemma 6.3. *If f and g are conformal mappings of D and $f(D)$, respectively, then*

$$\{g \circ f, z\} = \{g, f(z)\} \cdot f'(z)^2 + \{f, z\}, \quad z \in D. \tag{6.1}$$

Moreover, a conformal mapping of D is a Möbius transformation if and only if $\{f, z\} = 0$ on D.

Proof. A straightforward calculation gives us the first assertion. We have already seen that a Möbius transformation f satisfies $\{f, z\} = 0$ on D. Conversely, if $\{f, z\} = 0$ on D, it turns out that

$$\{f, z\} = (\log f'(z))'' - \frac{1}{2}\{(\log f'(z))'\}^2 = 0$$

on D. Solving this differential equation, we conclude that f is a Möbius transformation. □

6.1.3. Bers' Embedding and the Complex Structure of Teichmüller Space

For arbitrary $\mu \in B(H, \Gamma)_1$, we set

$$\varphi_\mu(z) = \{w_\mu, z\}, \qquad z \in H^*.$$

Then we have the following.

Lemma 6.4. *If $\gamma \in \Gamma$, then*

$$\varphi_\mu(\gamma(z))\gamma'(z)^2 = \varphi_\mu(z), \qquad z \in H^*. \tag{6.2}$$

Namely, φ_μ is a holomorphic automorphic form of weight -4 on H^* with respect to Γ, and it is regarded as a holomorphic quadratic differential on a Riemann surface H^*/Γ.

Moreover, for any two elements $\mu, \nu \in B(H, \Gamma)_1$, $[w_\mu] = [w_\nu]$ in $T_\beta(\Gamma)$ if and only if $\varphi_\mu = \varphi_\nu$ on H^*.

Proof. If $\gamma \in \Gamma$, then $\gamma_\mu = w_\mu \circ \gamma \circ (w_\mu)^{-1}$ is a Möbius transformation, because μ is in $B(H, \Gamma)_1$. Taking the Schwarzian derivative of $w_\mu \circ \gamma = \gamma_\mu \circ w_\mu$, by Lemma 6.3 we get $\{w_\mu, \gamma(z)\}\gamma'(z)^2 = \{w_\mu, z\}$ on H^*. Thus we have formula (6.2).

If $[w_\mu] = [w_\nu]$ in $T_\beta(\Gamma)$, then $w_\mu = w_\nu$ on H^*, which implies that $\varphi_\mu = \varphi_\nu$ on H^*. Conversely, assume that $\varphi_\mu = \varphi_\nu$ on H^*. Then defining a conformal mapping $F\colon w_\mu(H^*) \to w_\nu(H^*)$ by $F = w_\nu \circ (w_\mu)^{-1}$, again by Lemma 6.3 we see that

$$\varphi_\nu(z) = \{ F \circ w_\mu, z \} = \{ F, w_\mu(z) \} \cdot w'_\mu(z)^2 + \varphi_\mu(z)$$

on H^*. By the assumption that $\varphi_\mu = \varphi_\nu$ on H^*, we have $\{ F, z \} = 0$ on $w_\mu(H^*)$. Thus F must be a Möbius transformation. Since F leaves each of 0, 1, and ∞ fixed, we see that F is the identity. Consequently, $w_\mu = w_\nu$ on H^*, that is, $[w_\mu] = [w_\nu]$ in $T_\beta(\Gamma)$. □

Let $A_2(H^*, \Gamma)$ be the complex vector space of holomorphic automorphic forms of weight -4 on H^* with respect to Γ. Since it is identified with the vector space $A_2(H^*/\Gamma)$ of holomorphic quadratic differentials on H^*/Γ, the Riemann-Roch theorem shows that $A_2(H^*, \Gamma)$ is a $(3g-3)$-dimensional complex vector space.

Now, define a mapping $\mathcal{B}$ of $T_\beta(\Gamma)$ into $A_2(H^*, \Gamma)$ by $\mathcal{B}([w_\mu]) = \varphi_\mu$, where $\varphi_\mu = \{ w_\mu, z \}$, the Schwarzian derivative of w_μ on H^*. Then, by Lemma 6.4 this $\mathcal{B}$ is well-defined and injective, and that is called *Bers' embedding*. The mapping $\Phi\colon B(H, \Gamma)_1 \to A_2(H^*, \Gamma)$ given by $\Phi(\mu) = \mathcal{B} \circ \beta(\mu)$ is called *Bers' projection*.

In §2.2 of the previous chapter, $A_2(H^*, \Gamma)$ was considered as a complex Banach space with L^1-norm. In this chapter, in connection with the next subsection §1.4, we introduce on $A_2(H^*, \Gamma)$ the hyperbolic L^∞-norm by using the Poincaré metric $ds_{H^*}{}^2 = |dz|^2/(\operatorname{Im} z)^2$ on H^* as follows. By formula (6.2) and the invariance of the Poincaré metric under $PSL(2, \mathbf{R})$, every element $\varphi \in A_2(H^*, \Gamma)$ satisfies

$$(\operatorname{Im} \gamma(z))^2 |\varphi(\gamma(z))| = (\operatorname{Im} z)^2 |\varphi(z)|, \qquad z \in H^*,\ \gamma \in \Gamma.$$

Thus, $(\operatorname{Im} z)^2 |\varphi(z)|$ is regarded as a function on $R^* = H^*/\Gamma$. The *hyperbolic L^∞-norm* of φ in $A_2(H^*, \Gamma)$ is defined by

$$\|\varphi\|_\infty = \sup_{z \in H^*} (\operatorname{Im} z)^2 |\varphi(z)|.$$

Here, note that the supremum suffices to be taken over, not the whole H^*, but only a fundamental domain in H^* for Γ. In our case, R^* being compact, we can pick a relatively compact subset in H^* as such a domain (see Example 5 in §4.2 of Chapter 2). Therefore, $\|\varphi\|_\infty$ is finite for any $\varphi \in A_2(H^*, \Gamma)$, and hence $A_2(H^*, \Gamma)$ becomes a complex Banach space with this norm. Throughout this chapter, we assume that $A_2(H^*, \Gamma)$ is equipped with this norm.

Proposition 6.5. *Both Bers' projection $\Phi\colon B(H, \Gamma)_1 \to A_2(H^*, \Gamma)$ and Bers' embedding $\mathcal{B}\colon T_\beta(\Gamma) \to A_2(H^*, \Gamma)$ are continuous.*

Proof. Note that $\{ \varphi_n \}_{n=1}^\infty$ converges to φ in $A_2(H^*, \Gamma)$ if and only if $\{ \varphi_n \}_{n=1}^\infty$ converges to φ uniformly on compact sets on H^*. Hence, Proposition 4.36 implies that Φ is continuous. By the definition of topology of $T_\beta(\Gamma)$, it follows that $\mathcal{B}$ is also continuous. □

As stated in Theorem 5.15, the Teichmüller space $T(\Gamma)$ is homeomorphic to $\mathbf{R}^{6g-6}$, and so is $T_\beta(\Gamma)$. Thus Brouwer's theorem on invariance of domains implies that the image $T_B(\Gamma)$ of the continuous injection $\mathcal{B}\colon T_\beta(\Gamma) \to A_2(H^*, \Gamma)$ is a domain in $A_2(H^*, \Gamma)$, and $\mathcal{B}\colon T_\beta(\Gamma) \to T_B(\Gamma)$ is a homeomorphism. Since $A_2(H^*, \Gamma)$ is a $(3g-3)$-dimensional complex vector space, $T_B(\Gamma)$ inherits the complex manifold structure of $A_2(H^*, \Gamma)$. This $T_B(\Gamma)$ is also called the *Teichmüller space* of Γ. Under identification with $T_B(\Gamma)$, the Teichmüller spaces $T(\Gamma)$, $T_\beta(\Gamma)$, and $T(R)$ are also considered as $(3g-3)$-dimensional complex manifolds, where $R = H/\Gamma$. (The definition of a higher dimensional complex manifold is similar to that for the one-dimensional case. Also, see §2.3.)

6.1.4. Boundedness of $T_B(\Gamma)$

The following theorem shows that $T_B(\Gamma)$ is a bounded domain in $A_2(H^*, \Gamma)$.

Theorem 6.6. *The Teichmüller space $T_B(\Gamma)$ is contained in the open ball in $A_2(H^*, \Gamma)$ with center 0 and radius 3/2.*

This theorem is an immediate consequence of an inequality for univalent functions on H^*, which is due to Nehari and Kraus.

Lemma 6.7. (Nehari and Kraus) *Every univalent function on H^* satisfies the inequality*

$$\|\{f, z\}\|_\infty = \sup_{z \in H^*} (\operatorname{Im} z)^2 |\{f, z\}| \leqq \frac{3}{2}.$$

Proof. First, we consider a univalent function F on Δ^*, where $\Delta^* = \hat{\mathbf{C}} - \bar{\Delta}$, the exterior of the unit disk Δ. We shall see that if F is given by

$$\zeta = F(w) = w + b_o + \frac{b_1}{w} + \frac{b_2}{w^2} + \cdots, \tag{6.3}$$

then $|b_1| \leqq 1$. In fact, for arbitrary $r > 1$, let C'_r be the image of the circle $C_r = \{ w \in \mathbf{C} \mid |w| = r \}$ under F. Then Green's formula implies that the area A_r of the bounded domain surrounded by C'_r is given by

$$A_r = \frac{1}{2i} \int_{C_r} \overline{F(w)}\, dF(w).$$

Thus, noting $\bar{w} = r^2/w$ for $w = re^{i\theta}$, we have

$$A_r = \pi \left(r^2 - \sum_{n=1}^{\infty} n|b_n|^2 r^{-2n} \right) \geqq 0,$$

which implies that

$$\sum_{n=1}^{N} n|b_n|^2 r^{-2n} \leqq r^2$$

for any positive integer N. Letting $r \to 1$ and then letting $N \to \infty$, we obtain the inequality

$$\sum_{n=1}^{\infty} n|b_n|^2 \leqq 1.$$

This is the content of the so-called *Bieberbach's area theorem*. In particular, we have $|b_1| \leqq 1$.

Differentiating the series in (6.3) term by term, we obtain

$$\{F, w\} = -\frac{6}{w^4} b_1 + \sum_{n=5}^{\infty} \frac{c_n}{w^n}, \qquad w \in \Delta^* - \{\infty\}.$$

Hence, we get

$$\lim_{w \to \infty} |w^4\{F, w\}| = 6|b_1| \leqq 6.$$

Now, let f be an arbitrary univalent function on H^*. For a given point $z_o = x_o + iy_o \in H^*$, first suppose that $f(z_o) \neq \infty$. Taking a Möbius transformation $T\colon H^* \to \Delta^*$ defined by $T(z) = (z - \overline{z_o})/(z - z_o)$, we put

$$F(w) = \frac{2iy_o f'(z_o)}{f(T^{-1}(w)) - f(z_o)}, \qquad w \in \Delta^*.$$

Then F is a univalent function on Δ^*, and has an expansion as (6.3). From formula (6.1), we have $\{f, z\} = \{F, T(z)\} \cdot T'(z)^2$ on H^*. Thus noting $T(z_o) = \infty$ and $T'(z)^2 = -4\,y_o^2\,T(z)^4/(z - \overline{z_o})^4$, we conclude that

$$\begin{aligned} |\{f, z_o\}| &= \lim_{z \to z_o} |\{F, T(z)\} \cdot T'(z)^2| \\ &= \lim_{w \to \infty} |w^4\{F, w\}| \cdot \lim_{z \to z_o} \frac{4y_o^2}{|z - \overline{z_o}|^4} \leqq \frac{3}{2y_o^2}. \end{aligned}$$

Next, suppose that $f(z_o) = \infty$. Then by the relation $\{f, z_o\} = \{1/f, z_o\}$ and the above argument, we see again that $|\{f, z_o\}| \leqq 3/(2\,y_o^2)$. Since z_o is arbitrary, we complete the proof of Lemma 6.7. □

6.2. Invariance of Complex Structure of Teichmüller Space

Let us prove that the complex structure of $T(\Gamma)$ which is introduced in the preceding section is independent of the choice of the Fuchsian model Γ of a closed Riemann surface of genus $g\,(\geq 2)$.

6.2.1. Local Inverse of Bers' Embedding

To construct explicitly a holomorphic inverse of Bers' embedding $\mathcal{B}: T_\beta(\Gamma) \to T_B(\Gamma)$ in a neighborhood of the origin in $T_B(\Gamma)$, we associate each element $\varphi \in A_2(H^*, \Gamma)$ with an element $\mu_\varphi \in B(H, \Gamma)$ as follows. For an arbitrary element $\varphi \in A_2(H^*, \Gamma)$, we get an element $\psi \in A_2(H, \Gamma)$ defined by $\psi(z) = \overline{\varphi(\bar{z})}$ for any $z \in H$. Let $ds_H^2 = |dz|^2/(\operatorname{Im} z)^2$ be the Poincaré metric on H. Then $\overline{\psi(z)dz^2/ds_H^2} = (\operatorname{Im} z)^2\varphi(\bar{z})d\bar{z}/dz$ is a Beltrami differential on H/Γ. Thus setting

$$\mu_\varphi(z) = -2(\operatorname{Im} z)^2\varphi(\bar{z}), \quad z \in H,$$

we obtain an element $\mu_\varphi \in B(H, \Gamma)$. This μ_φ is called the *Bers' Beltrami differential* constructed from φ. It is also called a *harmonic Beltrami differential*, since it is a harmonic (-1, 1) form with respect to the Laplace-Beltrami operator induced by the Poincaré metric. Let $V = \{\varphi \in A_2(H^*, \Gamma) \mid \|\varphi\|_\infty < 1/2\}$, a neighborhood of the origin in $A_2(H^*, \Gamma)$. Then μ_φ belongs to $B(H, \Gamma)_1$, i.e., a Beltrami coefficient with respect to Γ for every $\varphi \in V$. Thus we define a mapping $\Psi: V \to T_\beta(\Gamma)$ by

$$\Psi(\varphi) = [w_{\mu_\varphi}],$$

which is continuous from Proposition 6.2.

Theorem 6.8. *Under the preceding situation, $U = \Psi(V)$ is an open neighborhood of the base point in $T_\beta(\Gamma)$, and $\Psi: V \to U$ is a right inverse of $\mathcal{B}: U \to V$, i.e., a holomorphic mapping with $\mathcal{B} \circ \Psi = id$.*

This is an immediate consequence of the following theorem due to Ahlfors and Weill [14].

Theorem 6.9. (Ahlfors and Weill) *For any element $\varphi \in A_2(H^*, \Gamma)$ with $\|\varphi\|_\infty < 1/2$, the harmonic Beltrami differential μ_φ constructed from φ satisfies $\mathcal{B}([w_{\mu_\varphi}]) = \varphi$.*

Proof. In order to construct a univalent function on H^* with Schwarzian derivative φ, take linearly independent holomorphic solutions η_1 and η_2 of the second-order ordinary differential equation

$$\eta'' + \frac{1}{2}\varphi\eta = 0. \tag{6.4}$$

Here, we assume that η_1 and η_2 satisfy the normalization condition

$$\begin{cases} \eta_1(-i) = \eta_2'(-i) = 0 \\ \eta_1'(-i) = \eta_2(-i) = 1. \end{cases} \tag{6.5}$$

Since $(\eta_1'\eta_2 - \eta_1\eta_2')' = 0$ on H^*, we have

$$\eta_1'\eta_2 - \eta_1\eta_2' = 1 \tag{6.6}$$

on H^*. Set $f(z) = \eta_1(z)/\eta_2(z)$ for any $z \in H^*$. From (6.6), we see that f is a locally biholomorphic mapping of H^* into $\hat{\mathbf{C}}$. A straightforward calculation gives $\{f, z\} = \varphi$ on H^*.

Now, we put

$$F(z) = \frac{\eta_1(\bar{z}) + (z-\bar{z})\eta_1'(\bar{z})}{\eta_2(\bar{z}) + (z-\bar{z})\eta_2'(\bar{z})}, \quad z \in H. \tag{6.7}$$

Then F is a real-analytic mapping of H into $\hat{\mathbf{C}}$, because its numerator and denominator do not vanish simultaneously from (6.6). By a simple computation, we see that $F_{\bar{z}}/F_z = \mu_\varphi$ on H. Since $\|\mu_\varphi\|_\infty < 1$, the Jacobian of F is positive on H, and hence F is locally diffeomorphic on H.

Next, we set

$$\hat{f}(z) = \begin{cases} F(z), & z \in H \\ f(z), & z \in H^*. \end{cases}$$

We need to prove that $\hat{f}$ extends to a quasiconformal mapping of $\hat{\mathbf{C}}$ onto itself in such a way that $w_{\mu_\varphi} = S \circ \hat{f}$ for some Möbius transformation S, which implies that $\{w_{\mu_\varphi}, z\} = \varphi$.

For this purpose, first suppose that φ is holomorphic in a neighborhood of $H^* \cup \hat{\mathbf{R}}$ in $\hat{\mathbf{C}}$ and $|\varphi(z)| = O(|z|^{-4})$ as $z \to \infty$. Then η_1 and η_2 are defined on a neighborhood of the real axis $\mathbf{R}$, and so are f and F. Since $f = F$ on $\mathbf{R}$, we obtain a continuous mapping $\hat{f}$ of $\mathbf{C}$ into $\hat{\mathbf{C}}$ by putting $\hat{f} = f$ on $\mathbf{R}$. This extended mapping $\hat{f}$ is locally homeomorphic on a neighborhood of $\mathbf{R}$.

In fact, for any point z on $\mathbf{R}$, choosing a small disk D with center z, we see that both $f\colon D \to f(D)$ and $F\colon D \to F(D)$ are homeomorphic, and $f = F$ on $D \cap \mathbf{R}$. Since both f and F are orientation-preserving, $f(D \cap H^*)$ and $F(D \cap H)$ do not intersect, which implies that $\hat{f}$ is injective on D. It is easy to see that $\hat{f}$ is an open mapping, and hence $\hat{f}\colon D \to \hat{f}(D)$ is a homeomorphism.

Moreover, it is proved that $\hat{f}$ also extends to a local homeomorphism on $\hat{\mathbf{C}}$ as follows. Since $|\varphi(z)| = O(|z|^{-4})$ as $z \to \infty$, η_1 and η_2 are expanded near ∞ in the form

$$\eta_1(z) = a_1 z + b_1 + O(|z|^{-1}),$$
$$\eta_2(z) = a_2 z + b_2 + O(|z|^{-1}),$$

where $a_1 b_2 - a_2 b_1 = 1$. Hence, $f = \eta_1/\eta_2$ is a univalent function in a neighborhood of ∞, and $f(\infty) = a_1/a_2$. On the other hand, we have

$$F(z) = \frac{a_1 z + b_1 + O(|z|^{-1})}{a_2 z + b_2 + O(|z|^{-1})} \quad (z \to \infty).$$

Thus F is an orientation-preserving diffeomorphism on a neighborhood of ∞ with $F(\infty) = a_1/a_2$. Therefore, putting $\hat{f}(\infty) = a_1/a_2$, we see that $\hat{f}$ is a local homeomorphism of $\hat{\mathbf{C}}$ into itself. Then Lemma 4.28 implies that $\hat{f}\colon \hat{\mathbf{C}} \to \hat{\mathbf{C}}$ is a homeomorphism.

Applying Painlevé's theorem to $w_{\mu_\varphi} \circ (\hat{f})^{-1}$, we see that there exists a Möbius transformation S with $w_{\mu_\varphi} = S \circ \hat{f}$.

To remove the hypothesis that φ is holomorphic on $\hat{\mathbf{R}}$ with zero of order at least 4 at ∞, we pick up a Möbius transformation T_n given by

$$T_n(z) = \frac{2nz - i}{iz + 2n}$$

for any positive integer n. Then $T_n^{-1}(H)$ is a relatively compact subdomain of H, and T_n converges uniformly on compact subsets of $\mathbf{C}$ to id as $n \to \infty$. Setting $\varphi_n = \varphi(T_n(z)) \cdot T_n'(z)^2$, we see that φ_n is a holomorphic function defined outside of $T_n^{-1}(H)$, and has a zero of order at least 4 at ∞. Noting $T_n(H^*) \subset H^*$, by Proposition 3.1, we get

$$\frac{1}{(\operatorname{Im} z)^2} \geqq \frac{|T_n'(z)|^2}{(\operatorname{Im} T_n(z))^2}, \quad z \in H^*.$$

Thus we have

$$\begin{aligned} \|\varphi_n\|_\infty &= \sup_{z \in H^*} (\operatorname{Im} z)^2 \, |\varphi_n(z)| \\ &\leqq \sup_{z \in H^*} (\operatorname{Im} T_n(z))^2 \, |\varphi(T_n(z))| \\ &\leqq \|\varphi\|_\infty < \frac{1}{2}. \end{aligned}$$

Now, putting $\mu_n = \mu_{\varphi_n}$, we get $\|\mu_n\|_\infty = 2\|\varphi_n\|_\infty \leqq 2\|\varphi\|_\infty < 1$. Hence, by the construction given, φ_n produces a quasiconformal mapping $\hat{f}_n$ of $\hat{\mathbf{C}}$ onto itself such that $w_{\mu_n} = S_n \circ \hat{f}_n$ for some Möbius transformation S_n.

On letting $g_n = w_{\mu_n} \circ (w_{\mu_\varphi})^{-1}$, from Theorem 4.10 we see that g_n has a maximal dilatation $\leqq \{(1 + 2\|\varphi\|_\infty)/(1 - 2\|\varphi\|_\infty)\}^2$. Thus we obtain

$$\|\mu_{g_n}\|_\infty \leqq 4\|\varphi\|_\infty / (1 + 4\|\varphi\|_\infty^2) < 1.$$

Then, (4.8) in Proposition 4.13 implies that $\mu_{g_n} \to 0$ uniformly on compact subsets of $\mathbf{C} - \mathbf{R}$. Hence, by the same argument as in the proof of Lemma 5.7, we see that $g_n \to id$ uniformly on compact subsets of $\mathbf{C}$. Consequently, $w_{\mu_n} \to w_{\mu_\varphi}$ uniformly on compact subsets of $\mathbf{C}$.

On the other hand, the normalized solutions $\eta_{1,n}$ and $\eta_{2,n}$ of the differential equation (6.4) for φ_n converge uniformly on compact subsets of H^* to η_1 and η_2, respectively. Thus by construction, $\hat{f}_n \to \hat{f}$ uniformly (with respect to the spherical metric) on compact subsets of $\mathbf{C} - \mathbf{R}$. In particular, we see that $S_n \to S$ in $Aut(\hat{\mathbf{C}})$, and $\hat{f} = S^{-1} \circ w_{\mu_\varphi}$ on $\mathbf{C} - \mathbf{R}$, which implies that $\hat{f}$ extends to a quasiconformal mapping $S^{-1} \circ w_{\mu_\varphi}$ on $\hat{\mathbf{C}}$. Moreover, we obtain

$$\mathcal{B}([w_{\mu_\varphi}]) = \{\, w_{\mu_\varphi}, z \,\} = \{\, S \circ f, z \,\} = \varphi \quad \text{on} \quad H^*.$$

This completes the proof of Theorem 6.9. □

Remark 1. Theorem 6.9 holds for any Fuchsian group Γ, since in the proof we did not use the hypothesis that H/Γ is compact.

Remark 2. We would like to explain a heuristic method to find a solution in the form (6.7) of the Beltrami equation

$$w_{\bar{z}} = \frac{1}{2}(z-\bar{z})^2 \varphi(\bar{z}) w_z$$

on H for any $\varphi \in A_2(H^*, \Gamma)$ with $\|\varphi\|_\infty < 1/2$. This is due to Shigeru Furuya, and the authors learned it from Kôtaro Oikawa.

Setting $z = y$ and $\bar{z} = x$ in the above Beltrami equation, we have a partial differential equation

$$w_x = \frac{1}{2}(y-x)^2 \varphi(x) w_y.$$

Denote by $w(x,y) = C$ with an arbitrary constant C a general solution of Riccati's differential equation

$$y' = -\frac{1}{2}(y-x)^2 \varphi(x).$$

Then this w gives a solution of the above partial differential equation.

Putting $u = y - x$, we get

$$\frac{u'+1}{u^2} = -\frac{1}{2}\varphi(x).$$

Thus, setting $u = -v/v'$, we obtain the second-order differential equation

$$v'' = -\frac{1}{2}\varphi(x) v.$$

Take linearly independent solutions η_1 and η_2 of this equation. Then we see that

$$w = \frac{\eta_1 + (y-x)\eta_1'}{\eta_2 + (y-x)\eta_2'},$$

and hence we obtain F in (6.7).

Corollary. *For every $\varphi \in V$, there exists an element $\mu \in B(H,\Gamma)_1$ such that w_μ is real-analytic on H and $\mathcal{B}^{-1}(\varphi) = [w_\mu]$.*

Moreover, every point $[S, f]$ of the Teichmüller space $T(R)$ of $R = H/\Gamma$ is represented by a real-analytic quasiconformal mapping g of R to S, i.e., $[S, f] = [S, g]$ in $T(R)$.

Proof. In the proof of Theorem 6.9, we saw that $w_{\mu_\varphi} = S \circ \hat{f}$, $\hat{f}$ is real-analytic on H, and $\mathcal{B}^{-1}(\varphi) = [w_{\mu_\varphi}]$, which shows the first assertion.

Let us prove the second statement. Denote by D the set of all points $[S, f] \in T(R)$ such that $[S, f]$ is represented by a real-analytic quasiconformal mapping of R to S. Let $[S_o, f_o]$ be an arbitrary point in D. From the first assertion, we find a neighborhood U of the base point in $T(S_o)$ so that every point of U is given by $[S, f]$ with some real-analytic quasiconformal mapping f. Then $\{ [S, f \circ f_o] \mid [S, f] \in U \}$ is a neighborhood of $[S_o, f_o]$ in $T(R)$ and contained in D. Thus D is an open subset of $T(R)$.

Next, let $\{ [S_n, f_n] \}_{n=1}^{\infty}$ be a sequence in D which converges to a point $[S, f] \in T(R)$. We may assume that each f_n is real-analytic. Since $[S_n, f_n \circ f^{-1}]$ converges to the base point of $T(S)$, the first assertion implies that there exists a real-analytic quasiconformal mapping g_n of S to S_n with $[S_n, f_n \circ f^{-1}] = [S_n, g_n]$ for any sufficiently large n. Thus $g_n^{-1} \circ f_n$ is real-analytic and $[S, f] = [S, g_n^{-1} \circ f_n]$ in $T(R)$. Thus, $[S, f]$ belongs to D, and hence D is a closed subset of $T(R)$. Since $T(R)$ is connected, D must be equal to $T(R)$. □

Remark 3. Theorem 6.9 states that Bers' projection Φ has a holomorphic local section $\varphi \mapsto \mu_\varphi$ on a neighborhood of the base point in $T_B(\Gamma)$, which is called the *Ahlfors-Weill section*. By using quasiconformal reflection, it is possible to construct a holomorphic local section of Φ on a neighborhood of any point in $T_B(\Gamma)$ (see Ahlfors [A-2], pp. 131-135, and Ahlfors [9]).

There are other approaches to this theorem which are due to Earle and due to Royden. For details, we refer to the book of Nag [A-80], pp. 242-249.

6.2.2. Differentiation of Bers' Projection

Now, we wish to define the derivative $\dot{\Phi}_\mu[\nu]$ of Φ in the direction ν at μ for arbitrary $\mu \in B(H, \Gamma)_1$ and $\nu \in B(H, \Gamma)$.

Let μ_t be an element in $B(H, \Gamma)_1$ defined for any complex number t in a neighborhood of the origin such that $\mu_t = \mu + t\nu + t\varepsilon(t)$, where ν belongs to $B(H, \Gamma)$, and $\|\varepsilon(t)\|_\infty \to 0$ as $t \to 0$. Then $\dot{\Phi}_\mu[\nu]$ is defined by

$$\dot{\Phi}_\mu[\nu] = \lim_{t \to \infty} \frac{1}{t} \left(\Phi(\mu_t) - \Phi(\mu) \right),$$

where the convergence is norm convergence with respect to the hyperbolic L^∞-norm in the preceding section. The existence and integral representation of $\dot{\Phi}_\mu[\nu]$ is given in Theorem 6.11.

First, we consider the case where $\mu = 0$.

Theorem 6.10. *For every $\nu \in B(H, \Gamma)$, the derivative $\dot{\Phi}_0[\nu]$ exists and is given by*

$$\dot{\Phi}_0[\nu](z) = -\frac{6}{\pi} \iint_H \frac{\nu(\zeta)}{(\zeta - z)^4} \, d\xi d\eta, \quad z \in H^*. \tag{6.8}$$

Proof. By Theorem 4.37, we have

$$w_{\mu_t}(z) = z + t\dot{w}[\nu](z) + o(t)$$

uniformly on compact subsets of $\mathbf{C}$ as $t \to 0$, where

$$\dot{w}[\nu](z) = -\frac{1}{\pi} \iint_H \nu(\zeta) \frac{z(z-1)}{\zeta(\zeta - 1)(\zeta - z)} \, d\xi d\eta. \tag{6.9}$$

Since w_{μ_t} is holomorphic on H^*, from Weierstrass' theorem on double series, we get

$$w'_{\mu_t} = 1 + t\dot{w}[\nu]' + o(t),$$
$$w''_{\mu_t} = t\dot{w}[\nu]'' + o(t),$$
$$w'''_{\mu_t} = t\dot{w}[\nu]''' + o(t)$$

uniformly on compact subsets of H^* as $t \to 0$. Thus we see that

$$\Phi(\mu_t) = \{\, w_{\mu_t}, z \,\} = t\dot{w}[\nu]''' + o(t)$$

uniformly on compact subsets of H^* as $t \to 0$. Since H^*/Γ is compact, it follows that $\dot{\Phi}_0[\nu]$ exists and is equal to $\dot{w}[\nu]'''$. Further, formula (6.9) provides (6.8). □

Theorem 6.11. *For every $\mu \in B(H,\Gamma)_1$ and $\nu \in B(H,\Gamma)$, the derivative $\dot{\Phi}_\mu[\nu]$ exists and is given by*

$$\dot{\Phi}_\mu[\nu](z) = \left[-\frac{6}{\pi}\iint_H \frac{\nu(\zeta)\left((w_\mu)_\zeta(\zeta)\right)^2}{\left(w_\mu(\zeta) - w_\mu(z)\right)^4}\, d\xi d\eta\right] w'_\mu(z)^2, \quad z \in H^*. \tag{6.10}$$

Proof. Set $f = w_\mu$, $g_t = w_{\mu_t} \circ w_\mu^{-1}$, and $\lambda_t = \mu_{g_t}$. Then we have

$$\lambda_t(\zeta) = \left(\frac{f_z}{\overline{f_z}}\, \frac{\mu_t - \mu}{1 - \bar{\mu}\mu_t}\right) \circ f^{-1}(\zeta), \quad \zeta \in f(H).$$

Thus, putting

$$\lambda(\zeta) = \left(\frac{f_z}{\overline{f_z}}\, \frac{\nu}{1 - |\mu|^2}\right) \circ f^{-1}(\zeta),$$

we get

$$\lambda_t = t\lambda + t\delta(t) \quad \text{on } f(H),$$

where $\|\delta(t)\|_\infty \to 0$ as $t \to 0$.

On the other hand, from the relation

$$\Phi(\mu_t)(z) = \{\, g_t \circ f, z \,\} = \{\, g_t, f(z) \,\} \cdot f'(z)^2 + \Phi(\mu)(z),$$

we obtain

$$\dot{\Phi}_\mu[\nu](z) = \left[\lim_{t \to 0} \frac{1}{t}\{\, g_t, f(z) \,\}\right] f'(z)^2, \quad z \in H^*.$$

Then by the same argument as in the proof of Theorem 6.10, we see that $\dot{\Phi}_\mu[\nu]$ exists and is represented in the form

$$\dot{\Phi}_\mu[\nu](z) = \left[-\frac{6}{\pi}\iint_{f(H)} \frac{\lambda(\zeta)}{(\zeta - f(z))^4}\, d\xi d\eta\right] f'(z)^2, \quad z \in H^*.$$

Therefore, by substituting $f(\zeta)$ for ζ in this integral, we obtain the integral formula (6.10). □

6.2.3. Invariance of Complex Structure of $T(\Gamma)$

To begin with, we review briefly complex manifolds of higher dimension. A complex-valued function f defined on a domain D of $\mathbf{C}^n$ is called *holomorphic* on D if for every $a = (a_1, \ldots, a_n) \in D$, it has a power series expansion

$$f(z) = \sum_{k_1,\ldots,k_n=0}^{\infty} c_{k_1\ldots k_n}(z_1 - a_1)^{k_1} \ldots (z_n - a_n)^{k_n}$$

which converges for all $z = (z_1, \ldots, z_n)$ in a neighborhood of a. By Cauchy's integral formula, it is easily seen that f is holomorphic in D provided that f is continuous in D and holomorphic in each variable separately. Further, Hartogs' theorem asserts that separate analyticity of f implies its continuity (see Bers [A-14], p.2). As in the case of one-dimensional manifolds in §1.1 of Chapter 1, we can define n-dimensional complex manifolds, holomorphic functions on a complex manifold, holomorphic and biholomorphic mappings between complex manifolds, and so on. For details, we refer to Griffiths and Harris [A-39], Chapter 0; and Morrow and Kodaira [A-77], Chapter 1.

Now, we shall prove that the complex structure of $T(\Gamma)$, which was defined in §1.3 by using Bers' embedding $\mathcal{B}\colon T_\beta(\Gamma) \to T_B(\Gamma)$, is independent of the choice of Γ.

Take a Fuchsian model Γ_1 of another closed Riemann surface R_1 of genus g such that each of points 0, 1, and ∞ is fixed by a suitable element in $\Gamma_1 - \{id\}$. Let ω be a lift of a quasiconformal mapping $f_1\colon R \to R_1$. We may assume that $\Gamma_1 = \omega\Gamma\omega^{-1}$. When we identify $T(R), T(R_1)$ with $T(\Gamma), T(\Gamma_1)$, respectively, the *translation of the base point* $[f_1]_*\colon T(R) \to T(R_1)$ induces a homeomorphism $[\omega]_*\colon T(\Gamma) \to T(\Gamma_1)$;

$$[w^\mu] \mapsto [\omega]_*([w^\mu]) = [w^\nu],$$

where α is a real Möbius transformation such that $\alpha \circ w^\mu \circ \omega^{-1}$ fixes each of 0, 1, and ∞, and such that $w^\nu = \alpha \circ w^\mu \circ \omega^{-1}$. Moreover, we define a homeomorphism $\langle\omega\rangle_*\colon T_\beta(\Gamma) \to T_\beta(\Gamma_1)$ by

$$[w_\mu] \mapsto \langle\omega\rangle_*([w_\mu]) = [w_\nu].$$

We also call $[\omega]_*$ or $\langle\omega\rangle_*$ a *translation of the base point.*

Let $\mathcal{B}_1$ be Bers' embedding of $T_\beta(\Gamma_1)$ into $A_2(H, \Gamma_1)$. Then $T_\beta(\Gamma_1)$ is realized as a bounded domain $T_B(\Gamma_1)$ of $A_2(H, \Gamma_1)$.

Theorem 6.12. *The following mappings are all biholomorphic:*

$$[f_1]_*\colon T(R) \to T(R_1),$$
$$[\omega]_*\colon T(\Gamma) \to T(\Gamma_1),$$
$$\langle\omega\rangle_*\colon T_\beta(\Gamma) \to T_\beta(\Gamma_1),$$
$$\mathcal{B}_1 \circ \langle\omega\rangle_* \circ \mathcal{B}^{-1}\colon T_B(\Gamma) \to T_B(\Gamma_1).$$

Proof. To see that $F = \mathcal{B}_1 \circ \langle\omega\rangle_* \circ \mathcal{B}^{-1}$ is biholomorphic, we need to prove that F is biholomorphic in a neighborhood of any point $\mathcal{B}([w_\mu]) \in T_B(\Gamma)$, since F is homeomorphic. Let $\Gamma^\mu = w^\mu \Gamma (w^\mu)^{-1}$ and $\mathcal{B}_\mu$ be Bers' embedding of $T_\mathcal{B}(\Gamma^\mu)$. If two mappings

$$F_1 = \mathcal{B} \circ \langle (w^\mu)^{-1} \rangle_* \circ \mathcal{B}_\mu^{-1} : T_B(\Gamma^\mu) \to T_B(\Gamma),$$
$$F_2 = \mathcal{B}_1 \circ \langle \omega \circ (w^\mu)^{-1} \rangle_* \circ \mathcal{B}_\mu^{-1} : T_B(\Gamma^\mu) \to T_B(\Gamma_1)$$

are biholomorphic in a neighborhood of the base point of $T_B(\Gamma^\mu)$, then $F = F_1 \circ F_2^{-1}$ is biholomorphic in a neighborhood of $\mathcal{B}([w_\mu]) \in T_B(\Gamma)$.

Thus, it is sufficient to give a proof that F is biholomorphic in a neighborhood of the base point of $T_B(\Gamma)$.

Take a neighborhood, say $V = \{ \varphi \in A_2(H^*, \Gamma) \mid \|\varphi\|_\infty < 1/2 \}$, of the base point of $T_B(\Gamma)$. For arbitrary $\varphi, \psi \in V$, we set $D = \{ t \in \mathbf{C} \mid \psi + t\varphi \in V \}$. Put $\varphi(t) = \psi + t\varphi$ and $\mu(t) = \mu_{\varphi(t)}$. From Theorem 6.9, we have $\mathcal{B}^{-1}(\varphi(t)) = [w_{\mu(t)}]$. Let $\lambda(t)$ be the Beltrami coefficient of $w^{\mu(t)} \circ \omega^{-1}$, which is given by

$$\lambda(t) = \left(\frac{\omega_z}{\overline{\omega_z}} \frac{\mu(t) - \mu_\omega}{1 - \overline{\mu_\omega}\, \mu(t)} \right) \circ \omega^{-1}$$

on H. By the construction, we get $F(\varphi(t)) = \{ w_{\lambda(t)}, z \}$. Since $\lambda(t)$ is holomorphic with respect to t, Theorem 6.11 implies that $F(\varphi(t))$ is holomorphic on D. Since φ and ψ are arbitrary, F is holomorphic on V. Since F is injective on $T_B(\Gamma)$, from the following lemma (Lemma 6.13) and the inverse mapping theorem, we see that F is biholomorphic on V.

By the definitions, the rest of this theorem is trivial. □

Lemma 6.13. *The Jacobian* $J_F = \det(\partial F_j / \partial z_k)_{1 \leqq j,k \leqq n}$ *of an injective holomorphic mapping* $F = (F_1, \ldots, F_n)$ *of a domain* D *in* $\mathbf{C}^n$ *into* $\mathbf{C}^n$ *vanishes at no points on* D.

Proof. We prove this assertion by induction for dimension n. First of all, clearly it holds for $n = 1$.

Given an integer $n > 1$, we assume that the assertion holds for any positive integer $\leqq n-1$. Let D_o be the set of all points in D where J_F vanishes. We want to prove that D_o is empty.

Suppose that D_o is non-empty. Further, assume that the Jacobi matrix of F is of rank r with $1 \leqq r \leqq n-1$ at some point $a \in D_o$. Then we may assume that $\det(\partial F_j / \partial z_k)_{1 \leqq j,k \leqq r}$ does not vanish at a. The inverse mapping theorem implies that $G(z) = (F_1(z), \ldots, F_r(z), z_{r+1}, \ldots, z_n)$ has the inverse mapping $H = (H_1, \ldots, H_n)$ in a neighborhood of a. Then we have $H_{r+1}(\zeta) = \zeta_{r+1}, \ldots, H_n(\zeta) = \zeta_n$, and $F \circ H(\zeta) = (\zeta_1, \ldots, \zeta_r, F_{r+1} \circ H(\zeta), \ldots, F_n \circ H(\zeta))$ in a neighborhood of $G(a)$. We set

$$W = \{ \zeta = (\zeta_1, \ldots, \zeta_n) \in \mathbf{C}^n \mid \zeta_1 = F_1(a), \ldots, \zeta_r = F_r(a) \},$$
$$W_1 = \{ w = (w_1, \ldots, w_n) \in \mathbf{C}^n \mid w_1 = F_1(a), \ldots, w_r = F_r(a) \}.$$

Then the restriction $F \circ H|_W$ is an injective holomorphic mapping of a neighborhood of $G(a)$ in W into W_1. By the hypothesis of induction, it follows that the Jacobi matrix of $F \circ H|_W$ is of rank $n-r$ at $G(a)$. Hence, the Jacobi matrix of $F \circ H$ is of rank n at $G(a)$. On the other hand, since G is biholomorphic in a neighborhood of a, the Jacobi matrix of F must be of rank n at a, which contradicts that F is of rank $\leqq n-1$. Thus every $\partial F_j/\partial z_k$ should vanish on D_o.

Now, assume that $D_o = D$. Then each F_j must be a constant, because all $\partial F_j/\partial z_k$ vanish on D. This contradicts that F is injective on D.

Next, assume that D_o is a proper subset of D. Since D_o is the set of zeros of the holomorphic function J_F on D, from Weierstrass' preparation theorem (Bers [A-14], p.40), we may assume that there exists a holomorphic function ψ on a domain U_1 in $\mathbf{C}^{n-1}$ such that $D_o \cap U = \{(\tau, \psi(\tau)) \mid \tau \in U_1\}$, where U is a suitable domain in $\mathbf{C}^n$. Let $\tilde{F}_j$ be the restriction of F_j to $D_o \cap U$, which is considered as a holomorphic function on U_1. Since all $\partial F_j/\partial z_k$ vanish on D_o, every $\partial \tilde{F}_j/\partial \tau_k$ vanishes on U_1. Hence, every $\tilde{F}_j$ must be a constant, which contradicts again that F is injective.

Therefore, D_o should be empty, and hence the lemma holds for n. □

Remark. As stated in §1.3, the complex structure on $T_\beta(\Gamma)$ was induced from the complex structure on $T_B(\Gamma)$, which is a bounded domain in the complex vector space $A_2(H^*, \Gamma)$. There, we needed to know that $T_\beta(\Gamma)$ is a real $(6g-6)$-dimensional topological manifold, since we used Brouwer's theorem on invariance of domains to see that Bers' embedding $\mathcal{B}\colon T_\beta(\Gamma) \to T_B(\Gamma)$ is homeomorphic.

However, to see that $T_\beta(\Gamma)$ has a complex manifold structure such that $\mathcal{B}$ is biholomorphic, the fact that $T_\beta(\Gamma)$ is a real $(6g-6)$-dimensional topological manifold is actually inessential. Namely, we take a coordinate neighborhood of any point $[w_\mu] \in T_\beta(\Gamma)$ as follows: setting $\Gamma^\mu = w^\mu \Gamma (w^\mu)^{-1}$, we have an injective mapping

$$F_\mu = \mathcal{B}_\mu \circ \langle w^\mu \rangle_* \colon T_\beta(\Gamma) \to T_B(\Gamma^\mu)$$

with $F_\mu([w_\mu]) = 0$, where $\mathcal{B}_\mu$ is Bers' embedding of $T_\beta(\Gamma^\mu)$ into $A_2(H^*, \Gamma^\mu)$. We set

$$V_\mu = \{\varphi \in A_2(H^*, \Gamma^\mu) \mid \|\varphi\|_\infty < 1/2\} \quad \text{and} \quad U_\mu = (F_\mu)^{-1}(V_\mu).$$

Then Theorem 6.8 asserts that U_μ is a neighborhood of $[w_\mu]$ in $T_\beta(\Gamma)$, and F_μ induces a homeomorphism of U_μ onto V_μ. Hence, we can take $(U_\mu, F_\mu|_{U_\mu})$ as a coordinate neighborhood around $[w_\mu]$. By Theorem 6.12, it follows that

$$F_\nu \circ (F_\mu)^{-1} \colon F_\mu(U_\mu \cap U_\nu) \to F_\nu(U_\mu \cap U_\nu)$$

is biholomorphic for every U_μ and U_ν with $U_\mu \cap U_\nu \neq \phi$. Therefore, $\{(U_\mu, F_\mu|_{U_\mu}) \mid [w_\mu] \in T_\beta(\Gamma)\}$ gives a complex manifold structure on $T_\beta(\Gamma)$. It is clear that Bers' embedding $\mathcal{B}$ is a biholomorphic mapping of $T_\beta(\Gamma)$ to $T_B(\Gamma)$. Hence, this complex structure on $T_\beta(\Gamma)$ is equivalent to the one induced in §1.3.

6.3. Teichmüller Modular Groups

We shall prove that the Teichmüller modular group $Mod(R)$ of a closed Riemann surface of genus $g\,(\geqq 2)$ acts properly discontinuously on the Teichmüller space $T(R)$ as a subgroup of the biholomorphic automorphism group $Aut(T(R))$ of $T(R)$.

6.3.1. Definition of Teichmüller Modular Groups

Let R be a closed Riemann surface of genus $g\,(\geqq 2)$. We define the *Teichmüller modular group* $Mod(R)$ of R as the factor group of the group of all quasiconformal self-mappings of R over the normal subgroup of those homotopic to the identity (cf. §3 of Chapter 1). The element of $Mod(R)$ defined by a quasiconformal self-mapping f_o of R is denoted by $[f_o]$. The action $[f_o]_*$ of an element $[f_o] \in Mod(R)$ on $T(R)$ is given by

$$[f_o]_*([S,f]) = [S, f\circ f_o^{-1}]$$

for every $[S,f] \in T(R)$ (see §1.3 of Chapter 5). We call such an $[f_o]_*$ a *Teichmüller modular transformation* of $T(R)$.

Let Γ be a Fuchsian model of R. By lifting, a quasiconformal self-mapping f_o of R corresponds to a quasiconformal self-mapping ω of the upper half-plane H with $\omega\Gamma\omega^{-1} = \Gamma$. Let ω_i be a lift of a quasiconformal self-mapping f_i of R with $\omega_i\Gamma(\omega_i)^{-1} = \Gamma$ for $i = 1, 2$. By the same argument as in the proof of Lemma 5.1, we see that $[f_1] = [f_2]$ in $Mod(R)$ if and only if $\omega_2 = \omega_1\circ\gamma_o$ holds on the real axis $\mathbf{R}$ for some $\gamma_o \in \Gamma$. With this in mind, two quasiconformal self-mappings ω_1 and ω_2 of H satisfying $\omega_i\Gamma\omega_i^{-1} = \Gamma$ $(i = 1, 2)$ are said to be *equivalent* if there exists an element γ_o of Γ such that $\omega_2 = \omega_1\circ\gamma_o$ on $\mathbf{R}$. Denote by $[\omega]$ the equivalence class of ω. The *Teichmüller modular group* $Mod(\Gamma)$ of Γ is the group of all these equivalence classes $[\omega]$. The action $[\omega]_*$ of an element $[\omega] \in Mod(\Gamma)$ on $T(\Gamma)$ is given by

$$[\omega]_*([w^\mu]) = [\alpha\circ w^\mu\circ\omega^{-1}]$$

for every $[w^\mu] \in T(\Gamma)$, where α is an element in $Aut(H)$ such that $\alpha\circ w^\mu\circ\omega^{-1}$ fixes each of 0, 1, and ∞. (cf. §2.3). Theorem 6.12 asserts that $[\omega]_*$ is a biholomorphic automorphism of $T(\Gamma)$. Furthermore, this $[\omega]_*$ induces a biholomorphic automorphism $\langle\omega\rangle_*$of $T_\beta(\Gamma)$ defined by

$$\langle\omega\rangle_*([w_\mu]) = [w_\nu]$$

for any $[w_\mu] \in T_\beta(\Gamma)$, where ν is the Beltrami coefficient of $\alpha\circ w^\mu\circ\omega^{-1}$. We use the same notation $\langle\omega\rangle_*$ for the biholomorphic automorphism of $T_B(\Gamma)$ instead of $\mathcal{B}\circ\langle\omega\rangle_*\circ\mathcal{B}^{-1}$, where $\mathcal{B}$ is Bers' embedding of $T_\beta(\Gamma)$. We also call $[\omega]_*$ or $\langle\omega\rangle_*$ a *Teichmüller modular transformation*. By the construction, it is obvious that $Mod(\Gamma)$ is isomorphic to $Mod(R)$.

By the identification of $T(\Gamma)$ and $T(R)$ (Proposition 5.3), the Teichmüller distance on $T(R)$ induces the Teichmüller distance on $T(\Gamma)$. Then Proposition 5.5 implies the following.

Theorem 6.14. *For every element $[\omega] \in Mod(\Gamma)$, the biholomorphic automorphism $[\omega]_*$ of $T(\Gamma)$ is an isometry with respect to the Teichmüller distance.*

6.3.2. Moduli Sets

To prove that $Mod(\Gamma)$ acts properly discontinuously on $T(\Gamma)$, we prepare some propositions.

Lemma 6.15. *For every compact subset K in the upper half-plane H and positive number M, there exist at most finitely many $\gamma \in \Gamma$ with $\min_{z\in K} \rho(z, \gamma(z)) \leqq M$, where ρ is the Poincaré distance on H.*

Proof. Suppose that there exists a sequence $\{ \gamma_n \}_{n=1}^{\infty}$ of mutually distinct elements in Γ such that every γ_n satisfies $\rho(z_n, \gamma_n(z_n)) \leqq M$ for some $z_n \in K$. Since K is compact and ρ is complete, taking a subsequence if necessary, we may assume that $z_n \to z_o \in K$ and $\gamma_n(z_n) \to w_o \in H$ as $n \to \infty$. Further, we may also assume that $\{ \gamma_n \}_{n=1}^{\infty}$ converges uniformly on compact sets to a holomorphic function f defined on H, because it is a normal family on H. Then we have $f(z_o) = w_o$. Since w_o is in H, Lemma 2.18 shows that f belongs to $Aut(H)$. Thus by Lemma 2.16, Γ is not a discrete subgroup of $Aut(H)$, a contradiction. □

Proposition 6.16. *The set of hyperbolic lengths of all closed geodesics on a closed Riemann surface R of genus g ($\geqq 2$) is discrete in $\mathbf{R}$.*

Moreover, for any positive ℓ, there exist at most finitely many closed geodesics on R with hyperbolic length ℓ.

Proof. Assume that there exists a sequence $\{ L_n \}_{n=1}^{\infty}$ of mutually distinct closed geodesics on R such that every L_n satisfies $\ell(L_n) \leq M$ for some positive number M, where $\ell(L_n)$ is the hyperbolic length of L_n. Choose a fundamental domain F for Γ which is relatively compact in H. Take an element $\gamma_n \in \Gamma$ so that γ_n covers L_n (see §1.3 of Chapter 3), and that $\bar{F} \cap A_{\gamma_n} \neq \phi$, where A_{γ_n} is the axis of γ_n. Then $\{ \gamma_n \}_{n=1}^{\infty}$ is a sequence of mutually distinct elements of Γ such that $\min_{z\in \bar{F}} \rho(z, \gamma_n(z)) = \ell(L_n) \leqq M$ for any n. This contradicts Lemma 6.15. □

Propositions 6.16 and 3.3 lead to the following corollary.

Corollary. *The set $\{ \mathrm{tr}^2(\gamma) \mid \gamma \in \Gamma \}$ is discrete in $\mathbf{R}$, where $\mathrm{tr}^2(\gamma)$ denotes the square of the trace of γ.*

As was shown in Theorem 3.12, a set of hyperbolic lengths of closed geodesics on R induces a normalized Fuchsian model Γ of R (cf. Theorem 3.12). Here, we shall show another such result, which is weaker than Theorem 3.12, but is verified more simply.

Proposition 6.17. *Let Γ be a Fuchsian model of a closed Riemann surface of genus g ($\geqq 2$). Let $\{\gamma_j\}_{j=1}^m$ be a system of generators for Γ such that γ_1 has the repelling fixed point 0 and the attractive fixed point ∞, and such that γ_2 has the repelling fixed point r with $r<0$ and the attractive fixed point 1. Then each γ_j is determined by the absolute values of traces of elements in the finite set*

$$\mathcal{G} = \{\gamma_1\circ\gamma_2, \gamma_j, \gamma_1^{\pm 1}\circ\gamma_k, \gamma_2^{\pm 1}\circ\gamma_k, (\gamma_1\circ\gamma_2)^{\pm 1}\circ\gamma_k\},$$

where $j=1,\dots,m$, and $k=3,\dots,m$.

Proof. First of all, we may assume that γ_1 has a matrix representation

$$A = \begin{bmatrix}\lambda & 0\\ 0 & \lambda^{-1}\end{bmatrix}, \quad \lambda > 1.$$

Thus γ_1 is determined by the absolute value of $\mathrm{tr}(\gamma_1)$.

Next, by the normalization condition we may assume that γ_2 has a matrix representation

$$B = \begin{bmatrix}a & b\\ c & d\end{bmatrix}, \quad a,b,c,d>0,\ ad-bc=1,\ a+b=c+d.$$

Thus we get

$$a+d = |\mathrm{tr}(\gamma_2)|, \quad a\lambda + d\lambda^{-1} = |\mathrm{tr}(\gamma_1\circ\gamma_2)|.$$

Hence both a and d are determined by the absolute values of traces of γ_1, γ_2, and $\gamma_1\circ\gamma_2$. Since the quadratic equation $\gamma_2(z)=z$ has a solution 1, we have

$$2c = a-d+\sqrt{(a+d)^2-4}, \quad b = c+d-a.$$

Consequently, both b and c are determined by a and d. Therefore, γ_2 is determined by the absolute values of traces of γ_1, γ_2, and $\gamma_1\circ\gamma_2$.

Now, for every $k=3,\dots,m$, the Möbius transformation γ_k has a matrix representation

$$C = \begin{bmatrix}p & q\\ r & s\end{bmatrix}, \quad p,q,r,s\in\mathbf{R},\ ps-qr=1.$$

Here, by Theorem 2.22 we may assume that

$$p+s>2.$$

Then, using the relation

$$\mathrm{tr}(A)\mathrm{tr}(C) = \mathrm{tr}(AC) + \mathrm{tr}(A^{-1}C), \tag{6.11}$$

we see that $\mathrm{tr}(AC)$ is determined by the absolute values of traces of γ_1, γ_k and $\gamma_1^{\pm}\circ\gamma_k$. In fact, since the left hand side of (6.11) is positive, we have $\mathrm{tr}(AC) = |\mathrm{tr}(AC)| = |\mathrm{tr}(\gamma_1\circ\gamma_k)|$ provided that $|\mathrm{tr}(AC)| \geqq |\mathrm{tr}(A^{-1}C)|$. If $|\mathrm{tr}(AC)| < |\mathrm{tr}(A^{-1}C)|$, then we obtain $\mathrm{tr}(A^{-1}C) = |\mathrm{tr}(A^{-1}C)| = |\mathrm{tr}(\gamma_1^{-1}\circ\gamma_k)|$, and hence $\mathrm{tr}(AC) = |\mathrm{tr}(\gamma_1)|\cdot|\mathrm{tr}(\gamma_k)| - |\mathrm{tr}(\gamma_1^{-1}\circ\gamma_k)|$.

In the same way, considering B instead of A, we see that $\mathrm{tr}(BC)$ is determined by the absolute values of traces of γ_2, γ_k, and $\gamma_2^{\pm}\circ\gamma_k$. Similarly, $\mathrm{tr}(ABC)$ is determined by the absolute values of traces of $\gamma_1\circ\gamma_2$, γ_k, and $(\gamma_1\circ\gamma_2)^{\pm}\circ\gamma_k$.

Solving the system of linear equations

$$\begin{aligned}\mathrm{tr}(C) &= p + s,\\ \mathrm{tr}(AC) &= p\lambda + s\lambda^{-1},\\ \mathrm{tr}(BC) &= ap + br + cq + ds,\\ \mathrm{tr}(ABC) &= ap\lambda + br\lambda + cq\lambda^{-1} + ds\lambda^{-1},\end{aligned}$$

we conclude that γ_k is determined by the absolute values of traces of γ_1, γ_2, $\gamma_1\circ\gamma_2$, γ_k, $\gamma_1^{\pm}\circ\gamma_k$, $\gamma_2^{\pm}\circ\gamma_k$, and $(\gamma_1\circ\gamma_2)^{\pm}\circ\gamma_k$. □

6.3.3. Discontinuity of Teichmüller Modular Groups

Theorem 6.18. *The Teichmüller modular group $Mod(\Gamma)$ acts properly discontinuously on $T(\Gamma)$ as a subgroup of the biholomorphic automorphism group $Aut(T(\Gamma))$.*

Proof. Recall that Γ has a system of generators satisfying the hypothesis in Proposition 6.17 (see Remark in §5 of Chapter 2).

Suppose that $Mod(\Gamma)$ does not act properly discontinuously on $T(\Gamma)$. Then there exists a sequence $\{f_n\}_{n=1}^{\infty}$ of mutually distinct elements in $Mod(\Gamma)$ such that, for a certain sequence $\{p_n\}_{n=1}^{\infty}$ in $T(\Gamma)$ which converges to a point $p_o \in T(\Gamma)$, the sequence $\{f_n(p_n)\}_{n=1}^{\infty}$ converges to a point $q_o \in T(\Gamma)$.

For each n, we set

$$g_n = f_n^{-1}, \quad h_n = g_{n+1}\circ f_n.$$

Since $T(\Gamma)$ is biholomorphic to a bounded domain $T_B(\Gamma)$ (Theorem 6.6), $\{f_n\}_{n=1}^{\infty}$ is a normal family. Thus, taking a subsequence if necessary, we may assume that $\{f_n\}_{n=1}^{\infty}$ converges uniformly on compact sets in $T(\Gamma)$ to a holomorphic mapping f_o. In particular, we have $f_o(p_o) = q_o$. Similarly, $\{g_n\}_{n=1}^{\infty}$ and $\{h_n\}_{n=1}^{\infty}$ converge uniformly on compact sets in $T(\Gamma)$ to holomorphic mappings g_o and h_o, respectively. From $g_n\circ f_n(p_n) = p_n$, we obtain $g_o(q_o) = p_o$. By $h_n(p_n) = g_{n+1}\circ f_n(p_n)$, we get $h_o(p_o) = g_o(q_o) = p_o$. Thus $\{h_n(p_o)\}_{n=1}^{\infty}$ converges to p_o. Hence, translating the base point of $T(\Gamma)$ into p_o, we may assume that $\{h_n[id]\}_{n=1}^{\infty}$ converges to [id].

For each n, we set $h_n = [\omega_n]_*$ with some quasiconformal self-mapping ω_n of H such that $\omega_n\Gamma\omega_n^{-1} = \Gamma$. Then we have

$$[\omega_n]_*([id]) = [\alpha_n\circ\omega_n^{-1}],$$

where $\alpha_n \in Aut(H)$ such that $\alpha_n\circ\omega_n^{-1}$ fixes each of 0, 1, and ∞. Since $\{[\alpha_n\circ\omega_n^{-1}]\}_{n=1}^{\infty}$ converges to $[id]$ in $T(\Gamma)$, we see that for every $\gamma \in \Gamma$, the sequence $\{(\alpha_n\circ\omega_n^{-1})\circ\gamma\circ(\alpha_n\circ\omega_n^{-1})^{-1}\}_{n=1}^{\infty}$ converges to γ. Thus we obtain

$$\lim_{n\to\infty} \mathrm{tr}^2(\omega_n^{-1}\circ\gamma\circ\omega_n) = \mathrm{tr}^2(\gamma), \quad \gamma \in \Gamma.$$

Since $\{\,\mathrm{tr}^2(\gamma) \mid \gamma \in \Gamma\,\}$ is discrete in $\mathbf{R}$ (the Corollary to Proposition 6.16), and since every $\omega_n\circ\gamma\circ\omega_n^{-1}$ belongs to Γ, we have

$$\mathrm{tr}^2(\omega_n^{-1}\circ\gamma\circ\omega_n) = \mathrm{tr}^2(\gamma), \quad \gamma \in \mathcal{G}$$

for every sufficiently large n, where $\mathcal{G}$ is the finite subset of Γ given in Proposition 6.17. Hence, Proposition 6.17 implies that for every sufficiently large n, there exists an element $\beta_n \in Aut(H)$ such that

$$\omega_n^{-1}\circ\gamma\circ\omega_n = \beta_n^{-1}\circ\gamma\circ\beta_n, \quad \gamma \in \Gamma.$$

This shows that β_n belongs to the normalizer $N(\Gamma)$ of Γ in $Aut(H)$, and $[\omega_n]_* = [\beta_n]_*$. Thus every such $[\omega_n]_*$ fixes the base point $[id]$ of $T(\Gamma)$.

By the definition, it is easy to see that the isotropy subgroup of $Mod(\Gamma)$ at $[id]$ is isomorphic to $N(\Gamma)/\Gamma$. On the other hand, it is well known that $N(\Gamma)/\Gamma$ is isomorphic to the biholomorphic automorphism group $Aut(H/\Gamma)$ of the closed Riemann surface H/Γ, and that $Aut(H/\Gamma)$ is a finite group (see the following Remark 1). Therefore, $\{\,[\omega_n]_*\,\}_{n=1}^{\infty}$ should be a finite set. This contradicts that $\{\,f_n\,\}_{n=1}^{\infty}$ consists of infinite elements. □

Remark 1. Every element $\alpha \in N(\Gamma)$ induces a biholomorphic automorphism $[\alpha]$ of H/Γ defined by $[\alpha]([z]) = [\alpha(z)]$ for any $[z] \in H/\Gamma$. It is easy to see that the mapping $\alpha \mapsto [\alpha]$ is a homomorphism of $N(\Gamma)$ onto $Aut(H/\Gamma)$ whose kernel is Γ. Thus $N(\Gamma)/\Gamma$ is isomorphic to $Aut(H/\Gamma)$.

H. A. Schwarz proved that $Aut(H/\Gamma)$ is a finite group. Further, A. Hurwitz showed that the number of $Aut(H/\Gamma)$ is not greater than $84(g-1)$. For these facts, we refer to Farkas and Kra [A-28], p. 242; Siegel [A-98], Vol.2, p. 91; Tsuji [A-108], p. 496; and Imayoshi [104].

Remark 2. In the proof of Theorem 6.18, we have used the fact that $T(\Gamma)$ is biholomorphic to a bounded domain. However, we can also verify Theorem 6.18 from Theorem 6.14, i.e., the fact that $[\omega]_*$ is an isometry with respect to the Teichmüller distance on $T(\Gamma)$ for all $[\omega] \in Mod(\Gamma)$ (see Gardiner [A-34], §8.5). Moreover, by using a theorem of Nielsen on topology of surfaces, we can show directly that $Mod(R)$ induces a discrete subgroup of the biholomorphic automorphism group of $T(R)$. The proof of this kind is in Nag [A-80], §7.1 of Chapter 2.

Now, we have the following fundamental theorem on the moduli space M_g.

Theorem 6.19. *The moduli space M_g of closed Riemann surfaces of genus g ($\geqq 2$) has a normal complex analytic space structure of dimension $3g-3$.*

This theorem is an immediate consequence of a theorem due to H. Cartan [48]. Namely, for a given discrete subgroup G of the biholomorphic automorphism group of a bounded domain D in $\mathbf{C}^n$, the quotient space D/G has a normal complex analytic space structure such that the projection $\pi: D \to D/G$ is holomorphic. For a proof of Cartan's theorem, we refer to Baily [A-10], §2 of Chapter 5.

It is known that every moduli space M_g $(g \geqq 2)$ actually has singular points. See Rauch [178], and [179], Theorem 4.

6.4 Royden's Theorems

H. L. Royden proved that the Teichmüller modular group $Mod(\Gamma)$ and the biholomorphic automorphism group $Aut(T(\Gamma))$ of the Teichmüller space $T(\Gamma)$ have the following relationship. We shall briefly explain his result. For details, see Royden [184], Gardiner [A-34], Earle and Kra [65], and Kra [123].

Theorem 6.20. (Royden)

$$Aut(T(\Gamma)) \cong \begin{cases} Mod(\Gamma)/\mathbf{Z}_2 & (g=2) \\ Mod(\Gamma) & (g>2). \end{cases}$$

We define a homomorphism i_* of $Mod(\Gamma)$ to $Aut(T(\Gamma))$ by $i_*([\omega]) = [\omega]_*$ for every $[\omega] \in Mod(\Gamma)$. For $g > 2$, it is known that there exists a closed Riemann surface R with $Aut(R) = \{ id \}$ (see Griffiths and Harris [A-39], p.276; Lewittes [130], and Rauch [179], Corollary to Proposition 7). Thus the kernel of i_* is equal to $\{ [id] \}$, and hence i_* is injective provided that $g > 2$. On the other hand, any closed Riemann surface R of genus two has a biholomorphic automorphism of order two, since R is hyperelliptic, that is, R is represented by a two-sheeted branched covering surface over $\hat{\mathbf{C}}$. From this fact, it follows that the kernel of i_* is isomorphic to $\mathbf{Z}_2$ provided that $g = 2$ (see Nag [A-80], pp. 125-127).

The surjectivity of i_* is proved by the following theorems (Theorems 6.21 and 6.22) which are also due to Royden.

To describe Royden's theorems, we introduce the Kobayashi distance, which is a generalization of the Poincaré distance ρ on the unit disc Δ to a general complex manifold.

Let M be a complex manifold. Given two points $p, q \in M$, we set

$$d_M^1(p,q) = \inf \rho(a,b),$$

where the infimum is taken over all points $a, b \in \Delta$ such that there exists a holomorphic mapping $f: \Delta \to M$ with $f(a) = p$ and $f(b) = q$. For any positive integer n, we put

$$d_M^n(p,q) = \inf \sum_{i=1}^{n} d_M^1(p_{i-1}, p_i),$$

where the infimum is taken over all points $p_o, \ldots, p_n \in M$ with $p_o = p$ and $p_n = q$. It is clear that

$$d_M^{n+1}(p,q) \leqq d_M^n(p,q), \quad p, q \in M$$

for all positive integers n. The *Kobayashi pseudo-distance* d_M on M is defined by

$$d_M(p,q) = \lim_{n\to\infty} d_M^n(p,q).$$

It is an easy matter to show that $d_M : M \times M \to \mathbf{R}$ is continuous and satisfies the axioms for pseudo-distance: $d_M(p,q) \geqq 0$, $d_M(p,q) = d_M(q,p)$, and $d_M(p,q) + d_M(q,r) \geqq d_M(p,r)$ for all $p,q,r \in M$. It is said that d_M is *nondegenerate* if $d_M(p,q) = 0$ is equivalent to $p = q$. Note that d_M is not always nondegenerate. For example, if $M = \mathbf{C}$, then obviously $d_M \equiv 0$. If d_M is nondegenerate, d_M is called the *Kobayashi distance* on M, and M is called a *hyperbolic complex manifold.* A hyperbolic complex manifold M is said to be *complete* if it is complete with respect to d_M.

The most important property of d_M is the *distance decreasing property*, the proof of which is trivial by the definition: let M and N be two complex manifolds and let $f : M \to N$ be a holomorphic mapping. Then it follows that

$$d_M(p,q) \geqq d_N(f(p), f(q)), \quad p,q \in M.$$

In particular, every biholomorphic mapping of a hyperbolic complex manifold M is an isometry with respect to d_M.

Theorem 6.21. (Royden) *Let $T(\Gamma)$ be the Teichmüller space of a Fuchsian model Γ of a closed Riemann surface of genus g ($\geqq 2$). Then the Teichmüller distance d on $T(\Gamma)$ is equal to the Kobayashi distance $d_{T(\Gamma)}$.*

Proof (an outline). We give a sketch of its proof.

In order to prove $d_{T(\Gamma)} = d$, it is sufficient to show that $d^1_{T(\Gamma)} = d$. In fact, if $d^1_{T(\Gamma)} = d$, then $d^1_{T(\Gamma)}$ satisfies the triangle inequality. Thus, by the definition we have $d^n_{T(\Gamma)} = d^1_{T(\Gamma)}$ for any positive integer n, and hence $d_{T(\Gamma)} = d$.

For any $[w^\mu] \in T(\Gamma)$, we put $\Gamma^\mu = w^\mu \Gamma (w^\mu)^{-1}$, and denote by d_μ the Teichmüller distance on $T(\Gamma^\mu)$. Recall that every $[w^\mu] \in T(\Gamma)$ induces a biholomorphic mapping $[w^\mu]_* : T(\Gamma) \to T(\Gamma^\mu)$ (cf. §3.1).

Now, to prove $d^1_{T(\Gamma)} = d$, it suffices to prove the equality

$$d^1_{T(\Gamma^\mu)}([id],[w^\lambda]) = d_\mu([id],[w^\lambda]), \quad [w^\lambda] \in T(\Gamma^\mu)$$

for any $[w^\mu] \in T(\Gamma)$. In fact, if this equality holds, then the relations

$$d([w^\mu],[w^\nu]) = d_\mu([id],[w^\nu \circ (w^\mu)^{-1}]),$$
$$d^1_{T(\Gamma)}([w^\mu],[w^\nu]) = d^1_{T(\Gamma^\mu)}([id],[w^\nu \circ (w^\mu)^{-1}])$$

imply that $d([w^\mu],[w^\nu]) = d^1_{T(\Gamma)}([w^\mu],[w^\nu])$ for all $[w^\mu],[w^\nu] \in T(\Gamma)$.

To simplify the notation, Γ^μ, d_μ, and w^λ in the above equality are rewitten as Γ, d, and w^μ, respectively. Then, we shall show that

$$d^1_{T(\Gamma)}([id],[w^\mu]) \leqq d([id],[w^\mu]), \quad [w^\mu] \in T(\Gamma),$$

as follows. From Teichmüller's existence theorem (Theorem 5.15), there exists an element $\varphi \in A_2(H, \Gamma)$ such that $[w^{\mu_o}] = [w^{\mu}]$ with $\mu_o = k\bar{\varphi}/|\varphi|$ for some k, $0 \leqq k < 1$. Then Teichmüller's uniqueness theorem (Theorem 5.9) yields

$$d([id], [w^{\mu}]) = \log \frac{1+k}{1-k}.$$

On the other hand, setting $\mu_\tau = \tau\bar{\varphi}/|\varphi|$, we see that the mapping $f_\varphi \colon \Delta \to T(\Gamma)$ given by $f_\varphi(\tau) = [w^{\mu_\tau}]$ is a holomorphic mapping with $f_\varphi(0) = [id]$, $f_\varphi(k) = [w^{\mu}]$. Thus we obtain

$$d^1_{T(\Gamma)}([id], [w^{\mu}]) \leqq \rho(0, k) = \log \frac{1+k}{1-k}.$$

Hence, we have

$$d^1_{T(\Gamma)}([id], [w^{\mu}]) \leqq d([id], [w^{\mu}]), \quad [w^{\mu}] \in T(\Gamma).$$

Finally, we need to prove

$$d^1_{T(\Gamma)}([id], [w^{\mu}]) \geqq d([id], [w^{\mu}]), \quad [w^{\mu}] \in T(\Gamma).$$

This is the most crucial part of the proof due to Royden. We briefly explain the proof of this inequality. For details, see Royden [184], and Gardiner [A-34], Chapter 7.

(i) For any point $[w^{\nu}] \in T(\Gamma)$ and $\lambda \in B(H, \Gamma)$, we set

$$F([w^{\nu}], \lambda) = \lim_{t \to 0, t > 0} \frac{d([w^{\nu}], [w^{\nu + t\lambda}])}{t}.$$

This F is considered as the infinitesimal metric on the tangent bundle of $T(\Gamma)$ induced from the Teichmüller metric d. It is proved that

$$F([w^{\nu}], \lambda) = 2 \sup \left| \operatorname{Re} \iint_{F^{\nu}} \left[\frac{(w^{\nu})_z}{\overline{(w^{\nu})_z}} \cdot \frac{\lambda}{1 - |\nu|^2} \right] \circ (w^{\nu})^{-1}(z)\, \varphi(z)\, dxdy \right|,$$

where F^{ν} is a fundamental domain for Γ^{ν} in H, and the supremum is taken over all $\varphi \in A_2(H, \Gamma^{\nu})$ with $\|\varphi\|_1 = 1$. It is verified that F is of class C^1 on the tangent bundle of $T(\Gamma)$ except at points of the zero section. (See also Earle [60].) It is also known that F is not of class C^{∞}.

(ii) For an arbitrary piecewise smooth path $C \colon [0, 1] \to T(\Gamma)$, we set

$$L(C) = \int_0^1 F(C(t), C'(t))\, dt.$$

For any two points $p, q \in T(\Gamma)$, we put

$$\tilde{d}(p, q) = \inf_{C} L(\gamma),$$

where the infimum is taken over all piecewise smooth paths C joining p and q in $T(\Gamma)$. Then it is shown that $\tilde{d} = d$.

(iii) It is verified that every holomorphic mapping $f\colon \Delta \to T(\Gamma)$ satisfies

$$F(f(\tau), f'(\tau)) \leqq \frac{2}{1-|\tau|^2}, \quad \tau \in \Delta.$$

To prove this inequality is essential, though we shall omit the details.

(iv) Take an arbitrary holomorphic mapping $f\colon \Delta \to T(\Gamma)$ with $f(a) = [id]$ and $f(b) = [w^\mu]$ for some points $a, b \in \Delta$. Then (ii) and (iii) imply that $d([id], [w^\mu]) = \tilde{d}([id], [w^\mu]) \leqq \rho(a, b)$. By the definition of $d^1_{T(\Gamma)}$, we get

$$d([id], [w^\mu]) \leqq d^1_{T(\Gamma)}([id], [w^\mu]), \quad [w^\mu] \in T(\Gamma).$$

This completes the proof of Theorem 6.21. □

Now, Theorem 6.21 asserts that every element $f \in Aut(T(\Gamma))$ is an isometry with respect to the Teichmüller distance on $T(\Gamma)$.

Take an element $f \in Aut(T(\Gamma))$. For every $p = [w^\mu] \in T(\Gamma)$, we set $q = f(p) = [w^\nu]$. The derivative $\dot{f}$ of f at p is a complex linear isometry of $T_p(T(\Gamma))$ to $T_q(T(\Gamma))$ with respect to the infinitesimal metric F, where $T_p(T(\Gamma))$ and $T_q(T(\Gamma))$ denote the holomorphic tangent spaces of $T(\Gamma)$ at p and q, respectively. Here we use the fact that the dual space of $T_p(T(\Gamma))$ is canonically isomorphic to the space $A_2(H, \Gamma^\mu)$ of holomorphic automorphic forms on H for Γ^μ, a fact which is proven in the next chapter (Theorem 7.5 and Proposition 7.8). Similarly, the dual space of $T_q(T(\Gamma))$ is identified with $A_2(H, \Gamma^\nu)$. Hence, $\dot{f}$ induces a complex linear isometry α of $A_2(H, \Gamma^\mu)$ to $A_2(H, \Gamma^\nu)$ with respect to the infinitesimal cometric induced by the Teichmüller distance d.

Here we know the following fact.

Theorem 6.22. (Royden) *Let α be a complex linear isometry of $A_2(H, \Gamma^\mu)$ to $A_2(H, \Gamma^\nu)$ with respect to the infinitesimal cometric induced by the Teichmüller distance d. Then there exists a biholomorphic mapping $h\colon H/\Gamma^\nu \to H/\Gamma^\mu$ and a complex number c with $|c| = 1$ such that $\alpha(\varphi) = c\varphi \circ \tilde{h} \cdot (\tilde{h}')^2$ for all $\varphi \in A_2(H, \Gamma^\mu)$, where $\tilde{h}$ is a lift of h to H.*

For a proof of this theorem, we refer to Royden [184], and Gardiner [A-34], Theorem 5 in Chapter 9.

Proof of the surjectivity of i_*. Now, we return to a proof of the surjectivity of i_*. From the previous observation and Theorem 6.22, for every $f \in Aut(T(\Gamma))$ and every point $p \in T(\Gamma)$ there exists an element $[\omega_p] \in Mod(\Gamma)$ with $[\omega_p]_*(p) = f(p)$. We need to show that $[\omega_p]$ can be chosen independently of p. Fix a point $q \in T(\Gamma)$ arbitrarily. Recall that $T(\Gamma)$ is biholomorphic to a bounded domain (Theorem 6.6), that the Teichmüller distance d is complete (Theorem 5.4), and that $Mod(\Gamma)$ acts properly discontinuously on $T(\Gamma)$ (Theorem 6.18). Then we can find a positive constant δ so that $d(p, [\omega]_*(p)) > 2\delta$ for any $p \in T(\Gamma)$ with $d(q, p) < \delta$, and for any $[\omega] \in Mod(\Gamma)$ with $[\omega]_*(p) \neq p$. Thus we have

$$\begin{aligned} d(p, [\omega_q]_*^{-1}\circ[\omega_p]_*(p)) &= d([\omega_q]_*(p), [\omega_p]_*(p)) \\ &\leqq d([\omega_q]_*(p), [\omega_q]_*(q)) + d([\omega_q]_*(q), [\omega_p]_*(p)) \\ &= d(p,q) + d(f(q), f(p)) \\ &= 2d(q,p) < 2\delta \end{aligned}$$

for all $p \in T(\Gamma)$ with $d(q,p) < \delta$. Hence, it follows that $[\omega_q]_*^{-1}\circ[\omega_p]_*(p) = p$, i.e., $[\omega_q]_*(p) = [\omega_p]_*(p) = f(p)$ for all $p \in T(\Gamma)$ with $d(q,p) < \delta$. Since $T(\Gamma)$ is connected, the uniqueness theorem for holomorphic functions implies that $[\omega_q]_* = f$ on $T(\Gamma)$. Thus we have proved Theorem 6.20.

Finally, from Theorems 5.4 and 6.21, we conclude that $T_B(\Gamma)$ is complete with respect to the Kobayashi distance. Thus $T_B(\Gamma)$ is pseudoconvex (see Kobayashi [A-54], Theorem 3.4 in Chapter V). By Oka's theorem (Bers [A-14], Theorem 40, p. 134), this is equivalent that $T_B(\Gamma)$ is a domain of holomorphy. Hence we have the following theorem.

Theorem 6.23. *The Teichmüller space $T(\Gamma)$ is a Stein manifold.*

6.5. Classification of Teichmüller Modular Transformations

Real Möbius transformations have been classified in Chapter 2 into three types, which are called elliptic, hyperbolic, and parabolic.

In this section, following Bers [38], we shall discuss a similar classification of Teichmüller modular transformations of T_g ($g \geq 2$), or equivalently, the action of elements of the mapping class group of a closed surface of genus g. Such a classification is also considered in the case of a (not necessarily closed) surface of analytically finite type. For full treatments, see Fathi, Laudenbach and Poénaru [A-29]. We also refer to Casson and Bleiler [A-19], Nielsen [A-87], and Kra [121].

As a consequence, we have Nielsen-Thurston's classification theorem of self-mappings of a closed surface (see Theorem 6.35 and Thurston [234]).

In this section, we always assume that R is an oriented closed (differentiable) surface of genus g (≥ 2), and that every mapping is homeomorphic and orientation-preserving. Further, we fix a Riemann surface whose underlying surface is R, and denote it by R_*. We also recall the classically well-known fact that two mappings of a closed surface are homotopic if and only if they are isotopic (cf. Epstein [69]).

6.5.1. Fundamental Extremal Problems

We can deal with the classification of real Möbius transformations relating to an extremal problem on hyperbolic translation length. More precisely, for every element $\gamma \in PSL(2, \mathbf{R})$, set

$$a(\gamma) = \inf_{z \in H} \rho(z, \gamma(z)),$$

where H is the upper half-plane, and ρ is the Poincaré distance on H. Then real Möbius transformations γ are classified as follows:

(i) γ is elliptic if $a(\gamma) = 0$ and there exists a point $z_\gamma \in H$ with $a(\gamma) = \rho(z_\gamma, \gamma(z_\gamma))$, i.e., z_γ is a fixed point of γ,
(ii) γ is parabolic if $a(\gamma) = 0$ but there exist no points $z_\gamma \in H$ with $a(\gamma) = \rho(z_\gamma, \gamma(z_\gamma))$, and
(iii) γ is hyperbolic if $a(\gamma) > 0$ (and then, there always exists a point $z_\gamma \in H$ with $a(\gamma) = \rho(z_\gamma, \gamma(z_\gamma))$).

Now, for Teichmüller modular transformations, we consider the following similar extremal problem.

Bers' extremal problem for Teichmüller modular transformations. For every Teichmüller modular transformation χ of $T(R_*)$, we set

$$a(\chi) = \inf_{p \in T(R_*)} d(p, \chi(p)),$$

where d is the Teichmüller distance on $T(R_*)$. Then find a point $p_\chi \in T(R_*)$ such that

$$a(\chi) = d(p_\chi, \chi(p_\chi)).$$

If there exists a solution $p_\chi \in T(R_*)$, then we call p_χ a *χ-minimal point*. We classify Teichmüller modular transformations χ into four types:

(i) χ is *elliptic* if $a(\chi) = 0$ and there exists a χ-minimal point (which should be a fixed point of χ),
(ii) χ is *parabolic* if $a(\chi) = 0$ but there exist no χ-minimal points,
(iii) χ is *hyperbolic* if $a(\chi) > 0$ and there exists a χ-minimal point, and
(iv) χ is *pseudo-hyperbolic* if $a(\chi) > 0$ but there exist no χ-minimal points.

Note that this classification is independent of the choice of the complex structure on R which is used to define the Teichmüller space $T(R_*)$.

Recall that every point $[S, f] \in T(R_*)$ is represented by $[R_\sigma, id]$ (see §1.4.1 of Chapter 1), where R_σ is a Riemann surface equipped with a complex structure σ on R, and id is the identity mapping of R_* onto R_σ. Hereafter, $[R_\sigma, id]$ is simply written as $[\sigma]$. Every Teichmüller modular transformation is represented by $[f]_*$ for a self-mapping f of R (see §3.1 of Chapter 6).

Now, to investigate χ-minimal points, we may consider the following version of this extremal problem.

Bers' extremal problem for complex structures. For every complex structure σ on R and every self-mapping f of R, considering f as a self-mapping of the Riemann surface R_σ, we denote by $K_\sigma(f)$ the maximal dilatation of f. Here, we put $K_\sigma(f) = \infty$ if f is not quasiconformal. Then find a complex structure σ_0 on R and a self-mapping f_0 of R homotopic to f such that

$$K_{\sigma_0}(f_0) \leq K_{\sigma_1}(f_1)$$

for every complex structure σ_1 on R and every self-mapping f_1 of R homotopic to f.

If there exists a solution, a pair (σ_0, f_0) as described, then we call σ_0 an *f-minimal complex structure* and $f_0\colon R_{\sigma_0} \to R_{\sigma_0}$ an *absolutely extremal self-mapping* of the Riemann surface R_{σ_0}. It is noted that $f_0\colon R_{\sigma_0} \to R_{\sigma_0}$ is a Teichmüller mapping, i.e., the unique extremal mapping in the family of all quasiconformal mappings $f_1 : R_{\sigma_0} \to R_{\sigma_0}$ which are homotopic to $f : R_{\sigma_0} \to R_{\sigma_0}$ (cf. Theorem 5.9).

Then we have the following assertion.

Proposition 6.24. *For a self-mapping f of R, a complex structure σ is f-minimal if and only if the point $[\sigma] \in T(R_*)$ corresponding to σ is $[f]_*$-minimal.*

Proof. Suppose that a complex structure σ is f-minimal. Let f_0 be the Teichmüller mapping (i.e., the unique extremal mapping) homotopic to f on the Riemann surface R_σ (cf. Theorem 5.9). By the definition, for every self-mapping f_1 of R homotopic to f and every complex structure σ_1 on R, we have

$$K_\sigma(f_0) \leq K_{\sigma_1}(f_1). \tag{6.12}$$

Let p_0 and p_1 be the points in $T(R_*)$ corresponding to σ and σ_1, respectively. Then by the definition of the action of f on $T(R_*)$ in §3.1, inequality (6.12) is equivalent to

$$d(p_0, [f]_*^{-1}(p_0)) \leq d(p_1, [f]_*^{-1}(p_1)).$$

Since $[f]_*$ is an isometry of $T(R_*)$ with respect to the Teichmüller distance, we have

$$d(p_0, [f]_*(p_0)) \leq d(p_1, [f]_*(p_1)).$$

Hence, p_0 is $[f]_*$-minimal.

The converse is shown similarly, and we conclude the assertion. □

6.5.2. Elliptic and Hyperbolic Transformations

For elliptic Teichmüller modular transformations, the following theorem is classically well-known.

Theorem 6.25. *A Teichmüller modular transformation χ is elliptic if and only if it is periodic.*

Proof. If χ has a fixed point, then there is a complex structure σ on R and a self-mapping f of R inducing χ such that f is a biholomorphic automorphism of R_σ. Since R is compact, it is well-known that f should be of finite order (see Remark 2 in §6.3), and hence is periodic.

Conversely, suppose that χ is periodic. Nielsen showed that χ has a fixed point in $T(R_*)$, whose proof we shall omit here. (Actually, it is shown that the action of every finite subgroup of $Mod(R_*)$ has a fixed point in $T(R_*)$. This is the affirmative solution for Nielsen's realization problem (cf. Notes of this chapter). See for instance Kerckhoff [112] or Wolpert [256].) □

Remark. A weaker version of Theorem 6.25 is easily shown. Namely, it is easy to prove that a self-mapping f of R is homotopic to a periodic self-mapping of R if and only if there exists a complex structure σ on R, and a self-mapping f_0 homotopic to f such that f_0 is holomorphic on R_σ.

Note that Proposition 6.24 implies the following theorem.

Theorem 6.26. *Let f be a self-mapping of R. Then there is an f-minimal complex structure if and only if the Teichmüller modular transformation $[f]_*$ corresponding to f is either elliptic or hyperbolic.*

Now, a finite non-empty set $\{C_1, \cdots, C_n\}$ of mutually disjoint simple closed curves on R is called *admissible* if every C_j is freely homotopic to none of $\{C_k, (C_k)^{-1}\}_{k\neq j}$, and is not homotopic to a point. We say that a self-mapping f of R is *reduced* by $\{C_1, \cdots, C_n\}$ if this set is admissible and

$$f(C_1 \cup \cdots \cup C_n) = C_1 \cup \cdots \cup C_n.$$

A self-mapping f of R is called *reducible* if it is homotopic to a reduced mapping, and *irreducible* if not. Then we have the following theorem.

Theorem 6.27. *If f is an irreducible self-mapping of R, then the Teichmüller modular transformation $[f]_*$ induced by f is either elliptic or hyperbolic.*

To prove this theorem, we prepare several lemmas. First, we start with the following fundamental one.

Lemma 6.28. (Wolpert) *Let f be a quasiconformal mapping of a Riemann surface S_1 onto another S_2, and C be a simple closed geodesic on S_1 with hyperbolic length ℓ_1. Then $f(C)$ is freely homotopic to a closed geodesic with length ℓ_2 such that*

$$\ell_2 \leq K(f)\ell_1, \tag{6.13}$$

where $K(f)$ is the maximal dilatation of f.

For the proof, see that of Lemma 3.1 in Wolpert [246].

Next, by the collar lemma (cf. Matelski [150]), we can easily show the following:

Lemma 6.29. *There is a constant $\delta_0 > 0$, which depends only on genus g, such that any two distinct simple closed geodesics on a closed Riemann surface of genus g are disjoint provided that hyperbolic lengths of them are less than δ_0.*

Thus we have the following:

Lemma 6.30. *Let C be a simple closed geodesic with hyperbolic length ℓ on a closed Riemann surface S of genus g (≥ 2). Then every irreducible self-mapping f of S satisfies*

$$K(f) \geq \left(\frac{\delta_0}{\ell}\right)^{1/(3g-3)}, \tag{6.14}$$

where δ_0 is the constant in Lemma 6.29.

Proof. Suppose that, for some f, inequality (6.14) does not hold, i.e.,

$$K(f)^{3g-3} \cdot \ell < \delta_0.$$

Set $C_0 = C$ and let C_j be the simple closed geodesic freely homotopic to $f^j(C)$ ($j = 1, \cdots, 3g-3$). By (6.13), the hyperbolic length of any one of $C_0, \cdots, C_{3g-3}$ is less than δ_0. Since not all of them are disjoint by Proposition 3.6, we may assume by Lemma 6.29 that there is a non-negative integer r such that $C_0, \cdots, C_r$ are disjoint and $C_{r+1} = C_0$ (as points sets).

However, we can deform f continuously to a mapping f' such that $f'(C_j) = C_{j+1}$ ($j = 0, \cdots, r-1$) and $f'(C_r) = C_0$ (as points sets). Thus f should be reducible, which contradicts the assumption. □

Finally, we recall the following Mumford's compactness theorem (cf. for instance Bers [30]).

Lemma 6.31. *Let $\{p_j\}_{j=1}^{\infty}$ be a sequence in $T(R_*)$, and S_j be the Riemann surface corresponding to p_j for every j. Suppose that there is a positive δ such that the hyperbolic length of any simple closed geodesic on each S_j is greater than δ. Then there exists a subsequence $\{p_{j_n}\}_{n=1}^{\infty}$ of $\{p_j\}_{j=1}^{\infty}$, and a sequence $\{\chi_n\}_{n=1}^{\infty}$ of Teichmüller modular transformations of $T(R_*)$ such that the sequence $\{\chi_n(p_{j_n})\}_{n=1}^{\infty}$ converges in $T(R_*)$.*

Proof of Theorem 6.27. Take a sequence $\{p_j\}_{j=1}^{\infty}$ in $T(R_*)$ such that

$$\lim_{j\to\infty} d(p_j, [f]_*(p_j)) = a([f]_*). \tag{6.15}$$

For every j, let σ_j be the complex structure corresponding to both p_j and $[f]_*(p_j)$, and let $h_j : R_{\sigma_j} \to R_{\sigma_j}$ be the Teichmüller mapping homotopic to f. By (6.15), we have

$$\lim_{j\to\infty} \log K(h_j) = a([f]_*).$$

In particular, there is a constant A such that $K(h_j) < A$ for every j. Since f is irreducible, every h_j is also irreducible. Hence, Lemma 6.30 implies that the hyperbolic length of any simple closed geodesic on each R_{σ_j} is greater than $\delta_0 A^{3-3g}$. By Lemma 6.31, we may assume, taking a subsequence if necessary, that there is a sequence $\{\chi_j\}_{j=1}^{\infty}$ of Teichmüller modular transformations such that the sequence $\{\chi_j(p_j)\}_{j=1}^{\infty}$ converges to a point $q \in T(R_*)$. We set $q_j = \chi_j(p_j)$ for every j. Since each χ_j is an isometry with respect to the Teichmüller metric, (6.15) gives

$$\lim_{j\to\infty} d(q_j, \chi_j \circ [f]_* \circ (\chi_j)^{-1}(q_j)) = a([f]_*). \tag{6.16}$$

Again taking a subsequence if necessary, we may assume that $\{\chi_j \circ [f]_* \circ (\chi_j)^{-1}(q_j)\}_{j=1}^{\infty}$ converges to a point $q' \in T(R_*)$, for $T(R_*)$ is finite-dimensional and is complete with respect to d. Since each $\chi_j \circ [f]_* \circ (\chi_j)^{-1}$ is an isometry, it is easy to see that

$$\lim_{j\to\infty} \chi_j \circ [f]_* \circ (\chi_j)^{-1}(q) = q'.$$

Hence, by Theorem 6.18 we may assume, taking a subsequence if necessary, that $\chi_j \circ [f]_* \circ (\chi_j)^{-1}$ has the same action on $T(R_*)$ for every sufficiently large j, say $j \geq j_0$.

Then (6.16) implies that

$$d(q, \chi_{j_0} \circ [f]_* \circ (\chi_{j_0})^{-1}(q)) = d((\chi_{j_0})^{-1}(q), [f]_*((\chi_{j_0})^{-1}(q))) = a([f]_*).$$

Thus there is an $[f]_*$-minimal point, which shows the assertion. □

6.5.3. Absolutely Extremal Mappings

Next, we characterize absolutely extremal mappings. We have seen that, when a self-mapping f is homotopic to a periodic mapping, then the corresponding absolutely extremal mapping is conformal (Theorem 6.25). Hence, we discuss the case that $[f]_*$ is of infinite order.

Here, note that the Teichmüller space $T(R_*)$ is a *straight line space* in the sense of Busemann (cf. Kravetz [128], and also see Masur [142]). In particular, any two distinct points $p_1, p_2 \in T(R_*)$ lie on a unique *straight line*, say L, which is an isometric image of $\mathbf{R}$ into $T(R_*)$ equipped with the Teichmüller metric, and contains all points p such that

$$d(p_1, p) + d(p, p_2) = d(p_1, p_2).$$

We also note the following elementary fact:

Theorem 6.32 *If a Teichmüller modular transformation χ is of infinite order, then a point $p \in T(R_*)$ is χ-minimal if and only if χ leaves a straight line through p invariant.*

Proof. Assume that p is χ-minimal. Since χ is of infinite order, three points p, $\chi(p)$, and $\chi^2(p)$ are distinct. Let p_1 and p_2 be the midpoints of the "segments" $[p, \chi(p)]$ and $[\chi(p), \chi^2(p)]$, respectively. Then it is easy to see that

$$d(p_1, \chi(p)) = d(\chi(p), p_2) = \frac{1}{2}a(\chi).$$

Hence, we conclude that $d(p_1, p_2) \leq a(\chi)$.

On the other hand, since $p_2 = \chi(p_1)$, we have $d(p_1, p_2) \geq a(\chi)$ by the definition of $a(\chi)$. Thus $\chi(p)$ is the midpoint of the segment $[p_1, p_2]$, which implies that the straight line, on which p and $\chi(p)$ (and hence also p_1, p_2, $\chi^2(p)$) lie, is invariant under χ.

Next, suppose that a straight line L through p is invariant under χ. Then, for any point $p' \in T(R_*)$ and any positive integer n, we can see, by using the triangle inequality, that

$$n \cdot d(p, \chi(p)) = d(p, \chi^n(p)) \leq 2d(p, p') + n \cdot d(p', \chi(p')).$$

Since n is arbitrary, we conclude that

$$d(p, \chi(p)) \leq d(p', \chi(p'))$$

for every $p' \in T(R_*)$, which implies that p is χ-minimal. □

Corollary. *For a non-periodic element $[f] \in Mod(R_*)$, the Teichmüller modular transformation $[f]_*$ is hyperbolic if and only if it leaves a straight line invariant.*

Remark 1. It follows from Thurston's theory that any Teichmüller modular transformation has at most one invariant straight line. See also Tanigawa [221].

Now, we obtain the following theorem.

Theorem 6.33. *Let S be a closed Riemann surface of genus g (≥ 2). Then a mapping $f \colon S \longrightarrow S$ is absolutely extremal if and only if it is either a conformal mapping or a Teichmüller mapping such that the mapping f^2 is also a Teichmüller mapping with $K(f^2) = K(f)^2$.*

Proof. We may assume that f is not conformal, and hence is a Teichmüller mapping. Put $S = R_\sigma$ with a suitable conformal structure σ. By Proposition 6.24, f is absolutely extremal if and only if the corresponding point $[\sigma] \in T(R_*)$ is $[f]_*$-minimal. By Theorem 6.32, this is equivalent to

$$\begin{aligned} &d([\sigma], [f]_*([\sigma])) + d([f]_*([\sigma]), ([f]_*)^2([\sigma])) \\ &= 2d([\sigma], [f]_*([\sigma])) \\ &= d([\sigma], ([f]_*)^2([\sigma])). \end{aligned} \tag{6.17}$$

Since f is a Teichmüller mapping, (6.17) is rewitten as

$$K(f)^2 = K(h), \tag{6.18}$$

where h is the Teichmüller mapping homotopic to f^2.

Now, assume that f is absolutely extremal. Since $K(f^2) \leq K(f)^2$ by Theorem 4.10 (iii), and since $K(h) \leq K(f^2)$, formula (6.18) implies that

$K(h) = K(f^2)$. Hence, the mapping f^2 is also a Teichmüller mapping with $K(f^2) = K(f)^2$. (Note that, by Theorem 5.9, we conclude that $h = f^2$.)

Conversely, if the mapping f^2 is also a Teichmüller mapping with $K(f^2) = K(f)^2$, then clearly (6.17) holds. Therefore, f is absolutely extremal. □

Remark 2. We can show further that the condition that the mapping f^2 is also a Teichmüller mapping with $K(f^2) = K(f)^2$ is equivalent to the condition that the initial and the terminal differential of f (cf. Proposition 5.19) coincide with each other up to a positive constant factor.

6.5.4. Reducible Mappings and Nielsen-Thurston's Theorem

For reducible mappings, we can show the following theorem.

Theorem 6.34. *Let f be a reducible self-mapping of R. If f is not homotopic to a periodic mapping, then $[f]_*$ is either parabolic or pseudo-hyperbolic.*

Thus, by Theorems 6.25, 6.26, 6.27, and 6.34, we conclude the following:

Corollary. *Let f be a self-mapping of R. An f-minimal complex structure exists if and only if f is either homotopic to a periodic mapping or irreducible.*

We shall omit the proof of Theorem 6.34. Instead, we shall explain the structure of a reducible mapping. (For this purpose, recall that the foregoing arguments still work even for the case of a surface of finite type.)

Let f be a reducible self-mapping f of R. Then we can deform f continuously to a completely reduced mapping, or more precisely, to a self-mapping f_0 of R which satisfies the following condition: there is an admissible set $\{C_1, \cdots, C_n\}$ of disjoint simple closed curves on R such that, for every component R' of $R - C_1 \cup \cdots \cup C_n$ and for the smallest positive integer N with $f^N(R') = R'$, the mapping $f^N|_{R'}$ is irreducible.

Let $\{R'_1, \cdots, R'_m\}$ be the components of $R - C_1 \cup \cdots \cup C_n$. As in the case of a closed surface, there is an $\left(f^{N_j}|_{R'_j}\right)$-minimal complex structure σ_j on R'_j and an absolutely extremal mapping $F_j\colon (R'_j)_{\sigma_j} \to (R'_j)_{\sigma_j}$ for every j ($j = 1, \cdots, m$), where N_j is the smallest positive integer such that $f^{N_j}(R'_j) = R'_j$. (We can further show that $(R'_j)_{\sigma_j}$ is a Riemann surface of analytically finite type.)

We can show that

$$a([f]_*) = \max\{K_{\sigma_1}(F_1), \cdots, K_{\sigma_m}(F_m)\},$$

but there exists no $[f]_*$-minimal point.

Hence, if all $K_{\sigma_j}(F_j)$ are equal to 1, or equivalently, if all F_j are conformal, then $[f]_*$ is parabolic. If not, then $[f]_*$ is pseudo-hyperbolic.

Example. Let $f\colon R \to R$ be the Dehn twist with respect to a simple closed curve C. Then f leaves $R - C$ invariant componentwise, and f is homotopic to the identity mapping on each component of $R - C$. Hence, we can see that $[f]_*$ is parabolic.

Finally, a Teichmüller mapping whose initial and terminal differentials coincide with each other up to a positive constant factor gives a so-called *pseudo-Anosov diffeomorphism*, when we forget the complex structure.

Conversely, for any given pseudo-Anosov diffeomorphism f of R, we can obtain a complex structure on R such that f becomes a Teichmüller mapping whose initial and terminal differentials coincide with each other up to a positive constant factor.

Therefore, Theorems 6.26, 6.33, and Remark 2 in §5.3 imply that f is homotopic to a pseudo-Anosov diffeomorphism if and only if $[f]_*$ is hyperbolic. Thus Theorems 6.27 and 6.34 imply the following theorem of Nielsen and Thurston.

Theorem 6.35. (Nielsen and Thurston) *A self-mapping of a closed surface of genus g (≥ 2), which is not homotopic to a periodic one, is homotopic to a pseudo-Anosov diffeomorphism or is reducible (but not both).*

Notes

For fundamental materials on several complex variables and complex manifolds, we refer to the books by Bers [A-14], Griffiths and Harris [A-39], Kodaira [A-57], and Morrow and Kodaira [A-77].

The complex structure of the Teichmüller space T_g of genus g ($\geqq 2$) was first introduced by Ahlfors [5], as shall be explained briefly in Appendix A. Alternative treatments of this subject are found in Jost [A-49], Fischer and Tromba [73], Nag and Verjovsky [164], and Saito [186].

We have rather explicit holomorphic coordinates on $T(\Gamma)$ which are called the *Maskit coordinates* (see Kra [125] and Maskit [139]).

The method to define the complex structure of T_g in this chapter is due to Bers [25]. This method is applicable to the Teichmüller space $T(\Gamma)$ of an arbitrary Fuchsian group Γ. It is known that $T(\Gamma)$ is finite dimensional if and only if the Riemann surface H/Γ is of analytically finite type (g, n), i.e., a Riemann surface which is obtained by removing n distinct points from a closed Riemann surface of genus g. In particular, Bers' embedding realizes $T(\Gamma)$ as a bounded domain $T_B(\Gamma)$ in $\mathbf{C}^{3g-3+n}$ provided that $2g-2+n > 0$. The boundary $\partial T_B(\Gamma)$ of $T_B(\Gamma)$ is closely related to the theory of Kleinian groups. For this subject, we refer to Abikoff [1], Bers [28], Maskit [138], and McMullen [155]. We also note that the compactification of T_g using Bers' embedding is different from Thurston's one. See Kerckhoff and Thurston [116] (cf. Bers [39] and Kerckhoff [111]).

Let $T_B(1)$ be the image of Bers' embedding of the universal Teichmüller space $T(1)$, where 1 denotes the trivial group. Gehring [83] proved that $T_B(1)$ coincides with the interior of the set $\mathcal{S}(1)$ consisting of Schwarzian derivatives of holomorphic univalent functions on H. It is also known that $\mathcal{S}(1) \not\subset \overline{T_B(1)}$ (see Astala [19], Gehring [84], and Thurston [232]). Shiga [200] showed that if Γ is a finitely generated Fuchsian group of the first kind, then $T_B(\Gamma)$ coincides with the interior $\mathcal{S}(\Gamma)$, where $\mathcal{S}(\Gamma) = \mathcal{S}(1) \cap A_2(H^*, \Gamma)$. When Γ is of the second kind, Sugawa [212] has shown recently that $\mathcal{S}(\Gamma) \not\subset \overline{T_B(1)}$. However, in the case where Γ is infinitely generated and of the first kind, it is unknown whether $\mathcal{S}(\Gamma) \subset \overline{T_B(\Gamma)}$ or not.

For an arbitrary Fuchsian group Γ, the Nehari-Kraus lemma (Lemma 6.7) implies that $T_B(\Gamma)$ is contained in the open ball with center 0 and radius 3/2 in $A_2(H^*, \Gamma)$. The infimum of radii of open balls with center 0 in $A_2(H^*, \Gamma)$ which includes $T_B(\Gamma)$ is studied by Nakanishi [166], Sekigawa [192], and Sekigawa and Yamamoto [193].

For connections with projective structures on Riemann surfaces, there are papers Gunning [88], Kra [120], and Shiga [202].

It was known by Fricke that the Teichmüller modular group $Mod(\Gamma)$ induces a discrete subgroup of $Aut(T(\Gamma))$ (see, for example, Fricke and Klein [A-33]). The proof in this chapter is due to Bers [31]. The modular group $Mod(\Gamma)$ is studied in Bers [39], Hejhal [100], Ivanov [108], Kerckhoff [111], McMullen [154], Mumford [161], and Wolpert [249]. For classification theory of Teichmüller modular transformations, we refer to Casson and Bleiler [A-19], Fathi, Laudenbach and Poénaru [A-29], Bers [38], Kra [121], Shiga [201], and Thurston [231]. Kerckhoff [112] and Wolpert [256] solved the Nielsen realization problem which asserts that the action of every finite subgroup of $Mod(\Gamma)$ has a fixed point in $T(\Gamma)$. See also Kerckhoff [113].

For Bers' fiber space over a Teichmüller space and the Teichmüller curve, we refer to Bers [31] and Earle [61]. The relation between Teichmüller spaces and holomorphic families of Riemann surfaces is found in Nag [A-80], Chapter 5, Earle [58], Earle and Fowler [64], and Imayoshi [102]. Their applications are treated in Griffiths [87], Imayoshi [103], [105], [106], Imayoshi and Shiga [107], and Riera [183]. For holomorphic sections over a holomorphic family of Reimann surfaces, see Hubbard [A-44], and Earle and Kra [66].

Many more details of the Kobayashi distance are found in Kobayashi [A-54], Lang [A-64], and Noguchi and Ochiai [A-88]. For a distance on $T(\Gamma)$ invariant under the action of $Mod(\Gamma)$ other than the Teichmüller distance, we have the Carathéodory distance. Earle [59] proved that $T(\Gamma)$ is complete with respect to the Carathéodory distance. The connection between the Teichmüller and Carathéodory distances is studied in Kra [122]. There are also invariant distances which are induced by the Bergman metric and the invariant metric on the Siegel upper half-space. For this subject, see Royden [185]. Even in the case where $T(\Gamma)$ is of infinite dimension, it is shown by Gardiner [81] that the Teichmüller distance also coincides with the Kobayashi distance. See also the book by Gardiner [A-34].

It was first proved by Bers and Ehrenpreis [41] that a finite dimensional Teichmüller space is a Stein manifold. There are alternative proofs in Shiga [199] and Wolpert [256].

The relation between the theory of Teichmüller spaces and the Kodaira-Spencer deformation theory of complex manifolds shall be discussed briefly in §2.4 of Chapter 7. We also refer to Wolpert [258], where the infinitesimal variation of a family of Riemann surfaces is studied from this point of view.

Chapter 7

Weil-Petersson Metric

Unless otherwise stated, a Fuchsian group Γ, considered in this chapter, is a Fuchsian model of a closed Riemann surface of genus g ($\geqq 2$). We also assume that each of 0, 1, and ∞ is fixed by an element in $\Gamma - \{id\}$.

As stated in §1.3 of Chapter 5, the Teichmüller distance on the Teichmüller space $T(\Gamma)$ measures a kind of magnitude of deformation of complex structures of Riemann surfaces, and $T(\Gamma)$ is complete with respect to the Teichmüller distance. We also saw in §3.4 of Chapter 6 that the Teichmüller distance is equal to the Kobayashi distance, which is defined complex analytically. However, the Finsler metric induced by the Teichmüller distance is not of class C^∞.

The purpose of this chapter is to introduce another natural metric on $T(\Gamma)$, which is called the Weil-Petersson metric, and to prove that it is a Kähler metric whose Ricci curvatures, scalar curvature, and holomorphic sectional curvatures are negative.

The first section is preliminary and devoted to studying the Petersson scalar product on the space $A_2(H, \Gamma)$ of holomorphic automorphic forms of weight -4 with respect to Γ on H, and related topics such as the reproducing formula for holomorphic automorphic forms, Poincaré series, and the Bergman projection. In Section 2, we see that $A_2(H, \Gamma)$ is the dual space of the holomorphic tangent space $T_0(T(\Gamma))$ of $T(\Gamma)$ at the base point. We also represent elements of $T_0(T(\Gamma))$ by harmonic Beltrami differentials. Then the Weil-Petersson metric on $T_0(T(\Gamma))$ is given by the dual metric of the Petersson scalar product on $A_2(H, \Gamma)$. In Section 3, we define the Weil-Petersson metric on $T(\Gamma)$ and verify that it is Kählerian and that its Ricci curvatures, scalar curvature, and holomorphic sectional curvatures are negative.

7.1. Petersson Scalar Product and Bergman Projection

7.1.1. Petersson Scalar Product and Reproducing Formula

First of all, we define a canonical Hermitian inner product on the space $A_2(H,\Gamma)$ of holomorphic automorphic forms with respect to Γ on H.

Let $ds_H^2 = \lambda_H(z)^2|dz|^2$ be the Poincaré metric on H, i.e., $\lambda_H(z) = (\operatorname{Im} z)^{-1}$. The area element $d\sigma = \lambda_H(z)^2 dxdy$ induced by ds_H^2 is regarded as an area element on the Riemann surface $R = H/\Gamma$.

For any elements $\varphi, \psi \in A_2(H,\Gamma)$, we set

$$\langle\varphi,\psi\rangle(z) = \lambda_H^{-4}\varphi(z)\overline{\psi(z)}, \quad z \in H.$$

Then $\langle\varphi,\psi\rangle$ is a function on H invariant under Γ, i.e., it is considered as a function on R.

Now, the Hermitian inner product $\langle\varphi,\psi\rangle_R$ is defined by

$$\langle\varphi,\psi\rangle_R = \iint_R \langle\varphi,\psi\rangle\, d\sigma = \iint_F \lambda_H(z)^{-2}\varphi(z)\overline{\psi(z)}\, dxdy,$$

where F is a fundamental domain for Γ in H. Since R is compact, it follows that $A_2(H,\Gamma)$ becomes a Hilbert space with this Hermitian inner product. We call this Hermitian inner product $\langle\cdot,\cdot\rangle_R$ the *Petersson scalar product* on $A_2(H,\Gamma)$.

Instead of a Fuchsian model Γ of R acting on the upper half-plane H, taking a Fuchsian model Γ' of R acting on the unit disk Δ, we define similarly, the Petersson scalar product on $A_2(\Delta,\Gamma')$. We use both Hilbert spaces $A_2(H,\Gamma)$ and $A_2(\Delta,\Gamma')$. In the same way, we define the Petersson scalar product on $A_2(H^*,\Gamma)$ of holomorphic automorphic forms with respect to Γ on the lower half-plane H^*.

Let $A_2(\Delta)$ be the complex vector space of holomorphic functions φ on Δ such that

$$||\varphi||^2 = \iint_\Delta \lambda(z)^{-2}\,|\varphi(z)|^2\, dxdy < \infty,$$

where $\lambda(z) = 2/(1-|z|^2)$, i.e., $ds^2 = \lambda(z)^2|dz|^2$ is the Poincaré metric on the unit disk Δ. We define the Petersson scalar product $\langle\,\cdot\,,\,\cdot\,\rangle_\Delta$ on $A_2(\Delta)$ by

$$\langle\varphi,\psi\rangle_\Delta = \iint_\Delta \lambda(z)^{-2}\,\varphi(z)\overline{\psi(z)}\, dxdy.$$

Then $A_2(\Delta)$ becomes a separable Hilbert space with this scalar product.

We put

$$\varphi_n(z) = \sqrt{\frac{2}{\pi}(n+1)(n+2)(n+3)}\; z^n \tag{7.1}$$

for any non-negative integer n. Then $\{\varphi_n\}_{n=0}^{\infty}$ is a complete orthonormal basis for $A_2(\Delta)$. In particular, every $\varphi \in A_2(\Delta)$ is written in the form

$$\varphi = \sum_{n=0}^{\infty}\langle\varphi,\varphi_n\rangle_\Delta\,\varphi_n.$$

We set

$$K_\Delta(z,\zeta) = \sum_{n=0}^{\infty} \overline{\varphi_n(z)}\,\varphi_n(\zeta), \quad z,\zeta \in \Delta. \tag{7.2}$$

Then $K_\Delta(z,\cdot)$ belongs to $A_2(\Delta)$ for any $z \in \Delta$, and satisfies the *reproducing formula*

$$\varphi(z) = \langle \varphi, K_\Delta(z,\cdot)\rangle_\Delta, \quad z \in \Delta,$$

for any $\varphi \in A_2(\Delta)$; that is, K_Δ is the *reproducing kernel* for $A_2(\Delta)$. From (7.1) and (7.2), we obtain

$$K_\Delta(z,\zeta) = \sum_{n=0}^{\infty} \frac{2}{\pi}(n+1)(n+2)(n+3)(\bar{z}\zeta)^n = \frac{12}{\pi(1-\bar{z}\zeta)^4}, \quad z,\zeta \in \Delta.$$

Similarly, we consider the Hilbert space $A_2(H)$ of holomorphic functions ψ on H such that

$$\|\psi\|^2 = \iint_H \lambda_H(z)^{-2}|\psi(z)|^2\,dxdy < \infty.$$

The Möbius transformation $T\colon H \to \Delta$ given by $T(z) = (z-i)/(z+i)$ induces an isomorphism $T^*: A_2(\Delta) \to A_2(H)$ defined by

$$T^*(\varphi) = (\varphi\circ T)\cdot(T')^2, \quad \varphi \in A_2(\Delta).$$

Hence the *reproducing kernel* K_H for $A_2(H)$ is given by

$$K_H(z,\zeta) = K_\Delta(T(z),T(\zeta))\,\overline{T'(z)^2}\,T'(\zeta)^2 = \frac{12}{\pi}\frac{1}{(\bar{z}-\zeta)^4}, \quad z,\zeta \in H. \tag{7.3}$$

Note that K_Δ and K_H are invariant under $Aut(\Delta)$ and $Aut(H)$, respectively. For example, K_H satisfies

$$K_H(z,\zeta) = K_H(\gamma(z),\gamma(\zeta))\,\overline{\gamma'(z)^2}\,\gamma'(\zeta)^2, \quad z,\zeta \in H, \tag{7.4}$$

for all $\gamma \in Aut(H)$.

Theorem 7.1. (Reproducing formula) *Every $\varphi \in A_2(H,\Gamma)$ satisfies*

$$\varphi(z) = \iint_H \lambda_H(\zeta)^{-2}\varphi(\zeta)\,\overline{K_H(z,\zeta)}\,d\xi d\eta, \quad z \in H. \tag{7.5}$$

Proof. It is sufficient to prove that

$$\psi(z) = \iint_\Delta \lambda(\zeta)^{-2}\psi(\zeta)\,\overline{K_\Delta(z,\zeta)}\,d\xi d\eta, \quad z \in \Delta, \tag{7.6}$$

for every $\psi \in A_2(\Delta,\Gamma')$, where $\Gamma' = T\Gamma T^{-1}$. Since $R = H/\Gamma$ is compact, $\lambda^{-2}|\psi|$ is bounded on Δ. Thus the integral in (7.6) converges absolutely for all z. By the mean value theorem for a holomorphic function, we have

$$\begin{aligned}\psi(0) &= \frac{3}{\pi}\iint_{\Delta}(1-|\zeta|^2)^2\psi(\zeta)\,d\xi d\eta \\ &= \iint_{\Delta}\lambda(\zeta)^{-2}\,\psi(\zeta)\,\overline{K_{\Delta}(0,\zeta)}\,d\xi d\eta.\end{aligned} \tag{7.7}$$

This asserts that formula (7.6) holds for $z=0$.

Now, for any $z\in\Delta$, choose an element $\gamma\in Aut(\Delta)$ with $\gamma(0)=z$. If we set $\psi_o=(\psi{\circ}\gamma)\cdot(\gamma')^2$, then ψ_o is an element of $A_2(\Delta,\gamma^{-1}\Gamma'\gamma)$, and satisfies $\psi_o(0)=\psi(z)\gamma'(0)^2$. Applying (7.7) to ψ_o and putting $\delta=\gamma^{-1}$, we obtain

$$\begin{aligned}\psi_o(0) &= \iint_{\Delta}\lambda(\delta(w))^{-2}\psi_o(\delta(w))\,\overline{K_{\Delta}(\delta(z),\delta(w))}\,|\delta'(w)|^2\,dudv \\ &= \left(\iint_{\Delta}\lambda(w)^{-2}\psi(w)\,\overline{K_{\Delta}(z,w)}\,dudv\right)\gamma'(0)^2.\end{aligned}$$

This shows that (7.6) holds for all $z\in\Delta$. □

7.1.2 Poincaré Series

Take a fundamental domain F for Γ' in Δ so that F is relatively compact in Δ. Then formula (7.6) is rewritten as

$$\begin{aligned}\psi(z) &= \sum_{\gamma\in\Gamma'}\iint_{\gamma(F)}\lambda(\zeta)^{-2}\psi(\zeta)\,\overline{K_{\Delta}(z,\zeta)}\,d\xi d\eta \\ &= \sum_{\gamma\in\Gamma'}\iint_{\gamma(F)}\lambda(\gamma(\zeta))^{-2}\psi(\gamma(\zeta))\,\overline{K_{\Delta}(z,\gamma(\zeta))}\,|\gamma'(\zeta)|^2\,d\xi d\eta \\ &= \sum_{\gamma\in\Gamma'}\left[\iint_{F}\lambda(\zeta)^{-2}\psi(\zeta)\,\overline{K_{\Delta}(\gamma^{-1}(z),\zeta)}\,d\xi d\eta\right](\gamma^{-1})'(z)^2 \\ &= \sum_{\gamma\in\Gamma'}\left[\iint_{F}\lambda(\zeta)^{-2}\psi(\zeta)\,\overline{K_{\Delta}(\gamma(z),\zeta)}\,d\xi d\eta\right]\gamma'(z)^2.\end{aligned} \tag{7.8}$$

Now, we set

$$f(z)=\iint_{F}\lambda(\zeta)^{-2}\psi(\zeta)\,\overline{K_{\Delta}(z,\zeta)}\,d\xi d\eta. \tag{7.9}$$

Since F is relatively compact in Δ, the function f is holomorphic on a neighborhood of the closure $\bar{\Delta}$ of Δ, and ψ is written as

$$\psi(z)=\sum_{\gamma\in\Gamma'}f(\gamma(z))\gamma'(z)^2,\quad z\in\Delta.$$

In general, for a holomorphic function f on Δ, we set

$$\Theta f(z)=\sum_{\gamma\in\Gamma'}f(\gamma(z))\gamma'(z)^2,\quad z\in\Delta.$$

We call this Θf the *Poincaré series* of f for Γ' of weight -4. Similarly, we define the Poincaré series of a holomorphic function on H for a Fuchsian group Γ acting on H.

Theorem 7.2. *Let f be an integrable holomorphic function on Δ, i.e.,*

$$\iint_{\Delta} |f(z)|\, dxdy < \infty.$$

Let Θf be the Poincaré series of f for a Fuchsian group Γ' acting on Δ. Then Θf converges absolutely and uniformly on compact sets in Δ, and belongs to $A_2(\Delta, \Gamma')$.

Proof. Denote by $B(z,r)$ the closed disk with center at z and radius r. Take any compact subset K in Δ. Since Γ' acts properly discontinuously on Δ, and since every element $\gamma \in \Gamma' - \{id\}$ has no fixed points in Δ, we can choose a sufficiently small radius r so that $\gamma(B(z,r)) \cap B(z,r) = \phi$ for any $z \in K$ and $\gamma \in \Gamma' - \{id\}$.

On the other hand, $|f(\gamma(z))\gamma'(z)^2|$ is a subharmonic function on Δ, and hence the mean value theorem for a subharmonic function yields that

$$\begin{aligned} |f(\gamma(z))\gamma'(z)^2| &\leqq \frac{1}{\pi r^2} \iint_{B(z,r)} |f(\gamma(\zeta))\gamma'(\zeta)^2|\, d\xi d\eta \\ &= \frac{1}{\pi r^2} \iint_{\gamma(B(z,r))} |f(\zeta)|\, d\xi d\eta. \end{aligned}$$

Thus we get

$$\begin{aligned} \sum_{\gamma \in \Gamma'} |f(\gamma(z))\gamma'(z)^2| &\leqq \frac{1}{\pi r^2} \sum_{\gamma \in \Gamma'} \iint_{\gamma(B(z,r))} |f(\zeta)|\, d\xi d\eta \\ &\leqq \frac{1}{\pi r^2} \iint_{\Delta} |f(\zeta)|\, d\xi d\eta < \infty, \quad z \in K. \end{aligned}$$

Hence, Θf converges absolutely and uniformly on K, which in turn implies that f is holomorphic on Δ.

Next, for any $\delta \in \Gamma'$ we have

$$\begin{aligned} \Theta f(\delta(z))\delta'(z)^2 &= \sum_{\gamma \in \Gamma'} f(\gamma \circ \delta(z))\gamma'(\delta(z))^2 \delta'(z)^2 \\ &= \sum_{\gamma \in \Gamma'} f(\gamma \circ \delta(z))(\gamma \circ \delta)'(z)^2 \\ &= \Theta f(z). \end{aligned}$$

Therefore, Θf belongs to $A_2(\Delta, \Gamma')$. □

From the observation preceding Theorem 7.2, we have the following corollaries.

Corollary 1. *For every $\varphi \in A_2(\Delta, \Gamma')$, there exists a holomorphic function f defined on a neighborhood of $\bar{\Delta}$ such that $\varphi = \Theta f$ on Δ.*

Corollary 2. *Every $\varphi \in A_2(H, \Gamma)$ is written in the form*

$$\varphi(z) = \sum_{\gamma \in \Gamma} \left[\iint_F \lambda_H(\zeta)^{-2} \varphi(\zeta) \overline{K_H(\gamma(z), \zeta)}\, d\xi d\eta \right] \gamma'(z)^2, \quad z \in H,$$

and the series on the right hand side converges absolutely and uniformly on compact sets in H, where F is a fundamental domain for Γ in H.

Example. We give an example of a Poincaré series. Let Γ be a group generated by $\gamma_0(z) = \lambda z$ with $\lambda > 1$, which is a Fuchsian group acting on H. Let f be an integrable holomorphic function on H given by $f(z) = (\lambda - 1)/(z(z-1)(z-\lambda))$. Then the Poincaré series Θf of f for Γ converges to $1/z^2$ absolutely and uniformly on compact sets of H (cf. the proof of Theorem 8.3).

7.1.3 Bergman Projection

We shall construct holomorphic quadratic differentials on the Riemann surface $R = H/\Gamma$ from measurable quadratic differentials on R by using the reproducing kernel K_H for $A_2(H)$.

A complex-valued function f on H is called a *measurable automorphic form* of weight -4 with respect to Γ on H if it is a measurable function on H, and satisfies

$$f(\gamma(z))\gamma'(z)^2 = f(z), \quad z \in H, \ \gamma \in \Gamma.$$

Let $L_2^\infty(H, \Gamma)$ be the set of all measurable automorphic forms of weight -4 with respect to Γ on H with

$$\|f\|_\infty = \operatorname*{ess.sup}_{z \in H} \lambda_H(z)^{-2} |f(z)| < \infty.$$

Then $L_2^\infty(H, \Gamma)$ becomes a Banach space with this norm. Since R is compact, $A_2(H, \Gamma)$ is a closed subspace of $L_2^\infty(H, \Gamma)$. Further, the Petersson scalar product $\langle \cdot, \cdot \rangle_R$ on $A_2(H, \Gamma)$ is extended to $L_2^\infty(H, \Gamma)$.

Theorem 7.3 *For an arbitrary $f \in L_2^\infty(H, \Gamma)$, set*

$$(\beta_2 f)(z) = \iint_H \lambda_H(\zeta)^{-2} f(\zeta) \overline{K_H(z, \zeta)}\, d\xi d\eta, \quad z \in H. \tag{7.10}$$

Then $\beta_2 f$ belongs to $A_2(H, \Gamma)$, and satisfies

$$\beta_2 f = \Theta g \quad \text{on } H, \tag{7.11}$$

where

$$g(z) = \iint_F \lambda_H(\zeta)^{-2} f(\zeta) \overline{K_H(z, \zeta)}\, d\xi d\eta$$

and F is a fundamental domain for Γ in H.

Proof. From $\|f\|_\infty < \infty$ and formula (7.3), we see that integral (7.10) converges absolutely. It is clear that $\beta_2 f$ is holomorphic on H. From formula (7.4), we conclude that $\beta_2 f$ belongs to $A_2(H, \Gamma)$. By an argument similar to that in the proof of the Corollary to Theorem 7.2, we get formula (7.11). □

It is easy to see that $\beta_2 \colon L_2^\infty(H, \Gamma) \to A_2(H, \Gamma)$ is a bounded linear operator. The reproducing formula in Theorem 7.2 implies that β_2 is the identity mapping on $A_2(H, \Gamma)$. We call β_2 the *Bergman projection* of $L_2^\infty(H, \Gamma)$ to $A_2(H, \Gamma)$.

Remark. We can also use the L^q-norm ($q \geqq 1$) instead of the L^∞-norm. Namely, let $L_2^q(H, \Gamma)$ be the Banach space consisting of all measurable automorphic forms f with respect to Γ on H such that

$$\|f\|_2^q = \iint_F \lambda_H(z)^{2-2q} |f(z)|^q \, dxdy < \infty.$$

Then we obtain the Bergman projection from $L_2^q(H, \Gamma)$ to $A_2(H, \Gamma)$ (equipped with L^q-norm). For details, we refer to Kra [A-58], Chapter 3, §§2 and 3.

The Bergman projection is a self-adjoint operator; that is, we have the following assertion.

Theorem 7.4. *Any two elements $f, g \in L_2^\infty(H, \Gamma)$ satisfy*

$$\langle \beta_2 f, g \rangle_R = \langle f, \beta_2 g \rangle_R.$$

Proof. Take a relatively compact fundamental domain F for Γ in H. By using formula (7.11), the transformation relations for λ_H, K_H, f, and g, and Fubini's theorem, we get

$$\begin{aligned}
&\langle \beta_2 f, g \rangle_R \\
&= \iint_F \lambda_H(z)^{-2} \left[\sum_{\gamma \in \Gamma} \iint_F \lambda_H(\zeta)^{-2} f(\zeta) \overline{K_H(\gamma(z), \zeta)} \, d\xi d\eta \right] \gamma'(z)^2 \, \overline{g(z)} \, dxdy \\
&= \iint_F \lambda_H(\zeta)^{-2} f(\zeta) \left[\sum_{\gamma \in \Gamma} \iint_F \lambda_H(z)^{-2} \overline{g(z)} \, \gamma'(z)^2 \, \overline{K_H(\gamma(z), \zeta)} \, dxdy \right] d\xi d\eta \\
&= \iint_F \lambda_H(\zeta)^{-2} f(\zeta) \\
&\qquad \times \overline{\left[\sum_{\gamma \in \Gamma} \iint_F \lambda_H(z)^{-2} g(z) \, \overline{K_H(\gamma^{-1}(\zeta), z)} \, dxdy \right] (\gamma^{-1})'(\zeta)^2} \, d\xi d\eta \\
&= \langle f, \beta_2 g \rangle_R .
\end{aligned}$$

□

Remark. In §2.2 of Chapter 6, the derivative $\dot{\Phi}_o[\mu]$ of Bers' projection Φ in the direction μ at the base point is given by

$$\dot{\Phi}_o[\mu](z) = -\frac{6}{\pi}\iint_H \frac{\mu(\zeta)}{(z-\zeta)^4}\,d\xi d\eta, \quad z \in H^*,$$

which belongs to $A_2(H^*, \Gamma)$. Thus we see that

$$\overline{\dot{\Phi}_o[\mu](\bar{z})} = -\frac{6}{\pi}\iint_H \frac{\overline{\mu(\zeta)}}{(z-\bar{\zeta})^4}\,d\xi d\eta$$

is an element of $A_2(H, \Gamma)$. By the definition of the Bergman projection, we obtain

$$\beta_2(\lambda_H^2 \bar{\mu})(z) = -2\overline{\dot{\Phi}_o[\mu](\bar{z})}, \quad z \in H. \tag{7.12}$$

7.2. Infinitesimal Theory of Teichmüller Spaces

We shall give an explicit representation for the tangent spaces of a Teichmüller space.

7.2.1. The Tangent Space at the Base Point

Denote by $T_o(T(\Gamma))$ the holomorphic tangent space of $T(\Gamma)$ at the base point. Identifying $T(\Gamma)$ with $T_B(\Gamma)$, we regard $T_o(T(\Gamma))$ as the holomorphic tangent space $T_o(T_B(\Gamma)) = A_2(H^*, \Gamma)$ of $T_B(\Gamma)$ at the base point. As was seen in §2 of Chapter 6, the derivative $\dot{\Phi}_o$ of Bers' projection Φ at the base point is a surjective linear mapping of the space $B(H, \Gamma)$ of Beltrami differentials for Γ on H to $T_o(T_B(\Gamma))$. Thus, setting $N(\Gamma) = \operatorname{Ker} \dot{\Phi}_o$, we have

$$T_o(T(\Gamma)) \cong B(H, \Gamma)/N(\Gamma). \tag{7.13}$$

Next, we shall identify canonically $T_o(T(\Gamma))$ with the dual space $A_2(H, \Gamma)^*$ of $A_2(H, \Gamma)$. Associate every $\mu \in B(H, \Gamma)$ with an element $\Lambda_\mu \in A_2(H, \Gamma)^*$, i.e., a linear functional on $A_2(H, \Gamma)$ defined by

$$\Lambda_\mu(\varphi) = (\mu, \varphi)_R = \iint_F \mu(z)\varphi(z)\,dxdy, \quad \varphi \in A_2(H, \Gamma).$$

Here, $R = H/\Gamma$ and F is a fundamental domain for Γ in H.

Theorem 7.5 *The mapping $\Lambda\colon B(H, \Gamma) \to A_2(H, \Gamma)^*$ given by $\Lambda(\mu) = \Lambda_\mu$ induces an isomorphism of $B(H, \Gamma)/N(\Gamma)$ onto $A_2(H, \Gamma)^*$. In particular,*

$$T_o(T(\Gamma)) \cong A_2(H, \Gamma)^*.$$

For the proof of this theorem, we need the following lemma due to Teichmüller.

Lemma 7.6. (Teichmüller) *An element $\mu \in B(H,\Gamma)$ belongs to $N(\Gamma)$ if and only if $\Lambda_\mu = 0$, i.e., $(\mu,\varphi)_R = 0$ for all $\varphi \in A_2(H,\Gamma)$.*

Proof of Theorem 7.5. Clearly, Λ is linear. Teichmüller's lemma asserts that $\mathrm{Ker}\Lambda = N(\Gamma)$. Every $f \in A_2(H,\Gamma)^*$ is written as $f = \langle \cdot, \psi \rangle_R$ for some $\psi \in A_2(H,\Gamma)$, where $\langle \cdot, \cdot \rangle_R$ denotes the Petersson scalar product on $A_2(H,\Gamma)$. Putting $\mu = \lambda_H^{-2}\bar{\psi}$, we see that $\mu \in B(H,\Gamma)$ and $\Lambda_\mu = f$. This shows that Λ is surjective. Hence, by the homomorphism theorem we conclude that Λ induces an isomorphism of $B(H,\Gamma)/N(\Gamma)$ onto $A_2(H,\Gamma)^*$. □

Proof of Lemma 7.6. We consider this lemma in the unit disk Δ instead of H. Take a Möbius transformation S given by $S(z) = -i(z+1)/(z-1)$, which sends Δ and Δ^* to H and H^*, respectively. Here, Δ^* is the exterior of Δ in $\hat{\mathbf{C}}$. We set $\Gamma' = S^{-1}\Gamma S$, a Fuchsian group acting on Δ and Δ^*. Then $\mu \in B(H,\Gamma)$ corresponds to $\nu \in B(\Delta,\Gamma')$ defined by

$$\nu(z) = \mu(S(z))\frac{\overline{S'(z)}}{S'(z)}, \quad z \in \Delta.$$

Further, $\dot{\Phi}_o[\mu] \in A_2(H^*,\Gamma)$ corresponds to $\Psi \in A_2(\Delta^*,\Gamma')$ given by

$$\Psi(z) = \dot{\Phi}_o[\mu](S(z))S'(z)^2, \quad z \in \Delta^*.$$

Now, formula (6.8) in Chapter 6 is rewritten in the form

$$\Psi(z) = -\frac{6}{\pi}\iint_\Delta \frac{\nu(w)}{(w-z)^4}\,dudv, \quad z \in \Delta^*.$$

Thus Ψ is expanded in the form

$$\Psi(z) = -\frac{1}{\pi}\sum_{n=3}^{\infty} a_n z^{-(n+1)}, \quad z \in \Delta^*,$$

where

$$a_n = n(n-1)(n-2)\iint_\Delta \nu(\zeta)\zeta^{n-3}\,d\xi d\eta.$$

Hence, $\Psi = 0$ if and only if $a_n = 0$ for all integers $n \geqq 3$, which is equivalent to the condition

$$\iint_\Delta \nu(\zeta)f(\zeta)\,d\xi d\eta = 0$$

for all holomorphic functions f in a neighborhood of $\bar{\Delta}$. By the same computation as that in (7.8), we obtain

$$\iint_\Delta \nu(\zeta)f(\zeta)\,d\xi d\eta = (\nu, \Theta f)_R.$$

Therefore, by Corollary 1 to Theorem 7.2, the lemma is proved. □

7.2.2. Harmonic Beltrami Differentials

In the proofs of the Ahlfors-Weill theorem (Theorem 6.9) and Theorem 7.5, so-called harmonic Beltrami differentials defined from elements of $A_2(H,\Gamma)$ played an important role. We use harmonic Beltrami differentials to represent tangent vectors of $T(\Gamma)$. Here, for every element $\varphi \in A_2(H,\Gamma)$, we define $\mu[\varphi] \in B(H,\Gamma)$ by

$$\mu[\varphi](z) = \lambda_H(z)^{-2}\overline{\varphi(z)}, \quad z \in H. \tag{7.14}$$

We call this $\mu[\varphi]$ the ***harmonic Beltrami differential*** induced by φ.

On the other hand, we associate every $\mu \in B(H,\Gamma)$ with an element $\varphi[\mu] \in A_2(H,\Gamma)$ given by

$$\varphi[\mu](z) = -2\overline{\dot{\Phi}_o[\mu](\bar{z})}, \quad z \in H. \tag{7.15}$$

As was stated in the Remark of §1.3, formula (7.12) yields

$$\varphi[\mu] = \beta_2(\lambda_H^2\bar{\mu}), \quad \mu \in B(H,\Gamma), \tag{7.16}$$

where β_2 is the Bergman projection of $L_2^\infty(H,\Gamma)$ to $A_2(H,\Gamma)$. From (7.14) and the reproducing formula, we have

$$\varphi[\mu[\varphi]] = \varphi, \quad \varphi \in A_2(H,\Gamma).$$

Next, for every $\mu \in B(H,\Gamma)$, we define an element $H[\mu] \in B(H,\Gamma)$ by

$$H[\mu](z) = \lambda_H(z)^{-2}\overline{\varphi[\mu](z)}. \tag{7.17}$$

This $H[\mu]$ is said to be the ***harmonic Beltrami differential*** induced by μ. Denote by $HB(H,\Gamma)$ the vector space of all these harmonic Beltrami differentials $H[\mu]$. In this way, we obtain a surjective linear mapping

$$H\colon B(H,\Gamma) \to HB(H,\Gamma).$$

Formulas (7.16) and (7.17) lead to

$$H[\mu] = \lambda_H^{-2}\overline{\beta_2(\lambda_H^2\bar{\mu})}, \quad \mu \in B(H,\Gamma). \tag{7.18}$$

From (7.15) and (7.17), we conclude that

$$\mathrm{Ker} H = N(\Gamma).$$

Further, by (7.14), (7.18) and the reproducing formula, we get

$$H[\mu[\varphi]] = \mu[\varphi], \quad \varphi \in A_2(H,\Gamma). \tag{7.19}$$

Moreover, (7.16), (7.17), and the reproducing formula imply that

$$\varphi[H[\mu]] = \varphi[\mu], \quad \mu \in B(H,\Gamma). \tag{7.20}$$

In particular, from (7.15) we obtain

$$\dot{\Phi}_o[H[\mu]] = \dot{\Phi}_o[\mu], \quad \mu \in B(H,\Gamma). \tag{7.21}$$

Using (7.17) and (7.20), we also have

$$H^2 = H. \tag{7.22}$$

With these preparations, we get the following assertion.

Theorem 7.7 *The space $B(H, \Gamma)$ of Beltrami differentials for Γ on H is the direct sum of $HB(H, \Gamma)$ and $N(\Gamma)$, i.e.,*

$$B(H, \Gamma) = HB(H, \Gamma) \oplus N(\Gamma). \tag{7.23}$$

The derivative $\dot{\Phi}_o$ of Bers' projection Φ at the base point induces the isomorphism $\dot{\Phi}_o \colon HB(H, \Gamma) \to T_o(T_B(\Gamma))$. In particular,

$$T_o(T(\Gamma)) \cong HB(H, \Gamma). \tag{7.24}$$

Moreover,

$$(\mu, \varphi)_R = (H[\mu], \varphi)_R, \quad \mu \in B(H, \Gamma), \ \varphi \in A_2(H, \Gamma). \tag{7.25}$$

Proof. Take an element $\mu \in N(\Gamma)$. Since $N(\Gamma) = \mathrm{Ker} H$, we have $H[\mu] = 0$. If $\mu \in HB(H, \Gamma)$, then by the definition there exists an element $\nu \in B(H, \Gamma)$ with $H[\nu] = \mu$. Hence, (7.22) leads to

$$\mu = H[\nu] = H^2[\nu] = H[\mu] = 0,$$

and we obtain

$$HB(H, \Gamma) \cap N(\Gamma) = \{0\}.$$

Every $\mu \in B(H, \Gamma)$ is decomposed into

$$\mu = H[\mu] + (\mu - H[\mu]),$$

and (7.22) implies that $\mu - H[\mu] \in \mathrm{Ker}\, H = N(\Gamma)$. Thus we have (7.23). It is obvious from (7.13) that $\dot{\Phi}_o \colon HB(H, \Gamma) \to T_o(T_B(\Gamma))$ is an isomorphism. Accordingly, we have (7.24). Further, Teichmüller's lemma (Lemma 7.6) yields (7.25). □

7.2.3. Tangent Space of $T(\Gamma)$ at a General Point

We shall give a representation of the holomorphic tangent space $T_p(T(\Gamma))$ of $T(\Gamma)$ at an arbitrary point $p = [w^\nu]$.

Let $T(\Gamma^\nu)$ be the Teichmüller space of $\Gamma^\nu = w^\nu \Gamma (w^\nu)^{-1}$. Then the translation mapping $[w^\nu]_*$ (§2.3 of Chapter 6) of $T(\Gamma)$ to $T(\Gamma^\nu)$ induces an isomorphism of $T_p(T(\Gamma))$ to $T_o(T(\Gamma^\nu))$. We give an explicit description of this isomorphism.

Defining

$$\kappa = F(\lambda) = \left(\frac{(w^\nu)_z}{\overline{(w^\nu)_z}} \frac{\lambda - \nu}{1 - \bar{\nu}\lambda} \right) \circ (w^\nu)^{-1},$$

we have

$$[w^\nu]_*([w^\lambda]) = [w^\kappa], \quad [w^\lambda] \in T(\Gamma).$$

Let $L^\nu : B(H,\Gamma) \to B(H,\Gamma^\nu)$ be an isomorphism given by

$$L^\nu[\mu] = \lim_{t\to 0} \frac{F(\nu + t\mu) - F(\nu)}{t} = \left(\frac{(w^\nu)_z}{\overline{(w^\nu)_z}} \frac{\mu}{1-|\nu|^2} \right) \circ (w^\nu)^{-1}. \tag{7.26}$$

Denote by H^ν the projection of $B(H,\Gamma^\nu)$ to the space $HB(H,\Gamma^\nu)$ of harmonic Beltrami differentials for Γ^ν on H, which is given by (7.17). Then we have the following assertion.

Proposition 7.8. *For every $p = [w^\nu] \in T(\Gamma)$, the mapping $H^\nu \circ L^\nu$ is an isomorphism of $B(H,\Gamma)/\mathrm{Ker}\,\dot{\Phi}_\nu \cong T_p(T(\Gamma))$ to $HB(H,\Gamma^\nu) \cong T_o(T(\Gamma^\nu))$, where $\dot{\Phi}_\nu$ is the derivative of Bers' projection at ν. Moreover, $H^\nu \circ L^\nu$ is independent of the choice of a representative w^ν of the point p.*

Proof. We consider $T_B(\Gamma)$ and $T_B(\Gamma^\nu)$ instead of $T(\Gamma)$ and $T(\Gamma^\nu)$, respectively. Denote by Ψ the translation mapping $\langle w^\nu\rangle_* : T_B(\Gamma) \to T_B(\Gamma^\nu)$. Let Φ and Φ^ν be Bers' projections of $B(H,\Gamma)_1$ and $B(H,\Gamma^\nu)_1$, respectively. Then we obtain the following commutative diagram:

$$\begin{array}{ccc} B(H,\Gamma)_1 & \xrightarrow{F} & B(H,\Gamma^\nu)_1 \\ {\scriptstyle\Phi}\downarrow & & \downarrow{\scriptstyle\Phi^\nu} \\ T_B(\Gamma) & \xrightarrow[\Psi]{} & T_B(\Gamma^\nu). \end{array}$$

Taking derivatives, we have the commutative diagram:

$$\begin{array}{ccc} B(H,\Gamma) & \xrightarrow{L^\nu} & B(H,\Gamma^\nu) \\ {\scriptstyle\dot{\Phi}_\nu}\downarrow & & \downarrow{\scriptstyle(\dot{\Phi}^\nu)_o} \\ T_p(T_B(\Gamma)) & \xrightarrow[\dot{\Psi}]{} & T_o(T_B(\Gamma^\nu)). \end{array}$$

Thus, noting that $\mathrm{Ker} H^\nu = \mathrm{Ker}(\dot{\Phi}^\nu)_o$, and that L^ν and $\dot{\Psi}$ are isomorphic, we see that

$$\mathrm{Ker}\,(H^\nu \circ L^\nu) = \mathrm{Ker}\,((\dot{\Phi}^\nu)_o \circ L^\nu) = \mathrm{Ker}\,(\dot{\Psi} \circ \dot{\Phi}_\nu) = \mathrm{Ker}\,\dot{\Phi}_\nu,$$

and that $H^\nu \circ L^\nu$ is surjective. Hence, the homomorphism theorem implies that $H^\nu \circ L^\nu : B(H,\Gamma)/\mathrm{Ker}\,\dot{\Phi}_\nu \to HB(H,\Gamma^\nu)$ is an isomorphism.

Further, from (7.15) and (7.17), we get

$$\begin{aligned}(H^\nu \circ L^\nu)[\mu](z) &= -2\lambda_H(z)^{-2}((\dot{\Phi}^\nu)_o \circ L^\nu[\mu])(\bar{z}) \\ &= -2\lambda_H(z)^{-2}(\dot{\Psi} \circ \dot{\Phi}_\nu[\mu])(\bar{z}), \quad z \in H.\end{aligned}$$

Since $\dot{\Phi}_\nu$ and $\dot{\Psi}$ are independent of the choice of a representative w^ν of p, we have the second assertion. □

Remark. In the rest of this chapter, we identify $T_p(T(\Gamma))$ and $T_o(T(\Gamma^\nu))$ with $B(H,\Gamma)/\mathrm{Ker}\,\dot{\Phi}_\nu$ and $HB(H,\Gamma^\nu)$, respectively. We also identify $T_p(T(\Gamma))$ with $HB(H,\Gamma^\nu)$ under the isomorphism $H^\nu \circ L^\nu$.

7.2.4. Connection with the Kodaira-Spencer Deformation Theory

The subject of this subsection is not needed for further development in this chapter. However, it is interesting by itself. We shall deal with the tangent space of $T(\Gamma)$ from the viewpoint of cohomology theory.

We recall the fundamental idea of Kodaira and Spencer on the deformation of complex structures. For details, we refer to Kodaira [A-57], and Morrow and Kodaira [A-77].

As was stated in §1.1 of Chapter 1, a Riemann surface R is obtained by patching domains $D_j = z_j(U_j)$ in the complex plane. The identification between D_j and D_k is given by a biholomorphic mapping z_{jk} of an open set $D_{kj} = z_k(U_j \cap U_k) \subset D_k$ onto $D_{jk} = z_j(U_j \cap U_k) \subset D_j$. A deformation R_t of R is considered to be the gluing of the same domains D_j via a different identification $f_{jk}(\,\cdot\,,t)$, where $f_{jk}(\,\cdot\,,t)$ is a biholomorphic mapping of D_{kj} onto D_{jk} with parameter $t = (t_1,\ldots,t_m)$ such that $f_{jk}(z_k,0) = z_{jk}(z_k)$. If all $f_{jk}(z_k,t)$ are C^∞ functions, we get a *differentiable family* $\{R_t\}_t$ of Riemann surfaces. In particular, if $f_{jk}(z_k,t)$ are holomorphic, we have a ***holomorphic family*** of Riemann surfaces. From here on, we consider a differentiable or holomorphic family of closed Riemann surfaces.

In order to know the actual dependence of the complex structure of R_t on the parameter t, we consider its infinitesimal deformation as follows. For simplicity, we assume that $m = 1$, that $\{U_j\}_j$ is a locally finite open covering of R, and that every D_j is an open disk in the complex plane. Take the differentiation of $f_{jk}(z_k,t)$ with respect to t at $t = 0$. This is regarded as a holomorphic vector field on $U_j \cap U_k$, which is written as

$$\theta_{jk} = \frac{\partial f_{jk}}{\partial t}(z_k,0)\frac{\partial}{\partial z_j}, \quad z_k = z_{kj}(z_j).$$

The relation $f_{jk}(f_{kj}(z_j,t),t) = z_j$ on $U_j \cap U_k$ gives

$$\theta_{jk} + \theta_{kj} = 0 \quad \text{on} \quad U_j \cap U_k.$$

Further, the relation $f_{j\ell}(z_\ell,t) = f_{jk}(f_{k\ell}(z_\ell,t),t)$ on $U_j \cap U_k \cap U_\ell$ yields

$$\theta_{jk} + \theta_{k\ell} + \theta_{\ell j} = 0 \quad \text{on} \quad U_j \cap U_k \cap U_\ell.$$

Thus it follows that $\theta = \{\theta_{jk}\}$ defines an element $[\theta]$ of the first cohomology group $H^1(R,\Theta)$ with coefficients in Θ, the sheaf of germs of holomorphic vector fields on R. For the cohomology theory, we refer to Gunning [A-40].

This $[\theta]$ represents in some sense the derivative of the complex structure of R_t with respect to t at $t = 0$, and is called an ***infinitesimal deformation*** of R. We call $H^1(R,\Theta)$ the ***space of infinitesimal deformations*** of R.

Theorem 7.9. *The space $H^1(R,\Theta)$ of infinitesimal deformations on R is identified with the tangent spcae $T_o(T(R))$ of the Teichmüller space $T(R)$ at the base point.*

Proof. Let κ be the canonical line bundle on R, i.e., its transition function κ_{jk} is given by $1/z'_{jk}$ on $U_j \cap U_k$. We adopt the notation:

$$\begin{aligned}
\mathcal{O}(\kappa^{-1}) &= \text{the sheaf of germs of holomorphic cross-sections of } \kappa^{-1},\\
\mathcal{E}^{0,0}(\kappa^{-1}) &= \text{the sheaf of germs of } C^\infty \text{ cross-sections of } \kappa^{-1},\\
\mathcal{E}^{0,1}(\kappa^{-1}) &= \text{the sheaf of germs of } C^\infty \text{ differential forms of type } (0,1)\\
&\qquad \text{with values which are cross-sections of } \kappa^{-1}, \text{ and}\\
\mathcal{O}^{1,0}(\kappa) &= \text{the sheaf of germs of holomorphic 1-forms}\\
&\qquad \text{with values which are cross-sections of } \kappa.
\end{aligned}$$

Then we have

$$\Theta = \mathcal{O}(\kappa^{-1}), \quad \text{and} \quad A_2(R) = H^0(R, \mathcal{O}^{1,0}(\kappa)).$$

We construct the isomorphism

$$\delta^* : \frac{H^0(R, \mathcal{E}^{0,1}(\kappa^{-1}))}{\bar\partial H^0(R, \mathcal{E}^{0,0}(\kappa^{-1}))} \to H^1(R,\Theta)$$

in the following way. (This is a consequence of Dolbeault's theorem.)

Note that an element of $H^0(R, \mathcal{E}^{0,1}(\kappa^{-1}))$ is a C^∞ differential form $\mu = \{\, \mu_j d\bar z_j / dz_j \,\}$ of type $(-1,1)$ on R, i.e., a C^∞ Beltrami differential on R. An element $v \in H^0(R, \mathcal{E}^{0,0}(\kappa^{-1}))$ is given by a C^∞ vector field $v = \{\, v_j \partial/\partial z_j \,\}$ on R, and $\bar\partial v$ is defined by $\{\, \mu_j d\bar z_j/dz_j \,\}$ with $\mu_j = \partial v_j/\partial \bar z_j$.

For any $\mu = \{\, \mu_j d\bar z_j/dz_j \,\}$, take a solution v_j of the differential equation $\partial v_j/\partial \bar z_j = \mu_j$ on U_j. Then $v_j \partial/\partial z_j$ is a C^∞ vector field on U_j. However, $\{\, v_j \partial/\partial z_j \,\}$ does not always define a global C^∞ vector field on R. A necessary and sufficient condition that $\{\, v_j\partial/\partial z_j \,\}$ gives a vector field on R is that every $v_k \partial/\partial z_k - v_j \partial/\partial z_j$ vanishes on $U_j \cap U_k$. We set

$$\theta_{jk} = v_k(z_k)\frac{\partial}{\partial z_k} - v_j(z_j)\frac{\partial}{\partial z_j} \quad \text{on} \quad U_j \cap U_k. \tag{7.27}$$

Then θ_{jk} is a holomorphic vector field on $U_j \cap U_k$, and $\theta = \{\, \theta_{jk} \,\}$ induces an element $[\theta]$ of $H^1(R,\Theta)$. This $[\theta]$ depends only on the equivalence class $[\mu]$ of μ in $H^0(R, \mathcal{E}^{0,1}(\kappa^{-1}))/\bar\partial H^0(R, \mathcal{E}^{0,0}(\kappa^{-1}))$. It is clear that the mapping δ^* defined by $\delta^*([\mu]) = [\theta]$ gives a homomorphism of $H^0(R, \mathcal{E}^{0,1}(\kappa^{-1}))/\bar\partial H^0(R, \mathcal{E}^{0,0}(\kappa^{-1}))$ to $H^1(R,\Theta)$.

The inverse mapping of δ^* is given as follows. Let $\{\, \rho_j \,\}$ be a partition of unity subordinate to the open covering $\{\, U_j \,\}$ of R. Take any element $[\theta] \in H^1(R,\Theta)$ which is represented by a cocycle $\theta = \{\, \theta_{jk} \,\}$ such that each $\theta_{jk} = a_{jk}(z_j)\partial/\partial z_j$ is a holomorphic vector field on $U_j \cap U_k$. We set $v_j \partial/\partial z_j = \sum_k \rho_k \theta_{kj}$, i.e., $v_j = \sum_k \rho_k(z_{kj}(z_j)) a_{kj}(z_{kj}(z_j))/z'_{kj}(z_j)$, which is a C^∞ vector field on U_j. Putting

$\mu_j = \partial v_j/\partial \bar{z}_j$, we get an element $\mu = \{\mu_j d\bar{z}_j/dz_j\} \in H^0(R, \mathcal{E}^{0,1}(\kappa^{-1}))$. Then the homomorphism of $H^1(R,\Theta)$ to $H^0(R,\mathcal{E}^{0,1}(\kappa^{-1}))/\bar{\partial}H^0(R,\mathcal{E}^{0,0}(\kappa^{-1}))$ sending $[\theta]$ to $[\mu]$ gives the inverse mapping of δ^*.

Next, we have a canonical isomorphism

$$\Lambda \circ (\delta^*)^{-1} : H^1(R,\Theta) \to A_2(R)^*,$$

where

$$\Lambda : H^0(R,\mathcal{E}^{0,1}(\kappa^{-1}))/\bar{\partial}H^0(R,\mathcal{E}^{0,0}(\kappa^{-1})) \to A_2(R)^*$$

is defined by

$$\Lambda[\mu](\varphi) = \iint_R \mu(z)\varphi(z)\,dxdy, \quad \mu \in H^0(R,\mathcal{E}^{0,1}(\kappa^{-1})),\ \varphi \in A_2(R).$$

(This is a consequence of Serre's duality theorem. See the proof of Theorem 7.5.)

Since $T_o(T(R))$ is isomorphic to $A_2(R)^*$ (Theorem 7.5), it follows that the infinitesimal deformation space $H^1(R,\Theta)$ is identified with the tangent space $T_o(T(R))$ of the Teichmüller space $T(R)$ of R at the base point. This completes the proof of Theorem 7.9. □

Now, we wish to see that $H^1(R,\Theta)$ is canonically identified with a subspace of the first Eichler cohomology group $H^1(\Gamma,\Pi_2)$, which is defined later, where Γ is a Fuchsian model of R, and Π_2 is the space of polynomials in one complex variable of degree at most two. Note that Π_2 is regarded as the space of holomorphic vector fields on the Riemann sphere $\hat{\mathbf{C}}$. Further, Π_2 is canonically identified with the Lie algebra $\mathfrak{sl}(2,\mathbf{C})$ of $SL(2,\mathbf{C})$, which is the tangent space of $SL(2,\mathbf{C})$ at the unit element.

Let (H,π,R) be the universal covering of R with covering transformation group Γ. We use the notation:

$$\mathcal{B}(H,\Gamma) = \text{the lift of } H^0(R,\mathcal{E}^{0,1}(\kappa^{-1})) \text{ under } \pi,$$
$$\mathcal{V}(H,\Gamma) = \text{the lift of } H^0(R,\mathcal{E}^{0,0}(\kappa^{-1})) \text{ under } \pi.$$

Then $\mathcal{B}(H,\Gamma)$ consists of smooth Beltrami differentials $\tilde{\mu}$ for Γ on H. An element of $\tilde{v} \in \mathcal{V}(H,\Gamma)$ is a C^∞ complex-valued function on H such that $\tilde{v}\circ\gamma/\gamma' = \tilde{v}$ for any $\gamma \in \Gamma$.

For every $\tilde{v} \in \mathcal{V}(H,\Gamma)$, we set $\bar{\partial}\tilde{v} = \partial\tilde{v}/\partial\bar{z}$, which belongs to $\mathcal{B}(H,\Gamma)$. Obviously, $H^0(R,\mathcal{E}^{0,1}(\kappa^{-1}))/\bar{\partial}H^0(R,\mathcal{E}^{0,0}(\kappa^{-1}))$ is isomorphic to $\mathcal{B}(H,\Gamma)/\bar{\partial}\mathcal{V}(H,\Gamma)$.

For any $\tilde{\mu} \in \mathcal{B}(H,\Gamma)$, we put

$$\hat{\mu}(z) = \begin{cases} \tilde{\mu}(z), & z \in H \\ 0, & z \in \mathbf{C} - H. \end{cases}$$

Let F be a continuous function on $\mathbf{C}$ which satisfies the differential equation

$$\frac{\partial F}{\partial \bar{z}} = \hat{\mu} \quad \text{on } \mathbf{C} \tag{7.28}$$

in the sense of distribution such that $F(z) = O(|z|^2)$ as $z \to \infty$. For example, we see from Lemma 4.20 and Theorem 4.37 that the function

$$F(z) = -\frac{z(z-1)}{\pi} \iint_H \frac{\tilde{\mu}(\zeta)}{\zeta(\zeta-1)(\zeta-z)}\, d\xi d\eta \tag{7.29}$$

gives a desired solution of (7.28). See also Lemma 1.4 of Chapter IV in Kra [A-58]. We call such an F an *potential* for $\hat{\mu}$.

For another potential G for $\hat{\mu}$, Weyl's lemma shows that $G - F$ is a holomorphic function on $\mathbf{C}$. Since $G(z) - F(z) = O(|z|^2)$ as $z \to \infty$, it follows that $G - F$ is a polynomial of degree at most two.

A potential F for $\hat{\mu}$ is not always contained in $\mathcal{V}(H, \Gamma)$. A necessary and sufficient condition that $F \in \mathcal{V}(H, \Gamma)$ is that $F \circ \gamma / \gamma' - F$ vanishes for any $\gamma \in \Gamma$.

Now, we set

$$\chi_F(\gamma) = \frac{F \circ \gamma}{\gamma'} - F, \quad \gamma \in \Gamma.$$

Since $F \circ \gamma / \gamma'$ is also a potential for $\hat{\mu}$, it follows that $\chi_F(\gamma) \in \Pi_2$. Hence, we obtain a mapping

$$\chi_F : \Gamma \to \Pi_2,$$

which satisfies the *cocycle condition*

$$\chi_F(\gamma_1 \circ \gamma_2) = (\gamma_2)_*(\chi_F(\gamma_1)) + \chi_F(\gamma_2), \quad \gamma_1, \gamma_2 \in \Gamma.$$

Here, Γ acts on Π_2 via

$$\gamma_*(P) = \frac{P \circ \gamma}{\gamma'}, \quad \gamma \in \Gamma,\ P \in \Pi_2.$$

If G is another potential for $\hat{\mu}$, then $P = G - F$ belongs to Π_2. Thus $\chi_G = \chi_F + \delta(P)$, where $\delta(P)$ is the *coboundary* of P which is given by the mapping of Γ into Π_2 such that $\delta(P)(\gamma) = \gamma_*(P) - P$.

For every $\tilde{\mu} \in \mathcal{B}(H, \Gamma)$, we have a differentiable family $\{ H_t/\Gamma_t \}$ with $H_t = f_t(H)$ and $\Gamma_t = \{ \gamma_t \mid \gamma_t = f_t \circ \gamma \circ f_t^{-1},\ \gamma \in \Gamma \}$. Here, f_t is the quasiconformal mapping $w_{t\hat{\mu}}$ of $\hat{\mathbf{C}}$ with Beltrami coefficient $t\hat{\mu}$, which is constructed in §1.1 of Chapter 6. This Γ_t corresponds to $\{ f_{jk}(\cdot, t) \}$. For every $\gamma \in \Gamma$, denote by $\dot{\gamma}$ the derivative of γ_t with respect to t at $t = 0$. From Theorem 4.37 in Chapter 4, the differentiation $\dot{f}[\tilde{\mu}]$ of f_t with respect to t at $t = 0$ is given by (7.29), that is, $\dot{f}[\tilde{\mu}]$ is a potential for $\hat{\mu}$. Since $f_t \circ \gamma = \gamma_t \circ f_t$ for every $\gamma \in \Gamma$, we have

$$\dot{f}[\tilde{\mu}] \circ \gamma = \dot{\gamma} + \dot{f}[\tilde{\mu}]\gamma'.$$

Thus we obtain

$$\chi_{\dot{f}[\tilde{\mu}]}(\gamma) = \frac{\dot{f}[\tilde{\mu}] \circ \gamma}{\gamma'} - \dot{f}[\tilde{\mu}] = \frac{\dot{\gamma}}{\gamma'}, \quad \gamma \in \Gamma.$$

This suggests that $\chi_{\dot{f}[\tilde{\mu}]}$ corresponds to θ defined by (7.27).

Now, we set

$$H^1(\Gamma, \Pi_2) = \frac{Z^1(\Gamma, \Pi_2)}{B^1(\Gamma, \Pi_2)},$$

where

$B^1(\Gamma, \Pi_2) = \delta(\Pi_2)$,
$Z^1(\Gamma, \Pi_2) = \{\chi \mid \chi\colon \Gamma \to \Pi_2,\ \chi(\gamma_1\gamma_2) = (\gamma_2)_*(\chi(\gamma_1)) + \chi(\gamma_2),\ \gamma_1, \gamma_2 \in \Gamma\}$.

This $H^1(\Gamma, \Pi_2)$ is called the ***first Eichler cohomology group*** of Γ.

Theorem 7.10. *Let β^* be the mapping*

$$\beta^* : \mathcal{B}(H, \Gamma)/\bar{\partial}\mathcal{V}(H, \Gamma) \to H^1(\Gamma, \Pi_2)$$

defined by

$$\beta^*([\tilde{\mu}]) = [\chi_F],$$

where F is a potential for $\hat{\mu}$. Then β^* is an injective homomorphism.

Proof. First, we show that β^* is well-defined. If $\tilde{\mu}, \tilde{\nu} \in \mathcal{B}(H, \Gamma)$ are equivalent, then there is an element $\tilde{v} \in \mathcal{V}(H, \Gamma)$ with $\tilde{\nu} = \tilde{\mu} + \partial\tilde{v}/\partial\bar{z}$. Let F and G be potentials for $\hat{\mu}$ and $\hat{\nu}$, respectively. We put

$$G_o(z) = \begin{cases} F(z) + \tilde{v}(z), & z \in H, \\ F(z), & z \in \mathbf{C} - H. \end{cases}$$

Since $\tilde{v}\circ\gamma = \tilde{v}\gamma'$ for every $\gamma \in \Gamma$, it is seen that for every $\zeta \in \mathbf{R}$, $\tilde{v}(z) \to 0$ as $z \to \zeta$ through H, and that $\tilde{v}(z) = O(|z|^2)$ as $z \to \infty$ through H. Thus G_o is a continuous function on $\mathbf{C}$, and satisfies $\partial G_o/\partial\bar{z} = \hat{\nu}$. Hence, G_o is also a potential for $\hat{\nu}$, which implies that $\chi_G = \chi_{G_o} + \delta(P)$ for some $P \in \Pi_2$. Noting that $\chi_{G_o} = \chi_F$, we have $\chi_G = \chi_F + \delta(P)$, and hence $[\chi_G] = [\chi_F]$ in $H^1(\Gamma, \Pi_2)$. This implies that β^* is well-defined. It is clear that β^* is a homomorphism.

Next, we verify that β^* is injective. Assume that $\beta^*([\tilde{\mu}]) = [\chi_F] = 0$. Then, there exists an element $P \in \Pi_2$ such that $\chi_F = \delta(P)$. Putting $\tilde{v} = F - P$, we see that $\tilde{v}$ is a potential for $\hat{\mu}$, and $\chi_{\tilde{v}} = 0$. Thus $\tilde{v}$ belongs to $\mathcal{V}(H, \Gamma)$, and $\bar{\partial}\tilde{v} = \tilde{\mu}$. This shows that $[\tilde{\mu}] = 0$ in $\mathcal{B}(H, \Gamma)/\bar{\partial}\mathcal{V}(H, \Gamma)$. Therefore, β^* is injective. □

We shall construct a canonical homomorphism

$$\beta\colon H^1(\Gamma, \Pi_2) \to \mathcal{B}(H, \Gamma)/\bar{\partial}\mathcal{V}(H, \Gamma).$$

(It is considered that β^* and β correspond to δ^* and $(\delta^*)^{-1}$ in the proof of Theorem 7.9, respectively.)

Choose a smooth function ρ on H which satisfies the following three conditions:

(i) $0 \leqq \rho \leqq 1$.
(ii) For each $z \in H$, there is a neighborhood U of z and a finite subset J of Γ such that $\rho = 0$ on $\gamma(U)$ for every $\gamma \in \Gamma - J$.
(iii) $\sum_{\gamma\in\Gamma} \rho\circ\gamma(z) = 1$ on H.

Such a ρ is called a ***partition of unity*** for Γ on H. For a proof of its existence, see Lemma 3.1 of Chapter V in Kra [A-58].

For any $[\chi] \in H^1(\Gamma, \Pi_2)$, we set

$$F_o(z) = -\sum_{\gamma \in \Gamma} \rho(\gamma(z))\chi(\gamma)(z), \quad z \in H. \tag{7.30}$$

Then F_o is a C^∞ function on H such that

$$\chi(\gamma) = F_o \circ \gamma/\gamma' - F_o, \quad \gamma \in \Gamma.$$

Thus $\mu = \partial F_o/\partial \bar{z}$ belongs to $\mathcal{B}(H,\Gamma)$. It is seen that the equivalence class $[\mu]$ in $\mathcal{B}(H,\Gamma)/\bar{\partial}\mathcal{V}(H,\Gamma)$ depends only on the equivalence class $[\chi] \in H^1(\Gamma, \Pi_2)$. Hence, we obtain a homomorphism β: $H^1(\Gamma, \Pi_2) \to \mathcal{B}(H,\Gamma)/\bar{\partial}\mathcal{V}(H,\Gamma)$ which sends $[\chi]$ to $[\mu]$.

From the construction of β and β^*, it is clear that $\beta \circ \beta^* = id$. It follows that β^* is injective (cf. Theorem 7.10), β is surjective, and

$$H^1(\Gamma, \Pi_2) = \operatorname{Im} \beta^* \oplus \operatorname{Ker} \beta.$$

Elements of $\operatorname{Im} \beta^*$ and $\operatorname{Ker} \beta$ are called *Bers cohomology classes* and *Eichler cohomology classes*, respectively.

Proposition 7.11. *Kerβ is identified with the space $A_2(H,\Gamma)$ of holomorphic quadratic forms on H for Γ.*

Proof. For any $[\chi] \in \operatorname{Ker} \beta$, the function F_o constructed by (7.30) satisfies $\partial F_o/\partial \bar{z} = \partial \tilde{v}/\partial \bar{z}$ for some $\tilde{v} \in \mathcal{V}(H,\Gamma)$. Thus, $f = F_o - \tilde{v}$ is a holomorphic function on H such that $\chi(\gamma) = f \circ \gamma/\gamma' - f$ for all $\gamma \in \Gamma$. It is proved that f''', the differentiation of f three times, is an element of $A_2(H,\Gamma)$, and depends only on the equivalence class $[\chi]$. We set $\alpha([\chi]) = f'''$. It is verified that α is an isomorphism of Kerβ onto $A_2(H,\Gamma)$. □

7.3. Weil-Petersson Metric

We shall define the Weil-Petersson metric on $T(\Gamma)$ and show that it is a Kähler metric.

7.3.1. Definition of the Weil-Petersson Metric

First of all, we give a Hermitian inner product on the holomorphic tangent space $T_o(T(\Gamma))$ of $T(\Gamma)$ at the base point. We identified $T_o(T(\Gamma))$ with $HB(H,\Gamma)$ (Theorem 7.7), and hence we start by giving a Hermitian inner product on $B(H,\Gamma)$. The inner product of two elements $\mu_1, \mu_2 \in B(H,\Gamma)$ is defined by

$$h(\mu_1, \mu_2) = \iint_F \lambda_H(z)^2 \mu_1(z)\, \overline{\mu_2(z)}\, dx dy, \tag{7.31}$$

where F is a relatively compact fundamental domain for Γ in H. Formulas (7.17) and (7.31) yield

$$h(H[\mu_1], H[\mu_2]) = \langle \varphi[\mu_2], \varphi[\mu_1] \rangle_R, \quad \mu_1, \mu_2 \in B(H,\Gamma).$$

Lemma 7.12. *For any* $\mu_1, \mu_2 \in B(H, \Gamma)$, *the following hold:*

$$h(H[\mu_1], H[\mu_2]) = h(H[\mu_1], \mu_2) = h(\mu_1, H[\mu_2]), \tag{7.32}$$

$$h(H[\mu_1], H[\mu_2]) = (H[\mu_1], \varphi[\mu_2])_R = (\mu_1, \varphi[\mu_2])_R. \tag{7.33}$$

Proof. First, (7.18) and (7.31) lead to

$$h(H[\mu_1], \mu_2) = \left\langle \lambda_H^2 \overline{\mu_2}, \lambda_H^2 \overline{H[\mu_1]} \right\rangle_R = \langle \lambda_H^2 \overline{\mu_2}, \beta_2(\lambda_H^2 \overline{\mu_1}) \rangle_R \,.$$

Since the Bergman projection β_2 is self-adjoint (Theorem 7.4), we have

$$h(H[\mu_1], \mu_2) = \langle \beta_2(\lambda_H^2 \overline{\mu_2}), \lambda_H^2 \overline{\mu_1} \rangle_R = h(\mu_1, H[\mu_2]).$$

In particular, using (7.22), we get

$$h(H[\mu_1], H[\mu_2]) = h(\mu_1, H^2[\mu_2]) = h(\mu_1, H[\mu_2]),$$

which shows (7.32). Moreover, from (7.17), (7.25), and (7.31), we obtain (7.33). □

Now, a Hermitian inner product on $T_o(T(\Gamma))$ is induced by h under the identification $T_o(T(\Gamma)) \cong HB(H, \Gamma)$.

Next, we define a Hermitian inner product on the tangent space $T_p(T(\Gamma))$ of $T(\Gamma)$ at an arbitrary point $p = [w^\nu]$ as follows: using the identification $T_p(T(\Gamma)) \cong HB(H, \Gamma^\nu)$ (see Remark in §2.3), an inner product on $T_p(T(\Gamma))$ is induced by the inner product h^ν on $B(H, \Gamma^\nu)$ which is given as (7.31). Actually, the inner product of elements $H^\nu \circ L^\nu [\mu_1]$ and $H^\nu \circ L^\nu [\mu_2]$ in $HB(H, \Gamma^\nu)$, which are also considered as elements in $T_p(T(\Gamma))$, is given by

$$h^\nu(H^\nu \circ L^\nu [\mu_1], H^\nu \circ L^\nu [\mu_2]), \quad \mu_1, \mu_2 \in B(H, \Gamma).$$

In this way, we have defined the Hermitian inner product on each tangent space $T_p(T(\Gamma))$ of $T(\Gamma)$.

Now, we study the dependence of the inner product with respect to p. For this purpose, it is sufficient to consider dependence in a neighborhood of the base point in $T(\Gamma)$.

Take a basis $\{ \varphi_j \}_{j=1}^{3g-3}$ for $A_2(H, \Gamma)$, and set

$$\nu_j = \mu[\varphi_j], \quad j = 1, \cdots, 3g-3.$$

Then $\{ \nu_j \}_{j=1}^{3g-3}$ is a basis for $HB(H, \Gamma)$. Put

$$\nu(t) = \sum_{j=1}^{3g-3} t^j \nu_j, \quad t = (t^1, \cdots, t^{3g-3}) \in D,$$

where D is an open neighborhood of the origin in $\mathbf{C}^{3g-3}$.

Choosing a sufficiently small D, we may assume by Theorem 6.8 that the mapping $\Phi_o : D \to T_B(\Gamma)$ defined by $\Phi_o(t) = \Phi(\nu(t))$ is a biholomorphic mapping of D onto an open neighborhood of the base point in $T_B(\Gamma)$. We set

$$\nu_j(t) = L^{\nu(t)}[\nu_j], \quad t \in D.$$

Proposition 7.8 shows that $\{ H^{\nu(t)}[\nu_j(t)] \}_{j=1}^{3g-3}$ is a basis of $HB(H, \Gamma^{\nu(t)})$ for every $t \in D$. Thus the inner product $h_{j\bar{k}}(t)$ of tangent vectors $\partial/\partial t_j, \partial/\partial t_k \in T_{p(t)}(T(\Gamma))$ with $p(t) = [w^{\nu(t)}]$ is given by

$$h_{j\bar{k}}(t) = h^{\nu(t)}(H^{\nu(t)}[\nu_j(t)], H^{\nu(t)}[\nu_k(t)]) = (\nu_j(t), \varphi[\nu_k(t)])_{R(t)}, \tag{7.34}$$

where $R(t) = H/\Gamma^{\nu(t)}$. Then we have the following theorem.

Theorem 7.13. *Under the above circumstances, each $h_{j\bar{k}}$ is of class C^∞ on D.*

A proof of this assertion is given in Remark 1 in §3.2.

The inner product on the tangent bundle of $T(\Gamma)$ is called the *Weil-Petersson metric* on $T(\Gamma)$, and is denoted by h_{WP}. Locally, it is written as

$$ds_{WP}{}^2 = 2 \sum_{j,k=1}^{3g-3} h_{j\bar{k}}(t)\, dt^j\, \overline{dt^k}. \tag{7.35}$$

By the definition of the Weil-Petersson metric, we can show the following assertion.

Theorem 7.14. *The Weil-Petersson metric h_{WP} on $T(\Gamma)$ is invariant under the action of the Teichmüller modular group $Mod(\Gamma)$.*

7.3.2. Kählerity of the Weil-Petersson Metric

Let g_{WP} be the Riemannian metric on $T(\Gamma)$ induced by the Weil-Petersson metric h_{WP}. The real tangent space of $T(\Gamma)$ at a point p is identified with the holomorphic tangent space $T_p(T(\Gamma))$ as follows: taking local coordinates $z_j = x_j + iy_j$ $(j = 1, \ldots, 3g-3)$ around p, the identification is given by the real isomorphism sending $(\partial/\partial x_j)_p, (\partial/\partial y_j)_p$ to $(\partial/\partial z_j)_p, i(\partial/\partial z_j)_p$, respectively, for every j. For details, see Griffiths and Harris [A-39], pp.16-17. Then g_{WP} is written in the form

$$g_{WP}(X, Y) = 2\,\mathrm{Re}\, h_{WP}(X, Y).$$

Here, X and Y on the left hand side are real tangent vectors, and those on the right hand side are regarded as holomorphic tangent vectors under the identification given before. We call g_{WP} the *Weil-Petersson Riemannian metric* on $T(\Gamma)$.

The *fundamental 2-form*, or the *Weil-Petersson form* ω_{WP} of the Weil-Petersson metric h_{WP} is defined by

$$\omega_{WP}(X,Y) = g_{WP}(iX,Y), \quad X,Y \in T_p(T(\Gamma)).$$

Here, iX means the real tangent vector corresponding to the holomorphic tangent vector iX under the preceding identification. Namely, if J denotes the almost complex structure on $T(\Gamma)$ which corresponds to the complex multiplication by i, then iX means JX.

This ω_{WP} is also written as

$$\omega_{WP}(X,Y) = -2\,\mathrm{Im}\, h_{WP}(X,Y).$$

Locally, ω_{WP} is represented by

$$\omega_{WP} = i\sum_{j,k=1}^{3g-3} h_{j\bar{k}}(t)\, dt^j \wedge \overline{dt^k}. \tag{7.36}$$

Note that ω_{WP} is a positive $(1,1)$-form on $T(\Gamma)$.

We say that h_{WP} is a *Kähler metric* if the Weil-Petersson form ω_{WP} is d-closed, i.e., $d\omega_{WP} = 0$, which is equivalent to the condition that

$$\frac{\partial h_{j\bar{k}}}{\partial t^\ell} = \frac{\partial h_{\ell\bar{k}}}{\partial t^j}, \quad j,k,\ell = 1,\ldots,3g-3 \tag{7.37}$$

on D, where $h_{j\bar{k}}$ are as in (7.32). This condition means that h_{WP} osculates to order two to the Euclidean metric on $\mathbf{C}^{3g-3}$ at every point $p \in T(\Gamma)$, that is, we can find local coordinates $z^1,\ldots,z^{3g-3}$ around p for which

$$ds_{WP}{}^2 = \sum_{j,k=1}^{3g-3} (\delta_{jk} + a_{jk}(z))\, dz^j\, d\overline{z^k},$$

where all partial derivatives of a_{jk} of order less than two vanish at p.

Now, we have the following theorem due to Ahlfors.

Theorem 7.15. (Ahlfors) *The Weil-Petersson metric is Kählerian.*

Proof. We follow Ahlfors [7]. By a translation of the base point, it is sufficient to prove formula (7.37) at the base point.

We put $f^t = w^{\nu(t)}$, $\Gamma^{\nu(t)} = w^{\nu(t)}\Gamma(w^{\nu(t)})^{-1}$, $R(t) = H/\Gamma^{\nu(t)}$, and $F(t) = f^t(F)$. We set

$$K(z,\zeta) = \frac{1}{(z-\zeta)^2}.$$

We also set

$$K_t(z,\zeta) = \frac{(f^t)_z(z)\cdot(f^t)_\zeta(\zeta)}{(f^t(z)-f^t(\zeta))^2}.$$

Since $f^t(\bar{z}) = \overline{f^t(z)}$ for all $z \in \mathbf{C}$, it is easy to see that

$$\overline{K_t(z,\zeta)} = K_t(\bar{z},\bar{\zeta}),$$
$$\overline{K_t(z,\bar{\zeta})} = K_t(\bar{z},\zeta) = K_t(\zeta,\bar{z})$$

for all $z, \zeta \in H \cup H^*$ and $t \in D$.

Now, from (7.10) and (7.16), we see that $h_{j\bar{k}}$ in (7.34) is written in the form

$$h_{j\bar{k}}(t) = \iint_{F(t)} \left\{ \frac{12}{\pi} \iint_H K(z,\bar{\zeta})^2 \, \overline{\nu_k(t)(\zeta)} \, d\xi d\eta \right\} \nu_j(t)(z) \, dx dy.$$

In this integral, substitute $f^t(z), f^t(\zeta)$ for z, ζ, respectively. Then, recalling $\nu_j(t) = L^{\nu(t)}[\nu_j]$ and (7.26), we get

$$h_{j\bar{k}}(t) = \iint_F \left\{ \frac{12}{\pi} \iint_H K_t(z,\bar{\zeta})^2 \, \overline{\nu_k(\zeta)} \, d\xi d\eta \right\} \nu_j(z) \, dx dy.$$

Thus from $h_{j\bar{k}} = \overline{h_{k\bar{j}}}$, we obtain

$$h_{j\bar{k}}(t) = \iint_F \left\{ \frac{12}{\pi} \iint_H K_t(\zeta,\bar{z})^2 \, \overline{\nu_k(z)} \, \nu_j(\zeta) \, d\xi d\eta \right\} dx dy. \tag{7.38}$$

Since $f^t(z)$ is of class C^∞ for (z,t) in $H \times D$ (cf. Remark 2 at the end of this subsection), a simple computation leads to

$$K_t(\zeta,\bar{z}) = \frac{\partial^2}{\partial\zeta\partial\bar{z}} \log(f^t(\zeta) - \overline{f^t(z)}), \quad \zeta, z \in H.$$

Using $\overline{f^t(z)} = f^t(\bar{z})$ and the Corollary to Theorem 4.37, we obtain

$$\begin{aligned}
&\frac{\partial}{\partial t^\ell} \log(f^t(\zeta) - \overline{f^t(z)}) \\
&= -\frac{1}{\pi} \iint_H \left[\frac{1}{(f^t(w) - f^t(\zeta))(f^t(w) - \overline{f^t(z)})} \right. \\
&\qquad \left. + \frac{1}{f^t(w)} - \frac{1}{f^t(w) - 1} \right] \left((f^t)_w(w)\right)^2 \nu_\ell(w) \, du dv \\
&= -\frac{1}{\pi} \iint_H \left[\frac{1}{(w - f^t(\zeta))(w - \overline{f^t(z)})} + \frac{1}{w} - \frac{1}{w-1} \right] L^{\nu(t)}[\nu_\ell](w) \, du dv.
\end{aligned}$$

Since this integral can be differentiated with respect to $\bar{z}$ under the integral sign, we have

$$\begin{aligned}
&\frac{\partial^2}{\partial\bar{z}\partial t^\ell} \log(f^t(\zeta) - \overline{f^t(z)}) \\
&= -\frac{1}{\pi} \iint_H \frac{\overline{(f^t)_z(z)}}{(w - f^t(\zeta))(w - \overline{f^t(z)})^2} L^{\nu(t)}[\nu_\ell](w) \, du dv.
\end{aligned}$$

By an argument similar to that in the proof of Proposition 4.23, we see that this integral can be differentiated with respect to ζ under the integral sign. Thus, it follows that

$$\frac{\partial K_t(\zeta,\bar{z})}{\partial t^\ell} = -\frac{1}{\pi} \iint_H \frac{(f^t)_\zeta(\zeta)\overline{(f^t)_z(z)}}{(w - f^t(\zeta))^2 (w - \overline{f^t(z)})^2} L^{\nu(t)}[\nu_\ell](w) \, du dv. \tag{7.39}$$

Here, this integral is defined as the Cauchy principal value.

From (7.38), (7.39), and the definition of $K_t(z,\zeta)$, we have

$$\begin{aligned}
&\frac{\partial h_{j\bar{k}}}{\partial t^\ell}(t) \\
&= \frac{24}{\pi}\iint_F \left\{\iint_H \frac{\partial K_t(\zeta,\bar{z})}{\partial t^\ell} K_t(\zeta,\bar{z})\,\overline{\nu_k(z)}\,\nu_j(\zeta)\,d\xi d\eta\right\} dxdy \\
&= -\frac{24}{\pi^2}\iint_{F(t)} \left\{\iint_H K(\zeta,\bar{z})T_\ell(z,\zeta)\,\overline{L^{\nu(t)}[\nu_k](z)}\,L^{\nu(t)}[\nu_j](\zeta)\,d\xi d\eta\right\} dxdy,
\end{aligned} \tag{7.40}$$

where

$$T_\ell(z,\zeta) = \iint_H K(w,\bar{z})K(w,\zeta)L^{\nu(t)}[\nu_\ell](w)\,dudv.$$

This integral is also defined as the Cauchy principal value. In this step, we differentiated (7.38) under the integral sign. To justify this procedure, we need to show that the integral in (7.40) converges absolutely and uniformly with respect to t. More precisely, putting $H(r) = \{\, z \in H \mid |z| \geqq r \,\}$ for any positive number r, we shall prove the integral

$$\begin{aligned}
&I(r) \\
&= \iint_{F(t)} \left\{\iint_{H(r)} \left|K(\zeta,\bar{z})\,T_\ell(z,\zeta)\,\overline{L^{\nu(t)}[\nu_k](z)}\,L^{\nu(t)}[\nu_j](\zeta)\right|\,d\xi d\eta\right\} dxdy
\end{aligned}$$

converges uniformly to 0 as $r \to \infty$ for all t. By Lemma 4.21, the above T_ℓ is estimated as

$$\begin{aligned}
\iint_H |T_\ell(z,\zeta)|^2\,d\xi d\eta &\leqq \pi^2 \iint_H |K(w,\bar{z})L^{\nu(t)}[\nu_\ell](w)|^2\,d\xi d\eta \\
&\leqq C^2\pi^2 \iint_H |K(w,\bar{z})|^2\,d\xi d\eta,
\end{aligned}$$

where C is a constant such that $\|L^{\nu(t)}[\nu_\ell]\|_\infty \leqq C$ for any t and ℓ. Thus we have

$$\begin{aligned}
&\iint_{H(r)} |K(\zeta,\bar{z})T_\ell(z,\zeta)|\,d\xi d\eta \\
&\leqq C\pi \left(\iint_{H(r)} |K(\zeta,\bar{z})|^2\,d\xi d\eta\right)^{1/2} \cdot \left(\iint_H |K(w,\bar{z})|^2\,dudv\right)^{1/2}.
\end{aligned}$$

If z stays in a compact set of H, the first factor on the right hand side tends uniformly to 0 as $r \to \infty$, and the second factor remains bounded, which leads to the desired conclusion.

Finally, by using again the fact that the linear operator T in Lemma 4.21 is isometric on $L^2(\mathbf{C})$, we can show that

$$\begin{aligned}
&\frac{\partial h_{j\bar{k}}}{\partial t^{\ell}}(t) \\
&= -\frac{24}{\pi^2}\iint_{F(t)}\left\{\iint_H K(\zeta,\bar{z})T_\ell(z,\zeta)L^{\nu(t)}[\nu_j](\zeta)\,d\xi d\eta\right\}\overline{L^{\nu(t)}[\nu_k](z)}\,dxdy \\
&= -\frac{24}{\pi^2}\iint_{F(t)}\left\{\iint_H K(w,\bar{z})T_j(z,w)L^{\nu(t)}[\nu_\ell](w)\,dudv\right\}\overline{L^{\nu(t)}[\nu_k](z)}\,dxdy \\
&= \frac{\partial h_{\ell\bar{k}}}{\partial t^j}(t).
\end{aligned}$$

This completes the proof of Theorem 7.15. □

Remark 1. From the proof of Theorem 7.15, we see that all $\partial h_{j\bar{k}}/\partial t^\ell, \partial h_{j\bar{k}}/\partial \overline{t^\ell}$ are continuous, that is, all $h_{j\bar{k}}$ are of class C^1. Repeating the same argument as that in this proof, we can show that each $h_{j\bar{k}}(t)$ is a C^∞-function of t (see Ahlfors [7], p. 172).

Furthermore, each $h_{j\bar{k}}$ is real-analytic. In fact, from Wolpert's formula (Theorem 8.6), the fundamental form ω_{WP} of the Weil-Petersson metric is written in the form

$$\omega_{WP} = \sum_{j=1}^{3g-3} d\tau_j \wedge d\ell_j.$$

By Lemma 3.9, all ℓ_j and τ_j are real-analytic functions on the Fricke space F_g. Since Fricke coordinates are real-analytic on the Teichmüller space $T(\Gamma)$, all ℓ_j and τ_j are real-analytic on $T(\Gamma)$. Thus we see that each $h_{j\bar{k}}(t)$ is a real-analytic function of t.

Remark 2. We give a proof of the assertion that the quasiconformal mapping $f^t(z)$ in the proof of Theorem 7.15 is of class C^∞ with respect to $(z,t) \in H \times D$.

For any $t \in D$, we define an element $\varphi \in A_2(H^*, \Gamma)$ by

$$\varphi(z) = -\frac{1}{2}\sum_{j=1}^{3g-3} t^j\, \overline{\varphi_j(\bar{z})}, \quad z \in H^*.$$

Then, φ induces a quasiconformal mapping F_t of H, which is defined by (6.7) of Chapter 6, and has a Beltrami coefficient $\nu(t)$. Since $F_t(z)$ is a real-analytic mapping of $(z,t) \in H \times D$, the inverse mapping theorem shows that its inverse mapping $F_t^{-1}(w)$ is also a real-analytic mapping of (w,t).

Next, for all $t, s \in D$, we set

$$\mu_{t,s}(z) = \begin{cases} \nu(t)(z), & z \in H \\ 0, & z \in \mathbf{R} \\ \overline{\nu(\bar{s})(\bar{z})}, & z \in H^*. \end{cases}$$

Let $f_{t,s}$ be the canonical $\mu_{t,s}$-qc mapping of $\hat{\mathbf{C}}$. Note that $f^t = f_{t,\bar{t}}$. By Proposition 4.36, it follows that $f_{t,s}(z)$ is a continuous mapping of (z,t,s). Putting $h_{t,s} = f_{t,s} \circ F_t^{-1}$, we see that $h_{t,s}(w)$ is a continuous mapping of (w,t,s). If we fix

t and s, then $h_{t,s}$ is a conformal mapping of $F_t(H)$, because $f_{t,s}$ and F_t have the same Beltrami coefficient $\nu(t)$ on H. Thus, applying Cauchy's integral formula to $h_{t,s}$, we see that all derivatives of $h_{t,s}(w)$ with respect to w are continuous functions of (w,t,s). Hence, all partial derivatives of $f_{t,s}(z)$ with respect to z and $\bar{z}$ are continuous functions of $(z,t,s) \in H \times D^2$.

On the other hand, the Corollary to Theorem 4.37 implies that for a fixed z, $f_{t,s}(z)$ is a holomorphic function of (t,s). Hence, applying Cauchy's integral formula to $f_{t,s}(z)$, we conclude that all partial derivatives of $f_{t,s}(z)$ with respect to t, s, z, and $\bar{z}$ are continuous functions of $(z,t,s) \in H \times D^2$. Therefore, $f_{t,s}(z)$ is a C^∞-mapping of $(z,t,s) \in H \times D^2$. In particular, $f^t(z) = f_{t,\bar{t}}(z)$ is a C^∞-mapping of $(z,t) \in H \times D$.

7.3.3. Alternative Proof of Kählerity of the Weil-Petersson Metric

We shall give another proof of Theorem 7.15, which is also due to Ahlfors [6]. Here, we use the fact that the first variation of the area element induced by the hyperbolic metric vanishes (Lemma 7.16 below), which is interesting by itself.

By a translation of the base point, it is sufficient to prove the relations in (7.37) at the base point. We use the notation in §3.2.

We set

$$\psi_j(t)(z) = \varphi[\nu_j(t)](f^t(z))\{(f^t)_z(z)\}^2, \quad t \in D,\ z \in H.$$

Then from (7.16), we have

$$\psi_j(t)(z) = \frac{12}{\pi} \iint_H K_t(z,\bar{\zeta})^2\, \overline{\nu_j(\zeta)}\, d\xi d\eta.$$

By an argument similar to that in the proof of Theorem 7.15, we see that $\psi_j(t)(z)$ is a C^∞-function of (z,t). From formula (7.38), we get

$$h_{j\bar{k}}(t) = \iint_F \nu_j(z)\psi_k(t)(z)\, dxdy. \tag{7.41}$$

Further, setting

$$\rho_t(z) = \left(|(f^t)_z(z)|^2 - |(f^t)_{\bar{z}}(z)|^2\right) \cdot \lambda_H(f^t(z))^2, \tag{7.42}$$

we see that formula (7.34) is rewritten in the form

$$h_{j\bar{k}}(t) = \iint_F \overline{\psi_j(t)(z)}\, \psi_k(t)(z)\, \frac{(1 - |\nu(t)(z)|^2)^2}{\rho_t(z)}\, dxdy. \tag{7.43}$$

We can differentiate (7.43) with respect to t_ℓ under the integral sign, because F is a relatively compact set in H, and the integrand is a C^∞-function of (z,t). Hence, we obtain

$$\begin{aligned}
&\frac{\partial h_{j\bar{k}}}{\partial t^{\ell}}(0) \\
&= \iint_F \left\{ \overline{\frac{\partial \psi_j(t)}{\partial t^{\ell}}}\bigg|_{t=0}(z)\varphi_k(z) + \overline{\varphi_j(z)}\,\frac{\partial \psi_k(t)}{\partial t^{\ell}}\bigg|_{t=0}(z) \right\} \lambda_H(z)^{-2}\,dxdy \qquad (7.44)\\
&\quad + \iint_F \overline{\varphi_j(z)}\,\varphi_k(z) \left\{ \frac{\partial}{\partial t^{\ell}} \left(\frac{1}{\rho_t(z)} \right) \bigg|_{t=0} \right\} dxdy.
\end{aligned}$$

Here, using Lemma 7.16 below, we see that the second term on the right hand side of (7.44) vanishes.

Next, differentiating (7.41) with respect to t^{ℓ}, we also have

$$\frac{\partial h_{j\bar{k}}}{\partial t^{\ell}}(0) = \iint_F \overline{\varphi_j(z)}\,\frac{\partial \psi_k(t)}{\partial t^{\ell}}\bigg|_{t=0}(z)\lambda_H(z)^{-2}\,dxdy. \qquad (7.45)$$

By (7.44) and (7.45), we obtain

$$\iint_F \overline{\frac{\partial \psi_j(t)}{\partial t^{\ell}}}\bigg|_{t=0}(z)\varphi_k(z)\lambda_H(z)^{-2}\,dxdy = 0. \qquad (7.46)$$

Similarly, taking the derivative of $h_{j\bar{k}}$ with respect to $\overline{t^{\ell}}$, we have

$$\iint_F \overline{\frac{\partial \psi_j(t)}{\partial \overline{t^{\ell}}}}\bigg|_{t=0}(z)\varphi_k(z)\lambda_H(z)^{-2}\,dxdy = 0. \qquad (7.47)$$

Therefore, from (7.44), (7.46), and (7.47), we get

$$\frac{\partial h_{j\bar{k}}}{\partial t^{\ell}}(0) = 0.$$

Since j, k, and ℓ are arbitrary, the relations in (7.37) hold at $t = 0$, which proves that h_{WP} is Kählerian. □

Lemma 7.16. *The function ρ_t defined in (7.42) satisfies*

$$\frac{\partial \rho_t}{\partial t^j}\bigg|_{t=0} = 0, \quad j = 1, \ldots, 3g-3.$$

Proof. We set $\nu = \nu_j$ and $f^s = w^{s\nu}$. Here, s is any sufficiently small real number. By the definition, we have

$$\begin{aligned}
\rho_s(z) &= \lambda_H(f^s(z))^2 \left(|(f^s)_z(z)|^2 - |(f^s)_{\bar{z}}(z)|^2 \right) \\
&= \frac{-4}{(f^s(z) - \overline{f^s(z)})^2}(1 - |s\nu|^2)|(f^s)_z(z)|^2.
\end{aligned}$$

Thus the derivative of ρ_s in the direction ν at the origin is written as

$$\begin{aligned}
\dot{\rho}(z) &= \left.\frac{\partial \rho_s}{\partial s}\right|_{s=0}(z) \\
&= \frac{8}{(z-\bar{z})^3}(\dot{f}(z)-\overline{\dot{f}(z)}) - \frac{4}{(z-\bar{z})^2}(\dot{f}_z(z)+\overline{\dot{f}_z(z)}) \\
&= -8\mathrm{Re}\left\{\frac{\dot{f}_z(z)}{(z-\bar{z})^2} - \frac{2\dot{f}(z)}{(z-\bar{z})^3}\right\} \\
&= -8\mathrm{Re}\frac{\partial}{\partial z}\left\{\frac{\dot{f}(z)}{(z-\bar{z})^2}\right\},
\end{aligned} \tag{7.48}$$

where $\dot{f}(z) = (\partial f^s(z)/\partial s)|_{s=0}$. Thus we need to obtain more information on $\dot{f}(z)$.

First, we see the followings:

(i) $\dot{f}_{\bar{z}} = \nu = \lambda_H^{-2}\bar{\varphi}$ on H, where $\varphi = \varphi_j$.
(ii) $\dot{f}$ is continuous on $\mathbf{C}$, $\dot{f}(0) = \dot{f}(1) = 0$, and $\lim_{z\to\infty}\dot{f}(z)/z^2 = 0$.
(iii) $\dot{f}$ takes real values on the real axis $\mathbf{R}$.

For the proof of these assertions, we refer to Lemma 4.20 and Theorem 4.37. See also Kra [A-58], Lemma 1.4 in Chapter IV, p.136.

Now, we determine $\dot{f}$ as follows. Since φ is an element in $A_2(H,\Gamma)$, it follows that $\lambda_H^{-2}\bar{\varphi}$ is bounded on H. Thus, integrating φ three times, we get a holomorphic function ψ on H satisfying the following four conditions:

(a) $\psi''' = \varphi$ on H.
(b) ψ is extended continuously to the real axis $\mathbf{R}$, and satisfies that $\psi(0) = \psi(1) = 0$.
(c) $(\mathrm{Im}\, z)\psi'(z) \to 0$ and $(\mathrm{Im}\, z)^2\psi''(z) \to 0$ as $\mathrm{Im}\, z \to 0$.
(d) $\psi(z)/z^2 \to 0$, $(\mathrm{Im}\, z)\psi'(z)/z \to 0$, and $(\mathrm{Im}\, z)^2\psi''(z) \to 0$ as $z \to \infty$.

Solving the differential equation in (i), i.e.,

$$\dot{f}_{\bar{z}}(z) = -\frac{(z-\bar{z})^2}{4}\overline{\varphi(z)},$$

we obtain

$$\dot{f}(z) = -\frac{(z-\bar{z})^2}{4}\overline{\psi''(z)} - \frac{z-\bar{z}}{2}\overline{\psi'(z)} - \frac{\overline{\psi(z)}}{2} + F(z),$$

where F is a holomorphic function on H.

Let us determine F. From (ii), (b), and (c), we see that F is extended continuously to $\mathbf{R}$, and $\dot{f} = -\bar{\psi}/2 + F$ on $\mathbf{R}$. Further, by (iii) it follows that $-\bar{\psi}/2 + F = -\psi/2 + \bar{F}$ on $\mathbf{R}$. Thus $h = F + \psi/2$ is a holomorphic function on H, which has a continuous extension such that h is real-valued on $\mathbf{R}$. Hence, Schwarz' reflection principle shows that h is extended holomorphically to $\mathbf{C}$. Then from (ii) and (d), this h should be a polynomial of degree at most 1. Consequently, by (ii) and (b), we conclude that $h = 0$. This implies that

$$\dot{f}(z) = -\frac{(z-\bar{z})^2}{4}\overline{\psi''(z)} - \frac{z-\bar{z}}{2}\overline{\psi'(z)} - \frac{\overline{\psi(z)}}{2} - \frac{\psi(z)}{2}.$$

Hence, it follows that

$$-8\frac{\partial}{\partial z}\left\{\frac{\dot{f}(z)}{(z-\bar{z})^2}\right\} = 4\frac{\psi'(z) - \overline{\psi'(z)}}{(z-\bar{z})^2} - 8\frac{\psi(z) + \overline{\psi(z)}}{(z-\bar{z})^3}$$

is purely imaginary. Therefore, from (7.48) we conclude that $\dot{\rho} = 0$. □

7.3.4. Curvatures of the Weil-Petersson Metric

We show that the Ricci curvatures, the scalar curvature, and the holomorphic sectional curvatures of the Weil-Petersson metric are negative.

First, recall some notions of complex differential geometry on Kähler manifolds (see Kobayashi and Nomizu [A-55], Chapter IX).

Let $ds^2 = 2\sum_{j,k=1}^N h_{j\bar{k}}\, dt^j d\bar{t}^k$ be a Kähler metric on a complex manifold M of dimension N. The Christoffel symbols $\Gamma^j_{k\ell}$ of the metric connection associated with the Kähler metric ds^2 are given by

$$\Gamma^j_{k\ell} = \sum_{m=1}^{N} h^{j\bar{m}} \frac{\partial h_{k\bar{m}}}{\partial t^\ell},$$

where $(h^{j\bar{k}})$ is the inverse matrix to $(h_{j\bar{k}})$. Then the *curvature tensors* $R^j_{k\ell\bar{m}}$ are given by

$$R^j_{k\ell\bar{m}} = -\frac{\partial \Gamma^j_{k\ell}}{\partial \bar{t}^m}.$$

The *Ricci curvature tensors* $R_{j\bar{k}}$ are defined by

$$R_{j\bar{k}} = \sum_{\ell=1}^{N} R^\ell_{j\ell\bar{k}}.$$

Further, the *Riemannian curvature tensors* $R_{j\bar{k}\ell\bar{m}}$ are given by

$$R_{j\bar{k}\ell\bar{m}} = -\sum_{n=1}^{N} h_{j\bar{n}} \overline{R^n_{km\bar{\ell}}}.$$

Let $V = \sum_{j=1}^N v^j\, (\partial/\partial t^j)_p$ be an element of the holomorphic tangent space $T_p(M)$ of M at p such that $\langle V, V\rangle_p = 1$, i.e., V is a holomorphic tangent vector at p with unit norm. The *Ricci curvature* $R_p(V)$ at p in the direction V is defined by

$$R_p(V) = \sum_{j,k=1}^{N} R_{j\bar{k}}\, v^j \overline{v^k}. \tag{7.49}$$

It is said that the Ricci curvature of ds^2 is *negative* at p if $R_p(V) < 0$ for any direction V. The *holomorphic sectional curvature* $K_p(V)$ at p with respect to V is defined by

$$K_p(V) = - \sum_{j,k,\ell,m=1}^{N} R_{j\bar{k}\ell\bar{m}}\, v^j\, \overline{v^k}\, v^\ell\, \overline{v^m}. \tag{7.50}$$

We say that the holomorphic sectional curvature of ds^2 is *negative* at p if $K_p(V) < 0$ for any direction V. The *scalar curvature* R is given by

$$R = \sum_{j=1}^{N} R_{j\bar{j}}. \tag{7.51}$$

Theorem 7.17. *The Ricci curvatures, the scalar curvature, and the holomorphic sectional curvatures of the Weil-Petersson metric are negative on $T(\Gamma)$.*

Proof. It is sufficient to verify this assertions at the base point of $T(\Gamma)$.

We use the same notation as in §§3.1, 2, and 3. For simplicity, we set $N = 3g-3$. For a complex variable $z = x + iy$, we put $d\sigma(z) = dxdy$. We write $d\sigma(z_1, \ldots, z_n) = d\sigma(z_1) \cdots d\sigma(z_n)$ for complex variables $z_1, \ldots, z_n$.

Take an orthonormal basis $\{\varphi_j\}_{j=1}^N$ for $A_2(\Gamma, H)$ with respect to the Petersson scalar product. Denote by $ds^2_{WP} = 2\sum_{j,k=1}^N h_{j\bar{k}}(t)dt^j d\bar{t}^k$ the Weil-Petersson metric defined in (7.35). Then, as was seen in the preceding subsection, it follows that

$$h_{j\bar{k}}(0) = \delta_{jk}, \quad \frac{\partial h_{j\bar{k}}}{\partial t^\ell}(0) = 0, \quad j, k, \ell = 1, \ldots, N. \tag{7.52}$$

Hence, from the definitions and (7.52), we obtain

$$\begin{aligned} R^j_{k\ell\bar{m}} &= -\frac{\partial^2 h_{k\bar{j}}}{\partial t^\ell \partial \bar{t}^m}(0), \\ R_{j\bar{k}\ell\bar{m}} &= \frac{\partial^2 h_{j\bar{k}}}{\partial t^\ell \partial \bar{t}^m}(0), \\ R_{j\bar{k}} &= -\sum_{\ell=1}^{N} \frac{\partial^2 h_{j\bar{\ell}}}{\partial t^\ell \partial \bar{t}^k}(0) \end{aligned} \tag{7.53}$$

at $t = 0$. Therefore, we only need to calculate $\partial^2 h_{j\bar{k}}/\partial t^\ell \partial \bar{t}^m$ at $t = 0$.

From (7.40), we have

$$\frac{\partial h_{j\bar{k}}}{\partial t^\ell}(t) = \frac{24}{\pi} \int \cdots \int_{F\times H} \frac{\partial K_t(z_1, \overline{z_2})}{\partial t^\ell} K_t(z_1, \overline{z_2})\, \nu_j(z_1)\, \overline{\nu_k(z_2)}\, d\sigma(z_2, z_1),$$

where F is a relatively compact fundamental domain for Γ in H, and (z_2, z_1) runs over $F \times H$. Furthermore, from (7.39) and (7.26), we obtain

$$\frac{\partial K_t(z_1, \overline{z_2})}{\partial t^\ell} = -\frac{1}{\pi} \iint_H K_t(w_1, z_1)\, K_t(w_1, \overline{z_2})\, \nu_\ell(w_1)\, d\sigma(w_1). \tag{7.54}$$

By a similar calculation, we have

$$\frac{\partial K_t(z_1,\overline{z_2})}{\partial \bar{t}^m} = -\frac{1}{\pi}\iint_H K_t(\overline{w_2},z_1)\,K_t(\overline{w_2},\overline{z_2})\,\overline{\nu_\ell(w_2)}\,d\sigma(w_2). \tag{7.55}$$

Thus we get

$$\frac{\partial h_{j\bar{k}}}{\partial t^\ell}(t) = -\frac{24}{\pi^2}\int\cdots\int_{F\times H^2} A\;\nu_j(z_1)\;\overline{\nu_k(z_2)}\;\nu_\ell(w_1)\,d\sigma(z_2,z_1,w_1), \tag{7.56}$$

where $A = K_t(w_1,z_1)\,K_t(w_1,\overline{z_2})\,K_t(z_1,\overline{z_2})$.

Hence, formulas (7.54), (7.55), and (7.56) yield

$$\frac{\partial^2 h_{j\bar{k}}}{\partial t^\ell \partial \bar{t}^m}(0) = \frac{24}{\pi^3}\int\cdots\int_{F\times H^3} B\;\nu_j(z_1)\;\overline{\nu_k(z_2)}\;\nu_\ell(w_1)\;\overline{\nu_m(w_2)}\,d\sigma(z_2,z_1,w_1,w_2),$$

where

$$\begin{aligned} B =& K(\overline{w_2},z_1)\,K(\overline{w_2},w_1)\,K(w_1,\overline{z_2})\,K(z_1,\overline{z_2})\\ &+K(w_1,z_1)\,K(\overline{w_2},w_1)\,K(\overline{w_2},\overline{z_2})\,K(z_1,\overline{z_2})\\ &+K(w_1,z_1)\,K(w_1,\overline{z_2})\,K(\overline{w_2},z_1)\,K(\overline{w_2},\overline{z_2}). \end{aligned}$$

Thus we have

$$\begin{aligned} \frac{\partial^2 h_{j\bar{k}}}{\partial t^\ell \partial \bar{t}^m}(0) &= \overline{\left(\frac{\partial^2 h_{k\bar{j}}}{\partial t^m \partial \bar{t}^\ell}(0)\right)}\\ &= \frac{24}{\pi^3}\int\cdots\int_{F\times H^3} B\;\nu_j(z_1)\;\overline{\nu_k(z_2)}\;\nu_\ell(w_1)\;\overline{\nu_m(w_2)}\,d\sigma(z_1,z_2,w_1,w_2). \end{aligned} \tag{7.57}$$

Now, using Green's formula and the residue theorem, we obtain

$$K(u,\bar{v}) = -\frac{1}{\pi}\iint_H K(\bar{z},u)\,K(z,\bar{v})\,d\sigma(z), \quad u,v\in H. \tag{7.58}$$

We apply this formula to $K(\overline{w_2},z_1)$, $K(w_1,\overline{z_2})$ in the first term, to $K(\overline{w_2},w_1)$, $K(z_1,\overline{z_2})$ in the second term, and to $K(w_1,\overline{z_2})$, $K(\overline{w_2},z_1)$ in the third term of B on the right hand side of (7.57). Then we find

$$\begin{aligned} &\frac{\partial^2 h_{j\bar{k}}}{\partial t^\ell \partial \bar{t}^m}(0)\\ &= \frac{24}{\pi^5}\int\cdots\int_{F\times H^5} C\;\nu_j(z_1)\;\overline{\nu_k(z_2)}\;\nu_\ell(w_1)\;\overline{\nu_m(w_2)}\,d\sigma(z_1,z_2,w_1,w_2,z,\zeta), \end{aligned} \tag{7.59}$$

where

$$\begin{aligned} C =& K(z,\overline{w_2})\,K(\bar{z},z_1)\,K(\overline{w_2},w_1)\,K(\bar{\zeta},w_1)\,K(\zeta,\overline{z_2})\,K(z_1,\overline{z_2})\\ &+K(w_1,z_1)\,K(z,\overline{w_2})\,K(\bar{z},w_1)\,K(\overline{w_2},\overline{z_2})\,K(\bar{\zeta},z_1)\,K(\zeta,\overline{z_2})\\ &+K(w_1,z_1)\,K(\bar{z},w_1)\,K(z,\overline{z_2})\,K(\zeta,\overline{w_2})\,K(\bar{\zeta},z_1)\,K(\overline{w_2},\overline{z_2}). \end{aligned}$$

By the transformation rules for $d\sigma(z)$, $K(z,\zeta)$, $K(z,\bar{\zeta})$, and Beltrami differentials with respect to real Möbius transformations, we see that the integration variables in (7.59) can be interchanged in an arbitrary manner. Thus (7.59) is rewritten as

$$\begin{aligned}&\frac{\partial^2 h_{j\bar{k}}}{\partial t^\ell \partial \bar{t}^m}(0)\\&= \frac{24}{\pi^5}\int\cdots\int_{F\times H^5} C\ \nu_j(z_1)\ \overline{\nu_k(z_2)}\ \nu_\ell(w_1)\ \overline{\nu_m(w_2)}\, d\sigma(\zeta, z, z_1, z_2, w_1, w_2),\end{aligned} \tag{7.60}$$

where ζ ranges over F.

We introduce the notation

$$\begin{aligned}L_{jk}(\bar{\zeta},\bar{z}) &= \int\cdots\int_{H^2} K(z_1,\bar{\zeta})\, K(z_1,w_1)\, K(w_1,\bar{z})\, \nu_j(z_1)\, \nu_k(w_1)\, d\sigma(z_1,w_1),\\ L_{j\bar{k}}(\zeta,\bar{z}) &= \int\cdots\int_{H^2} K(\overline{z_2},\zeta)\, K(\overline{z_2},w_2)\, K(w_2,\bar{z})\, \nu_j(z_2)\, \overline{\nu_k(w_2)}\, d\sigma(z_2,w_2).\end{aligned} \tag{7.61}$$

Then, from (7.60) we find

$$\frac{\partial^2 h_{j\bar{k}}}{\partial t^\ell \partial \bar{t}^m}(0) = \frac{24}{\pi^5}\int\cdots\int_{F\times H} D_{j\bar{k}\ell\bar{m}}\ d\sigma(\zeta, z), \tag{7.62}$$

where

$$D_{j\bar{k}\ell\bar{m}} = L_{j\bar{k}}(\zeta,\bar{z})\ \overline{L_{m\bar{\ell}}(\zeta,\bar{z})} + L_{j\ell}(\bar{\zeta},\bar{z})\ \overline{L_{km}(\bar{\zeta},\bar{z})} + L_{j\ell}(\bar{\zeta},\bar{z})\ \overline{L_{mk}(\bar{\zeta},\bar{z})}.$$

Since $L_{jk}(\bar{\zeta},\bar{z}) = L_{kj}(\bar{z},\bar{\zeta})$, and since $L_{jk}(\bar{\zeta},\, \bar{z}) = L_{jk}(\overline{\gamma(\zeta)},\, \overline{\gamma(z)})\cdot\overline{\gamma'(\zeta)}\cdot\overline{\gamma'(z)}$ for any $\gamma\in\Gamma$, we see from (7.62) that

$$\frac{\partial^2 h_{j\bar{k}}}{\partial t^\ell \partial \bar{t}^m}(0) = \frac{12}{\pi^5}\int\cdots\int_{F\times H} E_{j\bar{k}\ell\bar{m}}\ d\sigma(\zeta, z), \tag{7.63}$$

where

$$E_{j\bar{k}\ell\bar{m}} = (L_{j\ell}(\bar{\zeta},\bar{z}) + L_{\ell j}(\bar{\zeta},\bar{z}))(\overline{L_{km}(\bar{\zeta},\bar{z})} + \overline{L_{mk}(\bar{\zeta},\bar{z})}) + 2L_{j\bar{k}}(\zeta,\bar{z})\ \overline{L_{m\bar{\ell}}(\zeta,\bar{z})}.$$

Now, we show that the Ricci curvatures are negative at the base point. In fact, from (7.49), (7.53), and (7.63), for any element $V = \sum_{j=1}^N v^j\, (\partial/\partial t^j)_0 \in T_0(T(\Gamma))$ with unit norm, the Ricci curvature $R_0(V)$ at the base point in the direction V is given by

$$\begin{aligned}R_0(V) &= -\frac{12}{\pi^5}\sum_{\ell=1}^N\int\cdots\int_{F\times H}\sum_{j,k=1}^N E_{j\bar{\ell}\ell\bar{k}} v^j\, \overline{v^k}\ d\sigma(\zeta,z)\\ &= -\frac{12}{\pi^5}\sum_{\ell=1}^N\int\cdots\int_{F\times H} F_\ell\, d\sigma(\zeta,z) \leqq 0,\end{aligned} \tag{7.64}$$

where

$$F_\ell = \left| \sum_{j=1}^{N} (L_{j\ell}(\bar{\zeta}, \bar{z}) + L_{\ell j}(\bar{\zeta}, \bar{z})) v^j \right|^2 + 2 \left| \sum_{j=1}^{N} L_{j\bar{\ell}}(\zeta, \bar{z})\, v^j \right|^2.$$

If $R_0(V) = 0$, then (7.64) implies that

$$\sum_{j=1}^{N} L_{j\bar{\ell}}(\zeta, \bar{z}) v^j = 0, \quad (\zeta, z) \in F \times H,\ \ell = 1, \ldots, N. \tag{7.65}$$

Hence by (7.38), (7.58), and (7.61), and by changing variables, we find

$$\begin{aligned}
&\sum_{j=1}^{N} v^j \int \cdots \int_{F \times H^2} K(\zeta, \overline{z_2})\, K(\overline{z_2}, z_1)\, K(z_1, \bar{\zeta})\, \nu_j(z_1)\, \overline{\nu_\ell(z_2)}\, d\sigma(\zeta, z_1, z_2) \\
&= \sum_{j=1}^{N} v^j \int \cdots \int_{F \times H^2} K(z_1, \bar{\zeta})\, K(\zeta, \overline{z_2})\, K(\overline{z_2}, z_1)\, \nu_j(z_1)\, \overline{\nu_\ell(z_2)}\, d\sigma(z_1, \zeta, z_2) \\
&= -\pi \sum_{j=1}^{N} v^j \int \cdots \int_{F \times H} K(z_1, \overline{z_2})^2\, \nu_j(z_1)\, \overline{\nu_\ell(z_2)}\, d\sigma(z_1, z_2) \\
&= -\frac{1}{12} \sum_{j=1}^{N} v^j\, h_{j\bar{\ell}}(0) = 0
\end{aligned}$$

for all ℓ, which leads to $V = 0$. Thus we conclude that $R_0(V) < 0$ for all directions V. Further, by (7.51) it is clear that the scalar curvature of the Weil-Petersson metric is negative at the base point.

Finally, we show that the holomorphic sectional curvatures of the Weil-Petersson metric are negative at the base point. For any $V = \sum_{j=1}^{N} v^j\, (\partial/\partial t^j)_0 \in T_0(T(\Gamma))$ with unit norm, (7.50) implies that

$$\begin{aligned}
K_0(V) &= -\sum_{j,k,\ell,m=1}^{N} R_{j\bar{k}\ell\bar{m}}(0)\, v^j\, \overline{v^k}\, v^\ell\, \overline{v^m} \\
&= -\sum_{j,k,\ell,m=1}^{N} \frac{\partial^2 h_{j\bar{k}}}{\partial t^\ell \partial \bar{t}^m}(0)\, v^j\, \overline{v^k}\, v^\ell\, \overline{v^m} \\
&= -\frac{12}{\pi^5} \int \cdots \int_{F \times H} \sum_{j,k,\ell,m=1}^{N} E_{j\bar{k}\ell\bar{m}}\, v^j\, \overline{v^k}\, v^\ell\, \overline{v^m} d\sigma(\zeta, z) \\
&= -\frac{12}{\pi^5} \int \cdots \int_{F \times H} G\, d\sigma(\zeta, z) \le 0,
\end{aligned}$$

where

$$G = \left| \sum_{j,\ell=1}^{N} (L_{j\ell}(\bar{\zeta}, \bar{z}) + L_{\ell j}(\bar{\zeta}, \bar{z}))\, v^j\, v^\ell \right|^2 + 2 \left| \sum_{j,k=1}^{N} L_{j\bar{k}}(\zeta, \bar{z})\, v^j\, \overline{v^k} \right|^2.$$

By the same argument as in the case of the Ricci curvature, we can show that if $K_0(V) = 0$, then $V = 0$. Thus $K_0(V) < 0$, and hence the holomorphic sectional curvatures are negative at the base point. □

Remark. Wolpert [253] obtained the following estimates for the curvatures of the Weil-Petersson metric:

(i) the holomorphic sectional curvatures and Ricci curvatures are bounded above by $-1/2\pi(g-1)$, and
(ii) the scalar curvature is bounded above by $-3(3g-2)/4\pi$.

We also refer to Jost [A-49], Chapter 6; Tromba [235], and Wolf [244].

7.3.5. Weil-Petersson Metric of the Teichmüller Space of Genus 1

We define a metric on the Teichmüller space T_1 of genus 1 which corresponds to the Weil-Petersson metric h_{WP} on T_g with $g \geqq 2$.

As was seen in §2.2 of Chapter 1, the Teichmüller space T_1 is identified with the upper half-plane H. In fact, for every point $\tau \in H$, we denote by Γ_τ the lattice group generated by 1 and τ. The torus $R_\tau = \mathbf{C}/\Gamma_\tau$ has a marking Σ_τ associated with the generators 1 and τ. Then the above identification of H to T_1 is given by the correspondence sending τ to $[R_\tau, \Sigma_\tau]$.

Let $\lambda_\tau^2|dz|^2$ be a metric on a torus $R_\tau = \mathbf{C}/\Gamma_\tau$, where $\tau \in H$, and λ_τ is a positive constant. Here, we impose a normalization condition on $\lambda_\tau^2|dz|^2$ so that the area of R_τ measured by this metric is 1, i.e., we put $\lambda_\tau = 1/\sqrt{\operatorname{Im}\tau}$.

Now, for any $t \in \mathbf{C}$ with sufficiently small $|t|$, we consider a quasiconformal mapping $f_t\colon R_\tau \to R_{\tau+t}$ induced by a linear mapping

$$\tilde{f}_t(z) = \left(1 + \frac{t}{\tau - \bar{\tau}}\right) z + \frac{-t}{\tau - \bar{\tau}}\,\bar{z}, \quad z \in \mathbf{C}.$$

Since the Beltrami coefficient μ_t of $\tilde{f}_t$ is equal to $-t/(\tau - \bar{\tau} + t)$, the derivative of μ_t at τ is given by

$$\dot{\mu}_\tau = \lim_{t\to 0} \frac{\mu_t}{t} = \frac{i}{2}\frac{1}{\operatorname{Im}\tau} = \frac{-1}{\tau - \bar{\tau}}.$$

This $\dot{\mu}_\tau$ is regarded as a holomorphic tangent vector $\partial/\partial\tau$ on T_1 at $p = [R_\tau, \Sigma_\tau]$, and gives a basis for the tangent space $T_p(T_1)$.

We define the scalar product of $\partial/\partial\tau$ and itself by

$$h_{WP}\Big(\frac{\partial}{\partial\tau}, \frac{\partial}{\partial\tau}\Big) = \iint_{R_\tau} \frac{1}{|\tau - \bar{\tau}|^2}\,\lambda_\tau^2\,dxdy = \frac{1}{4(\operatorname{Im}\tau)^2}.$$

This metric is the desired metric on T_1, and is written as

$$ds_{WP}^2 = \frac{1}{2(\operatorname{Im}\tau)^2}|d\tau|^2,$$

which coincides with the Poincaré metric on H up to a constant factor. The fundamental form ω_{WP} of ds_{WP}^2 is given by

$$\omega_{WP} = \frac{i}{4(\operatorname{Im}\tau)^2} d\tau \wedge d\bar{\tau}.$$

On the other hand, we set

$$\ell = \frac{1}{\sqrt{\operatorname{Im}\tau}}, \quad t = -\frac{\operatorname{Re}\tau}{\sqrt{\operatorname{Im}\tau}}.$$

Then as indicated in Fig. 7.1, we see that ℓ, t are distances from the origin to $1, -\operatorname{Re}\tau$, respectively, measured by $\lambda_\tau^2|dz|^2$. If we set $\theta = 2\pi t/\ell$, then (ℓ, θ) gives coordinates on T_1, which correspond to Fenchel-Nielsen coordinates on T_g with $g \geqq 2$.

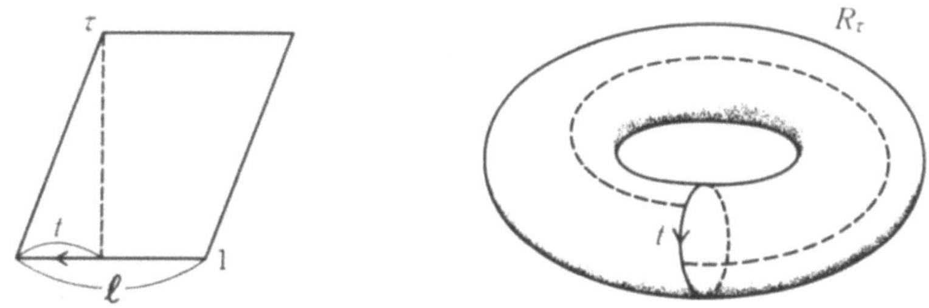

Fig. 7.1.

Furthermore, we have

$$\omega_{WP} = dt \wedge d\ell.$$

This formula corresponds to Wolpert's formula in Theorem 8.6 of Chapter 8 for T_g with $g \geqq 2$.

7.3.6. A Differential Geometric Interpretation of the Weil-Petersson Metric

As was stated in §5.2 of Chapter 1, we saw that the Teichmüller space $T(R)$ of a closed Riemann surface R is identified with a quotient space of the space $\mathcal{M}(R)$ of all Riemannian metrics on R by the group $Diff_0(R)$ of all orientation-preserving diffeomorphisms of R onto itself which are homotopic to the identity.

We see that a metric on $T(R)$ induced by a natural inner product on $\mathcal{M}(R)$ gives the Weil-Petersson metric on $T(R)$.

It is sufficient to consider the tangent space $T_o(T(R))$ of $T(R)$ at the base point. Take a Riemannian metric $ds^2 = \rho^2|dz|^2$ on R which determines the conformal structure of R (see §5.1 of Chapter 1). Here, (U, z) is a local coordinate on

R. The tangent space $\mathcal{T} = T_{ds^2}(\mathcal{M}(R))$ of $\mathcal{M}(R)$ at ds^2 consists of all symmetric tensors of degree 2 on R. Every element α of $\mathcal{T}$ is written as

$$\alpha = Adzd\bar{z} + Bdz^2 + \bar{B}d\bar{z}^2,$$

where A and B are smooth functions on U, A is real-valued, and B is complex-valued. This α corresponds to a real symmetric 2×2 matix

$$\tilde{\alpha} = \frac{1}{2}\begin{bmatrix} A+B+\bar{B} & i(B-\bar{B}) \\ i(B-\bar{B}) & A-B-\bar{B} \end{bmatrix}.$$

Now, the inner product of two elements

$$\alpha_j = A_j dzd\bar{z} + B_j dz^2 + \overline{B_j}\, d\bar{z}^2, \qquad j = 1, 2$$

in $\mathcal{T}$ is defined by

$$\langle \alpha_1, \alpha_2 \rangle_R = \iint_R \mathrm{tr}(\tilde{\alpha}_1\tilde{\alpha}_2)\rho^2\, dxdy.$$

Here, $\mathrm{tr}(\tilde{\alpha}_1\tilde{\alpha}_2)$ is the trace of the matrix $\tilde{\alpha}_1\tilde{\alpha}_2$ with respect to the metric ds^2, that is,

$$\mathrm{tr}(\tilde{\alpha}_1\tilde{\alpha}_2) = \frac{A_1A_2 + 2(B_1\overline{B_2} + \overline{B_1}B_2)}{2\rho^4}.$$

Thus we have

$$\langle \alpha_1, \alpha_2 \rangle_R = \frac{1}{2}\iint_R [A_1A_2 + 2(B_1\overline{B_2} + \overline{B_1}B_2)]\, \frac{dxdy}{\rho^2}.$$

Note that the following two types of elements in $\mathcal{T}$ correspond to zero vectors in $T_o(T(R))$:

(i) A vector induced by deformation of the scale factor ρ. This is an infinitesimal deformation $\psi\rho^2|dz|^2$ which is generated by a 1-parameter family $\{\rho^2 e^{t\psi}|dz|^2\}_{t\in\mathbf{R}}$ of conformal deformations of the metric ds^2, where ψ is a real-valued function on R.

(ii) A vector induced by diffeomorphisms of R to itself. Namely, this is an infinitesimal deformation

$$\rho^2\left(\frac{\partial\bar{a}}{\partial z}dz^2 + \frac{\partial a}{\partial\bar{z}}d\bar{z}^2\right)$$

induced by $\{f_t^*(ds^2)\}_{t\in\mathbf{R}}$ of deformations of ds^2, where $\{f_t\}_{t\in\mathbf{R}}$ is a 1-parameter family of transformations of R which is generated by a vector field $X = a(z)(\partial/\partial z)$ on R.

We shall obtain a condition on an element $\alpha = Adzd\bar{z} + Bdz^2 + \bar{B}d\bar{z}^2$ such that α is orthogonal to all elements of types (i) and (ii) in $\mathcal{T}$ with respect to the given inner product $\langle\cdot,\cdot\rangle_R$. First, in order that α satisfies

$$\langle \alpha, \beta \rangle_R = \frac{1}{2}\iint_R A\psi\, dxdy = 0$$

for any β of type (i), A should be 0. Next, in order that α satisfies

$$\langle \alpha, \beta \rangle_R = \iint_R \left(B \frac{\partial a}{\partial \bar{z}} + \bar{B} \frac{\partial \bar{a}}{\partial z} \right) dxdy = 0$$

for any element β of type (ii), from Green's formula we have

$$\iint_R \frac{\partial B}{\partial \bar{z}}\, a\, dxdy + \iint_R \frac{\partial \bar{B}}{\partial z}\, \bar{a}\, dxdy = 0.$$

Since $X = a(z)(\partial/\partial z)$ is arbitrary, we see that $\partial B/\partial \bar{z} = 0$, that is, B is holomorphic. Thus Bdz^2 belongs to the space $A_2(R)$ of holomorphic quadratic differentials on R.

Hence, an element α in $\mathcal{T}$ is orthogonal to all elements of types (i) and (ii) if and only if α is written as $\alpha = \psi + \bar{\psi}$ for some ψ in $A_2(R)$.

This observation shows that the subspace $\mathcal{T}_o = \{ \psi + \bar{\psi} \mid \psi \in A_2(R) \}$ of $\mathcal{T}$ corresponds to the holomorphic tangent space $T_o(T(R))$ of $T(R)$ at the base point. Moreover, the inner product of two elements $\alpha_j = \psi_j + \bar{\psi}_j$ $(j = 1, 2)$ in $\mathcal{T}_o$ is given by

$$\langle \alpha_1, \alpha_2 \rangle_R = 2\mathrm{Re} \iint_R \psi_1 \overline{\psi_2} \frac{dxdy}{\rho^2}.$$

In particular, taking the hyperbolic metric on R as ds^2, we see that this inner product coincides with the one on $T_o(T(R))$ which gives the Weil-Petersson Riemannian metric g_{WP} on $T(R)$.

Notes

A great deal of helpful introductory material on complex differential geometry is found in the books by Griffiths and Harris [A-39], Kobayashi and Nomizu [A-55], Kodaira [A-57], and Morrow and Kodaira [A-77].

Complete details of automorphic forms and Poincaré series in §1 are contained in the books by Kra [A-58], and Lehner [A-66]. Some of many papers on Poincaré series are by Kra [124], and Masumoto [140] and [141].

For a proof of Teichmüller's lemma (Lemma 7.6), see Lemma 1 of Chapter 6 in Ahlfors [A-2]. Our proof in §2 is due to Earle (see [A-41], §12 in Chapter 5, pp.155-158).

A great deal of the Eichler cohomology is found in the book by Kra [A-58]. For applications of the Eichler cohomology to quasiconformal stability of Kleinian groups, see Gardiner and Kra [82], Kra [120], Kra and Maskit [127], Matsuzaki [151], and Nakada [165].

The Weil-Petersson metric was first introduced by Weil [243]. Two proofs in this chapter for Kählerity of the Weil-Petersson metric are due to Ahlfors [6] and [7]. An alternative proof is given by Wolpert [253], which is based on the Maass calculus. See also Wolpert [248], [259], and [260]. Furthermore, we refer to Jost [A-49], Eells and Lemaire [67], Fisher and Tromba [72], Tromba [236], and Wolf [244], where Teichmüller spaces are studied from the viewpoint of harmonic maps. Moreover, see Takhtadzhyan [219] and [220], and Zograf and Takhtadzhyan [264], [265], and [266]. Tromba [235] showed that the sectional curvatures of the Weil-Petersson metric are also negative. It is also known that the Weil-Petersson metric is non-complete; proofs are found in Chu [49], Masur [143], and Wolpert [245].

For the subject in §3.6, we refer to Fischer and Tromba [72]. Such a differential geometric interpretation of the Weil-Petersson metric is closely related to the Polyakov integral in string theory (see, for example, Polyakov [176]). See also Nag and Verjovsky [164].

For the Weil-Petersson geometry on moduli spaces of higher dimensional complex manifolds, we refer to Besse [A-17], Fujiki and Schumacher [77], Koiso [118], [119], Schumacher [190], and Siu [208].

Chapter 8

Fenchel-Nielsen Deformations and the Weil-Petersson Metric

In this chapter, we shall give a beautiful representation, due to S. Wolpert, of the Weil-Petersson fundamental form on T_g ($g \geq 2$) by using Fenchel-Nielsen coordinates.

First, in Section 1, we describe Fenchel-Nielsen deformations by using quasiconformal mappings. Next, in Section 2, we calculate the tangent vector determined by a Fenchel-Nielsen deformation in T_g.

Finally, in Section 3, we prove a Wolpert's formula, namely, a simple representation of the Weil-Petersson fundamental form by Fenchel-Nielsen coordinates.

8.1. Fenchel-Nielsen Deformations

Let R be a closed Riemann surface of genus g (≥ 2). Fix an oriented simple closed geodesic C with respect to the hyperbolic metric ds_R^2. Recall that the *Fenchel-Nielsen deformation* of R with respect to C, which is simply called the *FN deformation* from here on, means the family $\{R_t \mid t \in \mathbf{R}\}$ of marked Riemann surfaces R_t obtained by cutting R along C, by twisting by hyperbolic length t and then by regluing the borders. See §2 of Chapter 3 and Fig. 8.1.

We represent such a deformation by using quasiconformal mappings.

Let ρ be the hyperbolic distance on R, the distance induced by ds_R^2. Set

$$W_C = \{p \in R \mid \rho(p, C) < a\}.$$

Here, we choose a positive constant a so small that W_C becomes a tubular neighborhood, i.e., a doubly connected neighborhood of C. Then we construct a quasiconformal mapping of R which represents the "twisting" along C by t in W_C, and is equal to id on $R - W_C$.

First, take a Fuchsian model Γ of R acting on the upper half-plane H. We may assume that 1 is a fixed point of some element of $\Gamma - \{id\}$, and that $\gamma_0(z) = \lambda z$ ($\lambda > 1$) belongs to Γ and covers C. Let $\tilde{W}_C$ be a lift of W_C on H (with respect to Γ) which contains the axis A_{γ_0} of γ_0. For a suitable θ_0 with $0 < \theta_0 < \pi/2$, we can express $\tilde{W}_C$ as

$$\tilde{W}_C = \left\{z \in H \;\middle|\; \frac{\pi}{2} - \theta_0 < \arg z < \frac{\pi}{2} + \theta_0\right\}.$$

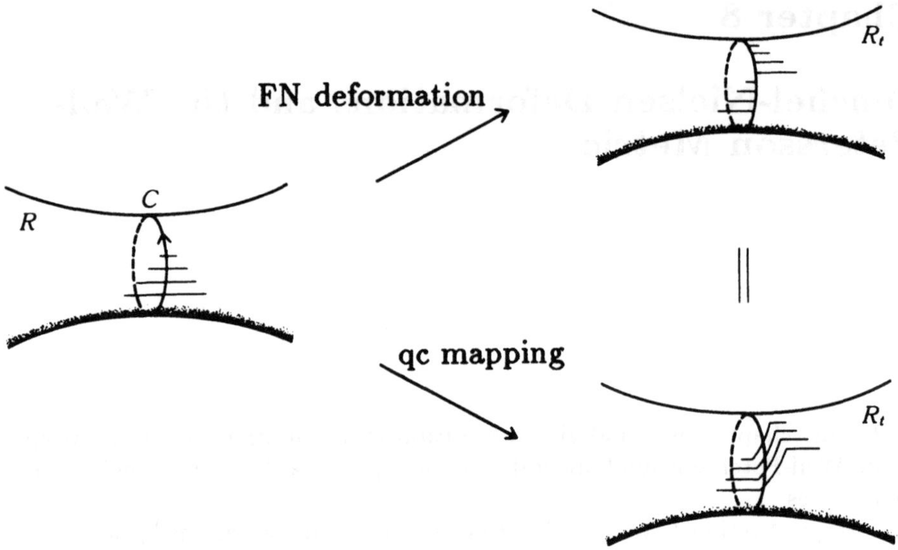

Fig. 8.1.

Next, for every $t \in \mathbf{R}$, we define a quasiconformal mapping w^t of H onto itself by

$$w^t(z) = \begin{cases} z, & 0 < \theta < \frac{\pi}{2} - \theta_0 \\ z \exp\left(\epsilon(\theta - \frac{\pi}{2} + \theta_0)\right), & \frac{\pi}{2} - \theta_0 \leq \theta \leq \frac{\pi}{2} + \theta_0 \\ z \exp(2\epsilon\theta_0), & \frac{\pi}{2} + \theta_0 < \theta < \pi. \end{cases}$$

Here, $\theta = \arg z$, and $\epsilon = -t/(2\theta_0)$. This w^t gives a surgery of H along the axis A_{γ_0}. Note that the sign of t is different from that in Wolpert's papers [247] and [251]. (See Fig. 8.2.)

Now, denote by τ_t the complex dilatation of w^t. A simple computation gives

$$\tau_t(z) = \frac{i\epsilon}{2 - i\epsilon}\chi_I(\theta)\frac{z}{\bar{z}}, \quad z \in H.$$

Here, χ_I is the characteristic function of $I = [\pi/2 - \theta_0, \pi/2 + \theta_0]$ on $\mathbf{R}$. Furthermore, τ_t satisfies

$$\tau_t \circ \gamma_0 \cdot (\overline{\gamma_0'}/\gamma_0') = \tau_t.$$

Thus, τ_t is a Beltrami coefficient with respect to the cyclic group $\langle\gamma_0\rangle$.

On the other hand, it is clear that the set Γ_C consisting of all elements in Γ which cover C is $\{\gamma \circ \gamma_0 \circ \gamma^{-1} \mid \gamma \in \Gamma\}$. By lifting this FN deformation to H, we have a family of self-mappings of H which give surgeries along the axes of all elements in Γ_C. Thus, we can construct a family of quasiconformal self-mappings of H which induces this FN deformation as follows. Denote by $\langle\gamma_0\rangle \setminus \Gamma$ the set of all right cosets of Γ with respect to $\langle\gamma_0\rangle$, and set

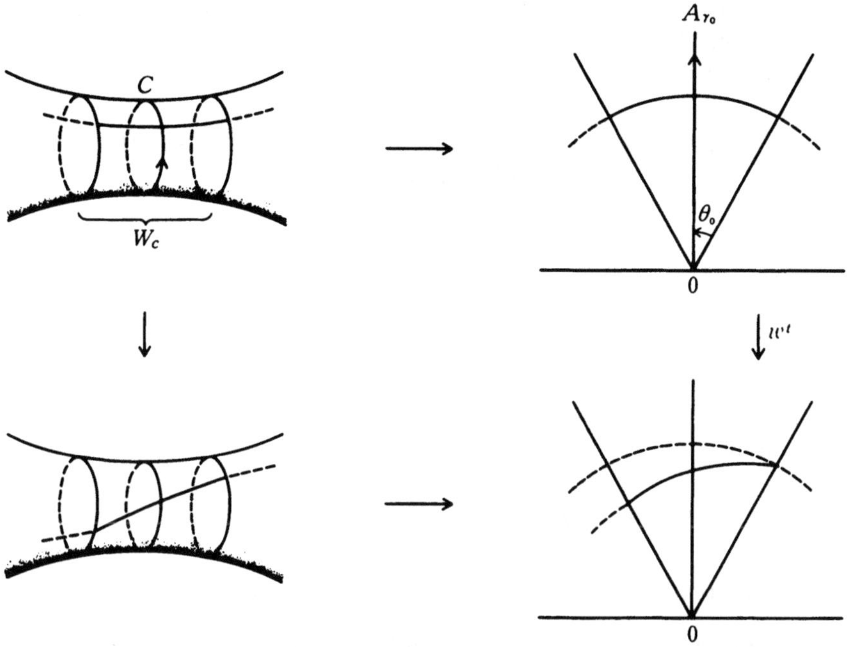

Fig. 8.2.

$$\mu_t = \sum_{\gamma \in \langle \gamma_0 \rangle \backslash \Gamma} (\tau_t \circ \gamma) \frac{\overline{\gamma'}}{\gamma'}.$$

It is clear from the definition that μ_t belongs to $B(H, \Gamma)_1$. Hence, for every t, the canonical μ_t-qc mapping w^{μ_t} of H exists, and determines a point $[w^{\mu_t}] \in T(\Gamma)$. Thus, we obtain a family $\{[w^{\mu_t}] \mid t \in \mathbf{R}\}$ in $T(\Gamma)$ which represents the FN deformation of R along C.

Remark. We can construct w^{μ_t} more directly and geometrically by using w^t, which "twists" along A_{γ_0} by t in H. Actually, "twist" inductively along $\gamma(A_{\gamma_0})$ by t on $\gamma(\tilde{W}_C)$ as described before for every $\langle \gamma_0 \rangle \gamma^{-1} \in \langle \gamma_0 \rangle \setminus \Gamma$, and then normalize the resulting mapping. For more details, see Wolpert [247], pp.503-505.

In particular, we have obtained a curve $\{\mu_t \mid t \in \mathbf{R}\}$ in $B(H, \Gamma)_1$, which represents the FN deformation of R along C. Next, we shall compute the tangent vector of this FN deformation in $T_B(\Gamma)$ considered as a real manifold.

For every μ_t, we set $w_t = w_{\mu_t}$, where w_{μ_t} is defined in §1.1 of Chapter 6. Then, as in the proof of Theorem 6.10, we can conclude that the tangent vector of the curve $\{\Phi(\mu_t) \mid t \in \mathbf{R}\}$ at the base point of $T_B(\Gamma)$ is equal to

$$\varphi_C = \frac{d^3}{dz^3} \left(\left. \frac{\partial w_t}{\partial t} \right|_{t=0} \right).$$

(Here, recall that such a tangent vector is considered as an element of $A_2(H^*, \Gamma)$.)

Furthermore, Theorem 4.37 gives an integral formula for $\dot{w} = (\partial w_t/\partial t)|_{t=0}$ as follows. We set

$$\nu(z) = \nu_0(z) + \sum_{\gamma \in \langle\gamma_0\rangle\backslash\Gamma,\, \gamma\notin\langle\gamma_0\rangle} \nu_0(\gamma(z)) \frac{\overline{\gamma'(z)}}{\gamma'(z)}, \quad z \in H, \tag{8.1}$$

where

$$\nu_0(z) = -\frac{i}{4\theta_0}\chi_I(\arg z)\frac{z}{\bar{z}}.$$

It is easy to see that

$$\lim_{t\to 0} \|\frac{\mu_t}{t} - \nu\|_\infty = 0.$$

Thus we obtain

$$\dot{w}(z) = -\frac{1}{\pi}\iint_H \nu(\zeta)\frac{z(z-1)}{\zeta(\zeta-1)(\zeta-z)}\,d\xi d\eta, \quad z \in \mathbf{C}. \tag{8.2}$$

As has been stated in the proof of Lemma 7.16, we know that

(i) $(\dot{w})_{\bar{z}} = \nu$ on $\mathbf{C}$ in the sense of distribution, and
(ii) $\dot{w}(0) = \dot{w}(1) = 0$, and $\dot{w}(z) = o(|z|^2)$ as $z \longrightarrow \infty$.

Conversely, these conditions (i) and (ii) characterize $\dot{w}$ in the class of continuous functions on $\mathbf{C}$, which can be easily shown by using Weyl's lemma (Lemma 4.6).

Now, to get a simpler representation of φ_C, we rewrite formula (8.2) as follows.

Lemma 8.1. *The derivative $\dot{w}$ is written as*

$$\dot{w}(z) = -z\left\{\int_0^{\arg z}\frac{\chi_I(t)}{2\theta_0}\,dt + \frac{i}{2\pi}\log z\right\} + \sum_{\gamma \in \langle\gamma_0\rangle\backslash\Gamma,\, \gamma\notin\langle\gamma_0\rangle} F_\gamma(z) \tag{8.3}$$

for every $z \in \mathbf{C}$. Here, $\arg z$ takes values in $[-\pi, \pi)$, and

$$F_\gamma(z) = -\frac{\gamma(z)}{\gamma'(z)}\left\{\int_0^{\arg\gamma(z)}\frac{\chi_I(t)}{2\theta_0}\,dt + \frac{i}{2\pi}\log\gamma(z)\right\} + \frac{i}{2\pi}P_\gamma(z),$$

where $P_\gamma(z)$ is a polynomial of degree at most two. This P_γ is uniquely determined by the conditions that $F_\gamma(0) = F_\gamma(1) = 0$ and that $F_\gamma(z) = o(|z|^2)$ as $z \longrightarrow \infty$.

Moreover, the series on the right hand side of (8.3) converges locally uniformly on $\mathbf{C}$.

Proof. First, substitute the right hand side of (8.1) for μ in (8.2). Then we have

$$\dot{w}(z) = -\frac{1}{\pi}\iint_H \nu_0(\zeta)R(\zeta,z)\,d\xi d\eta$$
$$- \sum_{\gamma\in<\gamma_0>\backslash\Gamma,\gamma\notin<\gamma_0>} \frac{1}{\pi}\iint_H \nu_0(\gamma(\zeta))\,\frac{\overline{\gamma'(\zeta)}}{\gamma'(\zeta)}\,R(\zeta,z)\,d\xi d\eta. \tag{8.4}$$

Here, we set

$$R(\zeta,z) = \frac{z(z-1)}{\zeta(\zeta-1)(\zeta-z)}.$$

Clearly, the series on the right hand side of (8.4) converges absolutely and locally uniformly on $\mathbf{C}$.

Denote by $I(z)$ the first term on the right hand side of (8.4). Then we have

$$\begin{aligned} I(z) =& \frac{i}{2\pi}\iint_H \frac{z(z-1)}{\zeta(\zeta-1)(\zeta-z)}\,\frac{\chi_I(\arg\zeta)}{2\theta_0}\,d\xi d\eta \\ =& \frac{iz}{2\pi}\int_0^\pi \frac{\chi_I(t)}{2\theta_0}\left\{\int_0^\infty \frac{(z-1)e^{it}dr}{(re^{it}-1)(re^{it}-z)}\right\}dt, \quad \zeta = re^{it}, \\ =& \frac{iz}{2\pi}\int_0^\pi \frac{\chi_I(t)}{2\theta_0}\left\{2\pi i\cdot\chi_{(0,\max\{0,\arg z\})}(t) - \log z\right\}dt \\ =& -z\left\{\int_0^{\arg z}\frac{\chi_I(t)}{2\theta_0}\,dt + \frac{i}{2\pi}\log z\right\}. \end{aligned}$$

Next, since $I_{\bar{z}} = \nu_0$ in the sense of distribution, we have

$$(I\circ\gamma)_{\bar{z}} = (\nu_0\circ\gamma)\cdot\overline{\gamma'}, \quad \gamma\in\Gamma.$$

Thus it is shown that

$$\frac{I\circ\gamma(z)}{\gamma'(z)} - \left\{-\frac{1}{\pi}\iint_H \nu_0(\gamma(\zeta))\,\frac{\overline{\gamma'(\zeta)}}{\gamma'(\zeta)}\,R(\zeta,z)\,d\xi d\eta\right\}$$

is holomorphic on $\mathbf{C}$, and is $O(|z|^2)$ as $z\to\infty$, which implies that this difference is a polynomial of degree at most two. We set this difference as $-iP_\gamma/2\pi$. Then the assertion follows. □

Theorem 8.2. *Under the foregoing circumstances, it follows that*

$$\varphi_C = \frac{i}{2\pi}\sum_{\gamma\in\langle\gamma_0\rangle\backslash\Gamma}\left(\frac{\gamma'}{\gamma}\right)^2,$$

which converges locally uniformly on H^*.

Proof. Lemma 8.1 gives

$$\dot{w} = -\frac{i}{2\pi}\sum_{\gamma\in\langle\gamma_0\rangle\backslash\Gamma}\left(\frac{\gamma}{\gamma'}\log\gamma - P_\gamma\right) \quad \text{on } H^*, \tag{8.5}$$

where $P_{id} = 0$.

On the other hand, a direct computation gives the following, Bol's equation:

$$\left(\frac{\gamma}{\gamma'}\log\gamma\right)''' = -\left(\frac{\gamma'}{\gamma}\right)^2.$$

Thus, differentiating both sides of (8.5) three times, we have the assertion. □

In general, for any simple closed geodesic C on R and any element $\gamma_0 \in \Gamma$ which covers C, we can construct a similar basic series as in Theorem 8.2. More precisely, let a and b be the two real fixed points of γ_0, and set

$$\omega_{\gamma_0}(z) = \frac{(a-b)^2}{(z-a)^2(z-b)^2}.$$

Then

$$\Theta_C = \sum_{\gamma \in \langle\gamma_0\rangle\backslash\Gamma} (\omega_{\gamma_0} \circ \gamma) \cdot (\gamma')^2$$

converges locally uniformly on H, and belongs to $A_2(H, \Gamma)$. We call this Θ_C the *Petersson series* for C. Using this series, Theorem 8.2 is reformulated as follows:

Corollary. *Let $N(\Gamma)$ be as in §2.1 of Chapter 7. Then*

$$\nu = -\frac{i}{\pi}\lambda_H^{-2}\overline{\Theta_C} \quad \mathit{mod}\ N(\Gamma).$$

Proof. Using the notation in §2 of Chapter 7, Theorem 8.2 and (7.15) in Chapter 7 imply that $\varphi[\nu] = \frac{i}{\pi}\Theta_C$. Hence, (7.17) in Chapter 7 gives

$$H[\nu] = -\frac{i}{\pi}\lambda_H^{-2}\overline{\Theta_C}.$$

Thus, the assertion follows by Theorem 7.7. □

8.2. A Variational Formula for Geodesic Length Functions

Fix a simple closed geodesic C on R arbitrarily. For every point $p = [S, f] \in T(R)$, let C_p be the simple closed geodesic on S freely homotopic to $f(C)$, and denote by $\ell_C(p)$ the hyperbolic length of C_p. Recall that ℓ_C is a real-analytic function on $T(R)$ (see Remark 1 in §3.2 of Chapter 7).

Here, we compute the variation of ℓ_C at the base point $p_0 = [R, id]$. More precisely, take $\mu \in B(H, \Gamma)$ arbitrarily, and let $w^{t\mu}$ be the canonical $t\mu$-qc mapping of H for every sufficiently small real t. Then $w^{t\mu}$ determines a point, say p_t, of $T(R)$ for such t. Under these circumstances, we compute the value

$$(d\ell_C)_{p_0}(\mu) = \frac{d}{dt}\ell_C(p_t)\bigg|_{t=0}.$$

This is essentially well-known (cf. Gardiner [79]).

Theorem 8.3.

$$(d\ell_C)_{p_0}(\mu) = \mathrm{Re}\left(\mu, \frac{2}{\pi}\Theta_C\right)_R.$$

Proof. As in §1, we may assume that $\gamma_0(z) = \lambda z$ ($\lambda > 1$) covers C. Then, for every t, a constant λ_t (> 1) is determined by the condition that

$$w^{t\mu} \circ \gamma_0 \circ (w^{t\mu})^{-1}(z) = \lambda_t z. \tag{8.6}$$

By the definition of ℓ_C, it follows that

$$\ell_C(p_t) = \log \lambda_t.$$

Next, we set

$$\nu(z) = \begin{cases} \mu(z), & z \in H \\ 0, & z \in \mathbf{R} \\ \overline{\mu(\bar{z})}, & z \in H^*. \end{cases}$$

Let $f^{t\nu}$ be the canonical $t\nu$-qc mapping of $\hat{\mathbf{C}}$. Then we know that $w^{t\mu} = f^{t\nu}$ on H. Theorem 4.37 implies that

$$\dot{W}(z) = \lim_{t\to 0} \frac{f^{t\nu}(z) - z}{t}, \quad z \in \mathbf{C},$$

has the integral representation:

$$\dot{W}(z) = -\frac{1}{\pi}\iint_{\mathbf{C}} \nu(\zeta) R(\zeta, z)\, d\xi d\eta. \tag{8.7}$$

On the other hand, (8.6) shows that

$$\dot{W}(\lambda z) = \frac{d\lambda_t}{dt}(0)\, z + \lambda \dot{W}(z).$$

Hence, we have the primary formula:

$$(d\ell_C)_{p_0}(\mu) = \frac{d\log\lambda_t}{dt}(0) = \frac{\dot{W}(\lambda z)}{\lambda z} - \frac{\dot{W}(z)}{z}, \quad z \neq 0.$$

Now, replace $\dot{W}$ in this equatity by the right hand side of (8.7), and rewrite the integrals as those on the domains $F_0 = \{z \in H \mid 1 < |z| < \lambda\}$ and $F_0^* = \{z \mid \bar{z} \in F_0\}$. Since $\nu(\lambda z) = \nu(z)$, we obtain

$$(d\ell_C)_{p_0}(\mu) = -\frac{1}{\pi}\iint_{F_0 \cup F_0^*} \nu(\zeta) \sum_{n=-\infty}^{\infty} \lambda^{2n} \left\{ \frac{R(\lambda^n\zeta, \lambda z)}{\lambda z} - \frac{R(\lambda^n\zeta, z)}{z} \right\} d\xi d\eta.$$

Furthermore, since

$$\lambda^{2n} R(\zeta, z) = -\frac{\lambda^n z}{\zeta}\left(\frac{\lambda^n}{\zeta-1} - \frac{\lambda^n}{\zeta - z}\right),$$

we see that

$$\begin{aligned}\sum_{n=-\infty}^{\infty} \lambda^{2n}\left\{\frac{R(\lambda^n\zeta, \lambda z)}{\lambda z} - \frac{R(\lambda^n \zeta, z)}{z}\right\} &= -\frac{1}{\zeta}\sum_{n=-\infty}^{\infty}\left\{\frac{\lambda^n}{\lambda^n\zeta - z} - \frac{\lambda^{n-1}}{\lambda^{n-1}\zeta - z}\right\}\\ &= -\frac{1}{\zeta}\lim_{n\to\infty}\frac{\lambda^n}{\lambda^n \zeta - z} = -\frac{1}{\zeta^2}.\end{aligned}$$

Hence, we have

$$\begin{aligned}(d\ell_C)_{p_0}(\mu) &= -\frac{1}{\pi}\iint_{F_0\cup F_0^*} \nu(\zeta)\left(-\frac{1}{\zeta^2}\right) d\xi d\eta\\ &=\frac{1}{\pi}\iint_{F_0}\left(\frac{\mu(\zeta)}{\zeta^2} + \frac{\overline{\mu(\zeta)}}{\bar{\zeta}^2}\right) d\xi d\eta\\ &=\frac{2}{\pi}\,\mathrm{Re}\iint_{F_0}\frac{\mu(\zeta)}{\zeta^2}\, d\xi d\eta.\end{aligned}$$

Finally, divide F_0 into domains $\{\gamma(F) \mid \gamma \in \Gamma\}$, where F is a fundamental domain for Γ. Then we conclude that

$$\begin{aligned}\iint_{F_0}\frac{\mu(\zeta)}{\zeta^2} d\xi d\eta &= \sum_{\gamma\in\langle\gamma_0\rangle\backslash\Gamma}\iint_F \mu(\zeta)\left(\frac{\gamma'(\zeta)}{\gamma(\zeta)}\right)^2 d\xi d\eta\\ &=\iint_F \mu(\zeta)\sum_{\gamma\in\langle\gamma_0\rangle\backslash\Gamma}\left(\frac{\gamma'(\zeta)}{\gamma(\zeta)}\right)^2 d\xi d\eta\\ &=(\mu, \Theta_C)_R.\end{aligned}$$

This completes the proof of Theorem 8.3. □

8.3. Wolpert's Formula

We have computed the tangent vector (represented by ν in §8.1) of the FN deformation with respect to C at the base point, where C is a given simple closed geodesic on R. By a translation of the base point, we can compute the tangent vector field on (the real manifold) $T(R)$ associated with the FN deformation with respect to C, which we denote by $\partial/\partial\tau_C$, and call the *FN vector field* for C. Namely, $\partial/\partial\tau_C$ is the vector field obtained by applying the FN deformation with respect to C with unit speed with respect to the hyperbolic length. Note that $\partial/\partial\tau_C$ is a real-analytic vector field.

On the other hand, from the geodesic length function ℓ_C, we have defined in §8.2 the real-analytic cotangent vector field $d\ell_C$.

We show the following *duality theorem.*

Theorem 8.4. *For every simple closed geodesic C on R,*

$$\left(i\frac{\partial}{\partial\tau_C}\right)^* = d\ell_C,$$

where i means the natural almost complex structure on the real manifold $T(R)$ (cf. §7.3.2), and $*$ is the operator given by taking the dual with respect to g_{WP}, i.e.,

$$g_{WP}\left(i\frac{\partial}{\partial\tau_C}, \cdot\right) = d\ell_C(\cdot).$$

Proof. By a translation of the base point, it suffices to show the former equality at the base point p_0 of $T(R)$.

First, from the Corollary to Theorem 8.2 and Lemma 7.12, we have

$$\begin{aligned} g_{WP}\left(i\frac{\partial}{\partial\tau_C}, \mu\right) &= 2\operatorname{Re} h_{WP}\left(i\frac{\partial}{\partial\tau_C}, \mu\right) \\ &= 2\operatorname{Re} h_{WP}\left(\frac{1}{\pi}\lambda_H{}^{-2}\overline{\Theta_C}, \mu\right) \\ &= 2\operatorname{Re}\left(\mu, \frac{1}{\pi}\Theta_C\right)_R \end{aligned}$$

for every $\mu \in HB(H,\Gamma)$.

On the other hand, Theorem 8.3 gives that the right hand side is equal to $(d\ell_C)_{p_0}(\mu)$. Hence we have the assertion. □

Corollary.

$$\omega_{WP}\left(\frac{\partial}{\partial\tau_C}, \cdot\right) = d\ell_C(\cdot).$$

Another corollary to Theorem 8.4 is the following *reciprocity theorem.*

Proposition 8.5. *For all simple closed geodesics C and C' on R,*

$$\frac{\partial\ell_{C'}}{\partial\tau_C} = -\frac{\partial\ell_C}{\partial\tau_{C'}}.$$

Proof. By the foregoing Corollary, we have

$$\begin{aligned} \frac{\partial\ell_{C'}}{\partial\tau_C} &= (d\ell_{C'})\left(\frac{\partial}{\partial\tau_C}\right) = \omega_{WP}\left(\frac{\partial}{\partial\tau_{C'}}, \frac{\partial}{\partial\tau_C}\right) \\ &= -\omega_{WP}\left(\frac{\partial}{\partial\tau_C}, \frac{\partial}{\partial\tau_{C'}}\right) = -\frac{\partial\ell_C}{\partial\tau_{C'}}. \end{aligned}$$

This shows the assertion. □

Finally, we have the following, *Wolpert's formula*, which is proved later.

Theorem 8.6. *Fix a system of decomposing curves $\mathcal{L} = \{C_j\}_{j=1}^{3g-3}$ on R arbitrarily. Denote by $\{\ell_{C_1}, \cdots, \ell_{C_{3g-3}}, \theta_{C_1}, \cdots, \theta_{C_{3g-3}}\}$ the Fenchel-Nielsen coordinates associated with $\mathcal{L}$. Set $\tau_{C_j} = (\ell_{C_j}/2\pi)\theta_{C_j}$ for every j. Then*

$$\omega_{WP} = \sum_{j=1}^{3g-3} d\tau_{C_j} \wedge d\ell_{C_j}.$$

Corollary. *For every simple closed geodesic C on R,*

$$\omega_{WP}\left(\frac{\partial}{\partial \ell_C}, \cdot\right) = -d\tau_C(\cdot).$$

Proof. Take a system of decomposing curves which contains C, and apply Theorem 8.6. □

Now, to prove Theorem 8.6, we take a base of tangent vector fields

$$\{X_1, \cdots, X_{6g-6}\} = \left\{\frac{\partial}{\partial \ell_{C_1}}, \cdots, \frac{\partial}{\partial \ell_{C_{3g-3}}}, \frac{\partial}{\partial \tau_{C_1}}, \cdots, \frac{\partial}{\partial \tau_{C_{3g-3}}}\right\}$$

on $T(R)$, where $T(R)$ is considered as a real manifold. Further, we set

$$\{x_1, \cdots, x_{6g-6}\} = \{\ell_{C_1}, \cdots, \ell_{C_{3g-3}}, \tau_{C_1}, \cdots, \tau_{C_{3g-3}}\}.$$

Then ω_{WP} is written in the form

$$\omega_{WP} = \sum_{1 \leq j < i \leq 6g-6} a_{ij}\, dx_i \wedge dx_j.$$

First, we shall show that every a_{ij} is invariant under the FN deformations with respect to C_k for every k.

Lemma 8.7. *For every i, j, and k,*

$$\frac{\partial a_{ij}}{\partial \tau_{C_k}} = 0 \quad on \quad T(R).$$

Proof. To prove the assertion, we use some basic notions and results from the differential geometry. See for instance Matsushima [A-72].

Let $I(X)$ be the interior product with respect to X. Then the Corollary to Theorem 8.4 gives

$$d\left(I\left(\frac{\partial}{\partial\tau_{C_k}}\right)\omega_{WP}\right)=d\left(\omega_{WP}\left(\frac{\partial}{\partial\tau_{C_k}},\cdot\right)\right)=d(d\ell_{C_k})=0.$$

Let L_X be the Lie derivative with respect to X. Then, H. Cartan's formula gives

$$L_{(\partial/\partial\tau_{C_k})}\omega_{WP}=d\left(I\left(\frac{\partial}{\partial\tau_{C_k}}\right)\omega_{WP}\right)+I\left(\frac{\partial}{\partial\tau_{C_k}}\right)d\omega_{WP}.$$

Since ω_{WP} is Kählerian, i.e., $d\omega_{WP}=0$ by Theorem 7.15, we conclude that

$$L_{(\partial/\partial\tau_{C_k})}\omega_{WP}=0. \tag{8.8}$$

Finally, from the definition of the Lie derivatives, we have

$$L_X\omega_{WP}(Y,Z)=X\omega_{WP}(Y,Z)-\omega_{WP}([X,Y],Z)-\omega_{WP}(Y,[X,Z]) \tag{8.9}$$

for every set of vector fields X,Y, and Z. Note that, if we take as X,Y, and Z vector fields in $\{X_j\}_{j=1}^{6g-6}$, then $[X,Y]=[X,Z]=0$. In particular, setting $X=\partial/\partial\tau_{C_k}$, $Y=X_i$, and $Z=X_j$, from (8.8) and (8.9) we obtain

$$\frac{\partial}{\partial\tau_{C_k}}\omega_{WP}(X_i,X_j)=\frac{\partial a_{ij}}{\partial\tau_{C_k}}=0$$

as desired. □

Lemma 8.7 implies that, when we change R to another R_0 by applying the FN deformation with respect to any C_j, the representation $\omega_{WP}=\sum a_{ij}dx_i\wedge dx_j$ is unchanged.

On the other hand, after suitable FN deformations, we can generate an R_0 admitting an anti-conformal reflection. In fact, let $\mathcal{P}=\{P_k\}_{k=1}^{2g-2}$ be the pants decomposition of R corresponding to the given $\mathcal{L}$. Then every pair of pants P_k has the reflection J_k of P_k (cf. the Corollary to Theorem 3.5). Moreover, the set F_{J_k} of fixed points of J_k divides each boundary component, say $L_{k,j}$ $(j=1,2,3)$ of R_k, into two geodesics with the same hyperbolic length. Thus, applying the FN deformation for every $L_{k,j}$ $(\in\mathcal{L})$, we find a Riemann surface R_0 satisfying the following condition: for every $L_j\in\mathcal{L}$, let $P_{j,1}$ and $P_{j,2}$ be pants mutually adjacent along L_j, and $J_{j,\ell}$ $(\ell=1,2)$ be the reflection of $P_{j,\ell}$ for each ℓ. Then the sets of fixed points $F_{J_{j,\ell}}\cap L_j$ of $J_{j,\ell}$ on L_j $(\ell=1,2)$ are coincident with each other.

Then we see that each reflection J_k becomes a restriction of a reflection J of R_0, an anti-conformal self-mapping of R_0 of order two, and that $F\cap P_k=F_{J_k}$ for every k, where F is the set of fixed points of J. Note that F consists of a finite number of simple closed geodesics on R_0. (See Fig. 8.3, where we omit some part of the sets of fixed points, to avoid unnecessary complexity.)

In the rest of this discussion, we may assume without loss of generality that R admits a reflection J. Then J induces a mapping, say $\mathcal{J}$, of $T(R)$ onto itself as follows. For every point $[S,f]\in T(R)$, let Γ_S be a Fuchsian model of S acting on H. Denote by S^* the mirror image H^*/Γ_S of S, where H^* is the lower half-plane. Define an anti-holomorphic canonical mapping

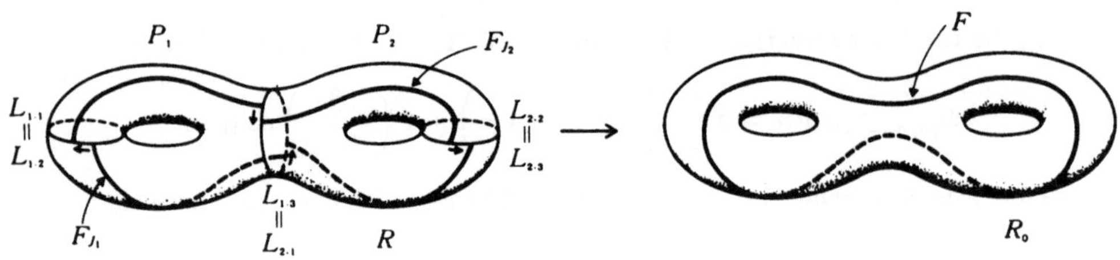

Fig. 8.3.

$$j_S : S \longrightarrow S^*$$

by setting

$$j_S([z]) = [\bar{z}], \quad [z] \in H/\Gamma_S.$$

Then we have a mapping

$$\mathcal{J} : T(R) \longrightarrow T(R)$$

given by

$$\mathcal{J}([S, f]) = [S^*, j_S \circ f \circ J], \quad [S, f] \in T(R).$$

This $\mathcal{J}$ is an anti-holomorphic automorphism of $T(R)$ which fixes the base point $[R, id]$. We can easily show from the definitions of $\mathcal{J}$ and the Weil-Petersson metric that g_{WP} is invariant under $\mathcal{J}$. Furthermore, we have the following:

Lemma 8.8. *Denote by $\mathcal{J}^*$ the pull-back operator induced by $\mathcal{J}$. Then*

$$\mathcal{J}^*(d\ell_{C_j}) = d\ell_{C_j}, \tag{8.10}$$

$$\mathcal{J}^*(d\tau_{C_j}) = -d\tau_{C_j} + \frac{n_j}{2} d\ell_{C_j}, \quad n_j \in \mathbf{Z} \tag{8.11}$$

for every i $(i = 1, \ldots, 3g-3)$. Further, ω_{WP} satisfies

$$\mathcal{J}^*(\omega_{WP}) = -\omega_{WP}. \tag{8.12}$$

Proof. The assertion (8.10) follows by taking the derivative of both sides of $\ell_{C_j} \circ \mathcal{J} = \ell_{J(C_j)} = \ell_{C_j}$.

Next, for every j, we see that, though $J(C_j) = C_j$ as points sets, the orientation of C_j at p is the converse of that at $\mathcal{J}(p)$ for every $p \in T(R)$. It is clear that τ_{C_j} is determined modulo $\ell_{C_j}/2$. Hence, we have

$$\tau_{C_j} \circ \mathcal{J} = -\tau_{C_j} + \frac{n_j}{2} \ell_j$$

with a suitable integer n_j, which implies (8.11).

Finally, let X and Y be arbitrary tangent vector fields on $T(R)$. Then we obtain

$$\begin{aligned}(\mathcal{J}^*\omega_{WP})(X,Y) =&\omega_{WP}(\mathcal{J}_*X,\mathcal{J}_*Y)\\ =&g_{WP}(i\mathcal{J}_*X,\mathcal{J}_*Y).\end{aligned}$$

Since $\mathcal{J}$ is anti-holomorphic, we see that $i\mathcal{J}_*X = -\mathcal{J}_*(iX)$. Since g_{WP} is invariant under $\mathcal{J}$, we conclude that

$$\begin{aligned}(\mathcal{J}^*\omega_{WP})(X,Y) =& - g_{WP}(\mathcal{J}_*(iX),\mathcal{J}_*Y)\\ =& - g_{WP}(iX,Y)\\ =& - \omega_{WP}(X,Y).\end{aligned}$$

This implies (8.12). □

Proof of Theorem 8.6. As before, it suffices to show the formula at the base point. Hence, all relations below are to be considered at the base point, which is a fixed point of $\mathcal{J}$.

First, by the Corollary to Theorem 8.4, we have

$$a_{3g-3+j,k} = \omega_{WP}\left(\frac{\partial}{\partial\tau_{C_j}},\frac{\partial}{\partial\ell_{C_k}}\right) = \frac{\partial\ell_{C_j}}{\partial\ell_{C_k}} = \delta_{jk}$$

for all j,k with $1\le j,k\le 3g-3$.

Next, taking the dual in (8.10) and (8.11), we obtain

$$\mathcal{J}_*\left(\frac{\partial}{\partial\ell_{C_j}}\right) = \frac{\partial}{\partial\ell_{C_j}} + \frac{n_j}{2}\frac{\partial}{\partial\tau_{C_j}},$$

and

$$\mathcal{J}_*\left(\frac{\partial}{\partial\tau_{C_j}}\right) = -\frac{\partial}{\partial\tau_{C_j}}$$

for every j ($j=1,\cdots,3g-3$). Hence by (8.12), we conclude that

$$\begin{aligned}a_{3g-3+j,3g-3+k} =&\omega_{WP}\left(\frac{\partial}{\partial\tau_{C_j}},\frac{\partial}{\partial\tau_{C_k}}\right)\\ =&\omega_{WP}\left(-\mathcal{J}_*\left(\frac{\partial}{\partial\tau_{C_j}}\right),-\mathcal{J}_*\left(\frac{\partial}{\partial\tau_{C_k}}\right)\right)\\ =&(\mathcal{J}^*\omega_{WP})\left(\frac{\partial}{\partial\tau_{C_j}},\frac{\partial}{\partial\tau_{C_k}}\right)\\ =&-\omega_{WP}\left(\frac{\partial}{\partial\tau_{C_j}},\frac{\partial}{\partial\tau_{C_k}}\right)\\ =&-a_{3g-3+j,3g-3+k}.\end{aligned}$$

Thus, we have $a_{3g-3+j,3g-3+k}=0$ for all j,k with $1\le k<j\le 3g-3$.

Similarly, for all such j,k, we see that

$$\begin{aligned} a_{jk} &= \omega_{WP}\left(\frac{\partial}{\partial \ell_{C_j}}, \frac{\partial}{\partial \ell_{C_k}}\right) \\ &= -\,\omega_{WP}\left(\frac{\partial}{\partial \ell_{C_j}} + \frac{n_j}{2}\frac{\partial}{\partial \tau_{C_j}}, \frac{\partial}{\partial \ell_{C_k}} + \frac{n_k}{2}\frac{\partial}{\partial \tau_{C_k}}\right) \\ &= -\,a_{jk}. \end{aligned}$$

Hence, we get $a_{jk} = 0$ for all j, k with $1 \leq k < j \leq 3g-3$.

Therefore, we have proved that

$$\omega_{WP} = \sum_{j=1}^{3g-3} dx_{3g-3+j} \wedge dx_j,$$

as desired. □

Notes

This chapter follows Wolpert's papers [247] and [251]. We remark again that the sign of the parameter t of the FN deformation is different from that in these papers.

For some applications of Wolpert's formula, see Appendix B.5.

Several potential functions of the fundamental form ω_{WP} of the Weil-Petersson metric have been obtained, for instance, in Tromba [236], Wolf [244], Wolpert [254], and Zograf and Takhtadzhyan [264], [265], [266]. See also Jost [A-49], and Takhtadzhyan [219].

Appendix A

Classical Variations on Riemann Surfaces

We have discussed several ways to introduce global coordinates on the Teichmüller space $T(R)$ of a given Riemann surface R. However, for the local theory, we need only to formulate "small deformation" of the given surface R.

Before quasiconformal mappings began to be used popularly, such "small deformation" had been investigated as classical variations of Riemann surfaces. A typical classical variation is Schiffer's interior variation, which is considered as a generalization of the more classical and fundamental Hadamard's variation (cf. Appendix of Courant [A-23], and Schiffer and Spencer [A-94]).

The complex structure of the Teichmüller space $T(R)$ of a closed Riemann surface R was firstly introduced by L. Ahlfors, by considering "small deformations" induced by Teichmüller mappings. This is also considered as another generalization of Hadamard's variation.

We explain briefly Schiffer's interior variation and Ahlfors' argument to introduce the complex structure of the Teichmüller space T_g of genus g (≥ 2) in A.1 and A.2, respectively.

Finally, in A.3, we shall discuss degeneration of Riemann surfaces, another type of fundamental classical variation.

A.1. Schiffer's Interior Variation

Roughly speaking, Hadamard's variation is a deformation of the boundary, but Schiffer's interior variation is a deformation in the interior of one (or more) local disk(s). We shall explain Schiffer's interior variation in a simple but typical case.

Let R be an arbitrary Riemann surface. Fix a coordinate neighborhood (U, z) around a given point $p \in R$. Assume that $z(p) = 0$ and $z(U) = \{ z \in \mathbf{C} \mid |z| < 2 \}$. Set $B = \{ z \in \mathbf{C} \mid |z| < 1 \}$, and $D_p = z^{-1}(B)$. Denote by C_p the relative boundary of D_p.

We construct a family $\{R_\epsilon\}$ of Riemann surfaces depending on a complex parameter ϵ (with sufficiently small $|\epsilon|$), by deforming the complex structure of R only on D_p. For this purpose, we consider the mapping

$$z_\epsilon = z + \frac{\epsilon}{z}$$

on U for every ϵ. When $|\epsilon|$ is sufficiently small, $z_\epsilon(C_p)$ is a simple closed curve (actually an ellipse in this case) in the z_ϵ-plane, which is denoted by C_ϵ, and z_ϵ gives a conformal mapping of a suitable neighborhood A_ϵ of C_p (see Fig. A.1).

Now, delete D_p from R, and paste the domain D_ϵ in the z_ϵ-plane surrounded by C_ϵ. More precisely, set $V_\epsilon = D_\epsilon \cup z_\epsilon(A_\epsilon)$, and glue $(R - D_p) \cup A_\epsilon$ and V_ϵ by identifying $z_\epsilon(A_\epsilon)$ and A_ϵ under the mapping z_ϵ. Then we have a family $\{R_\epsilon\}$ of Riemann surfaces depending on the complex parameter ϵ, which is a special case of Schiffer's interior variation. Here, note that, considering V_ϵ as a subdomain of R_ϵ, we can take z_ϵ as a local coordinate on V_ϵ.

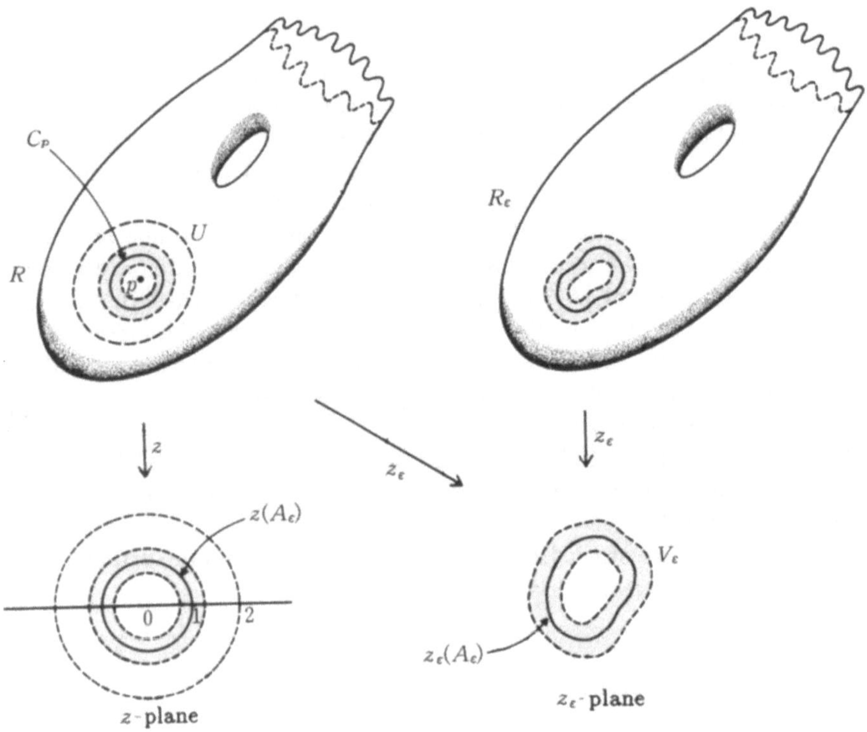

Fig. A.1.

When several mutually disjoint points, say $p_1, \cdots, p_n$, are given on R, we take a coordinate neighborhood (U_j, z_j) for every p_j so that

$$\begin{aligned}
&z_j(p_j) = 0, \\
&z_j(U_j) = \{ z \in \mathbf{C} \mid |z| < 2 \}, \quad j = 1, \cdots, n, \\
&\overline{U}_j \cap \overline{U}_k = \phi, \quad j \neq k.
\end{aligned}$$

Set $D_j = z_j^{-1}(\{z \in \mathbf{C} \mid |z| < 1\})$. For any complex number ϵ_j with sufficiently small $|\epsilon_j|$, consider the mapping

$$z_{j,\epsilon} = z_j(p) + \frac{\epsilon_j}{z_j(p)}, \quad p \in D_j.$$

As before, we can construct a family $\{R_\epsilon\}$ of Riemann surfaces depending on complex parameters $\epsilon = (\epsilon_1, \cdots, \epsilon_n)$.

Now, the following theorem is fundamental. (For instance, see Patt [174].)

Theorem A.1. *Let R be a closed Riemann surface of genus g (≥ 2). Let $3g-3$ mutually distinct points $\{p_j^*\}_{j=1}^{3g-3}$ on R, and a neighborhood V_j of p_j^* for every j, be given arbitrarily. Then there exists a set of points $\{p_j\}_{j=1}^{3g-3}$ with $p_j \in V_j$ which satisfies the following condition: let $\{R_\epsilon\}$ be the family of Riemann surfaces constructed as before from $\{p_j\}_{j=1}^{3g-3}$ and ϵ in a sufficiently small neighborhood $W(\subset \mathbf{C}^{3g-3})$ of $\epsilon = 0$. Then the inverse of the mapping*

$$\epsilon \longmapsto [R_\epsilon]$$

of W into $T(R) = T_g$ gives a system of holomorphic local coordinates around the base point of $T(R)$. (Here, R_ϵ is equipped with the naturally defined marking, which is henceforth omitted.)

Moreover, the inverse of the mapping $\epsilon \longmapsto [R_\epsilon]$ for $\{p_j\}_{j=1}^{3g-3}$ gives a system of holomorphic local coordinates if and only if any element in $A_2(R)$ vanishing at every p_j is identically zero.

This classical result is easily shown by representing Schiffer's interior variation by a quasiconformal deformation as follows. See also Gardiner [79], and Nag [A-80], Chapter 4.

To give a proof of Theorem A.1, we put

$$f_{j,\epsilon}(z_j) = z_j + \epsilon_j \bar{z}_j$$

for every j $(= 1, \cdots, 3g-3)$. Each $f_{j,\epsilon}$ is of class C^∞ on $B_j = \{ z \in \mathbf{C} \mid |z_j| < 1 \}$, continuous on $\overline{B_j}$, and coincident with $z_{j,\epsilon}$ on the boundary of B_j. We set

$$f_\epsilon(p) = \begin{cases} (z_{j,\epsilon})^{-1} \circ f_{j,\epsilon} \circ z_j(p), & p \in D_j,\ j = 1, \cdots, 3g-3 \\ p, & p \in R - \bigcup_{j=1}^{3g-3} D_j. \end{cases}$$

A simple computation shows that f_ϵ is a quasiconformal mapping of R to R_ϵ with Beltrami coefficient

$$\mu_\epsilon(p) = \begin{cases} \epsilon_j \, d\overline{z_j}/dz_j, & p \in D_j,\ j = 1, \cdots, 3g-3 \\ 0, & p \in R - \bigcup_{j=1}^{3g-3} D_j \end{cases}$$

for every ϵ sufficiently near to the origin.

Proof of Theorem A.1. Fix a sufficiently small neighborhood W of the origin in $\mathbf{C}^{3g-3}$. Define a mapping $F\colon W \longrightarrow B(R)_1$ by $F(\epsilon) = \mu_\epsilon$. Then F is a holomorphic mapping on W, and $(\partial F/\partial \epsilon_j)(0) \in B(R)$ is given by

$$\mu_j(p) = \begin{cases} d\overline{z_j}/dz_j, & p \in D_j \\ 0, & p \in R - D_j. \end{cases}$$

On the other hand, by using the mean value property we obtain

$$\iint_R \psi \cdot \mu_j = \iint_{B_j} \psi(z_j)\, dz_j \wedge d\overline{z_j} = -2\pi i \psi(p_j).$$

Here, writing $\psi = \psi_j(z_j)dz_j^2$ on B_j, we set $\psi(p_j) = \psi_j(0)$.

Now, to prove that F is biholomorphic in a neighborhood of the origin, it suffices by the implicit function theorem that $\{\mu_j\}_{j=1}^{3g-3}$ gives a basis of the tangent space $T_0(T(R))$. Since $T_0(T(R))$ is identified with $A_2(R)^*$ by Theorem 7.5, this is equivalent to assert that any complex vector $(c_1, \cdots, c_{3g-3})$ satisfying

$$\left(\sum_{j=1}^{3g-3} c_j \mu_j, \psi \right)_R = 0, \quad \psi \in A_2(R),$$

should be zero.

The second assertion is seen easily by linear algebra. As $\{p_j\}_{j=1}^{3g-3}$ in the first assertion, we can choose a set of points such that $\det(\varphi_k(p_j)) \neq 0$, where $\{\varphi_k\}_{k=1}^{3g-3}$ is a base for $A_2(R)$. □

A.2. Period Matrices as Moduli

As stated before, the first introduction of the complex structure of T_g ($g \geq 2$) was based on investigations of period matrices. We shall show Rauch's variational formula for the period matrices, following Ahlfors [5], and explain how to get local coordinates of T_g by using this formula.

First, we recall some fundamental terminology. Let R be a closed Riemann surface of genus g (≥ 2). Fix a set of $2g$ simple closed curves on R which induces a *canonical homology base*, i.e., a canonical base of the first homology group $H_1(R, \mathbf{Z})$. In this section, we use the notation $\{A_j, B_j\}_{j=1}^g$ for this set (see Fig. A.2).

For every $[S, f] \in T(R) = T_g$, we have a set $\{f(A_j), f(B_j)\}_{j=1}^g$ of simple closed curves on S which induces a canonical homology base on S. We denote this set by the same notation $\{A_j, B_j\}_{j=1}^g$.

Next, on every S, there exists uniquely a set $\{\theta_j\}_{j=1}^g$ of holomorphic Abelian differentials, i.e., holomorphic 1-forms on S such that

$$\int_{A_k} \theta_j = \delta_{jk}, \quad j, k = 1, \cdots, g.$$

We call this $\{\theta_j\}_{j=1}^g$ the *canonical base of the space of holomorphic Abelian differentials* on S with respect to $\{A_j, B_j\}_{j=1}^g$. Further, we put

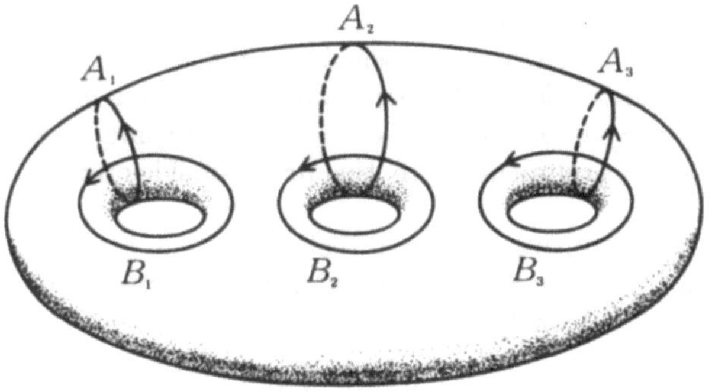

Fig. A.2.

$$\pi_{jk} = \int_{B_k} \theta_j, \quad j,k = 1, \cdots, g.$$

We call this $g \times g$ matrix $\Pi(S) = (\pi_{jk})$ the *canonical period matrix* of S with respect to $\{A_j, B_j\}_{j=1}^g$. Recall that the classical period relation implies that $\Pi(S)$ is symmetric, and that the imaginary part of $\Pi(S)$ is positive definite. Hence, we obtain a mapping

$$\Pi : T(R) \to \mathfrak{S}_g$$

defined by

$$\Pi([S, f]) = \Pi(S), \quad [S, f] \in T(R).$$

Here, $\mathfrak{S}_g$ ($\subset \mathbf{C}^{g(g+1)/2}$) is the *Siegel upper half-space*. (For more details, see for instance, Ahlfors and Sario [A-6], and Farkas and Kra [A-28].)

We give a proof of the following theorem.

Theorem A.2. *The mapping $\Pi : T_g \longrightarrow \mathfrak{S}_g$ is holomorphic.*

Moreover, when $g > 2$, then the derivative $d\Pi$ of Π has the maximal rank $3g - 3$ at any point of T_g except for those corresponding to hyperelliptic surfaces.

When $g = 2$, $d\Pi$ has the maximal rank 3 at every point of T_2.

To prove the first assertion, we show complex differentiability of Π. By a translation of the base point, it suffices to prove this at the base point.

Proposition A.3. (Rauch's variational formula) *For every $\mu \in B(R)$ of class C^∞, the derivative $(d\pi_{jk})_0[\mu]$ of every π_{jk} in the direction μ at the base point exists, and equals*

$$(d\pi_{jk})_0[\mu] = \iint_R \theta_j \theta_k \cdot \mu.$$

(Here, the product $\theta_j \theta_k$ is regarded as a quadratic differential.)

Proof of Theorem A.2. From Proposition A.3 and Hartogs' theorem (cf. Bers [A-14]), we can see the first assertion.

Assume that $g > 2$. Then the classical theorem of M. Noether gives that, if S is a non-hyperelliptic closed Riemann surface of genus g, we can find a base of $A_2(S)$ among the set of products of two holomorphic Abelian differentials. Hence, by the same argument as in the proof of Theorem A.1, we have the second assertion.

Finally, when $g = 2$, we can see directly that the set of products of holomorphic Abelian differentials spans $A_2(R)$ for every closed Riemann surface S of genus two. Thus, we conclude the third assertion. □

Furthermore, Theorem A.2 implies the following:

Corollary. *The complex structure of T_g introduced in Chapter 6 is the unique one under the condition that the canonical period matrix moves holomorphically on T_g.*

Remark. Besides $\{\pi_{jk}\}$, Ahlfors considered integrals of holomorphic Abelian differentials along suitable 1-chains, and succeeded in introducing a system of holomorphic local coordinates at every point of T_g (cf. Ahlfors [5]). This was the first introduction of the standard complex structure of T_g.

Proof of Proposition A.3. Fix a smooth Beltrami differential $\mu \in B(R)$. For every complex number ϵ with $\|\epsilon\mu\|_\infty < 1$, let $f_\epsilon : R \longrightarrow R_\epsilon$ be a quasiconformal mapping with complex dilatation $\epsilon\mu$. Let $\{\theta_{j,\epsilon}\}_{j=1}^g$ be the canonical base of holomorphic Abelian differentials on R_ϵ with respect to $\{A_j, B_j\}_{j=1}^g$.

Fix j arbitrarily. We set

$$\omega = (f_\epsilon)^*(\theta_{j,\epsilon}) - \theta_{j,0},$$

where $(f_\epsilon)^*(\theta_{j,\epsilon})$ is the pull-back of $\theta_{j,\epsilon}$ by f_ϵ. Then ω is a square integrable closed differential on R, and we have

$$\int_{A_k} \omega = \int_{A_k} \theta_{j,\epsilon} - \int_{A_k} \theta_{j,0} = 0, \quad k = 1, \cdots, g.$$

Hence, the period relation implies that

$$(\omega, \omega^*) = \iint_R -\omega \wedge \bar{\omega} = 0. \tag{A.1}$$

Let z_ϵ be a generic local parameter on R_ϵ. We write

$$\theta_{j,\epsilon} = a_{j,\epsilon}(z_\epsilon)dz_\epsilon$$

with holomorphic function $a_{j,\epsilon}$. Letting z be a generic local parameter on R, we decompose ω as

$$\omega = \omega_1 + \omega_2; \tag{A.2}$$

$$\omega_1 = (a_{j,\epsilon} \circ f_\epsilon(z) \cdot (f_\epsilon)_z(z) - a_{j,0}(z))\, dz,$$
$$\omega_2 = a_{j,\epsilon} \circ f_\epsilon(z) \cdot (f_\epsilon)_{\bar z}(z) d\bar z.$$

A simple computation shows that (A.1) is equivalent to

$$||\omega_1||^2 = ||\omega_2||^2, \tag{A.3}$$

where $||\omega_j||^2 = (\omega_j, \omega_j)$ for each j.

Next, we set

$$\hat\omega_2 = a_{j,\epsilon} \circ f_\epsilon(z) \cdot (f_\epsilon)_z(z) dz.$$

Then we have

$$\omega_2 = \epsilon\mu\,\hat\omega_2.$$

By (A.3) and the triangle inequality, we can show from (A.2) that

$$||\hat\omega_2|| \le \frac{||\theta_{j,0}||}{1 - ||\epsilon\mu||_\infty}. \tag{A.4}$$

In particular, by (A.3),

$$||\omega_1|| \le \frac{||\epsilon\mu||_\infty \cdot ||\theta_{j,0}||}{1 - ||\epsilon\mu||_\infty}. \tag{A.5}$$

Finally, again by the period relation, we have

$$\pi_{jk}(\epsilon) - \pi_{jk}(0) = -(\theta_{k,0}, {}^*\bar\omega) = \iint_R \theta_{k,0} \wedge \omega_2.$$

We express this difference as

$$\pi_{jk}(\epsilon) - \pi_{jk}(0) = \epsilon \iint_R \theta_{k,0}\theta_{j,0} \cdot \mu + \eta_{jk}.$$

From (A.5), we obtain

$$\begin{aligned} |\eta_{jk}| \le & \left| \iint_R \theta_{k,0}\omega_1 \cdot \epsilon\mu \right| \\ \le & \frac{|\epsilon|^2 ||\mu||_\infty^2}{1 - ||\epsilon\mu||_\infty} \cdot ||\theta_{j,0}|| \cdot ||\theta_{k,0}|| = O(|\epsilon|^2). \end{aligned}$$

Thus the assertion follows. □

A.3. Degeneration of Riemann Surfaces

As another classical and fundamental variation of Riemann surfaces, so-called degenerations of surfaces, have been deeply investigated. This variation is essentially different from those due to quasiconformal mappings, and gives a prototype of sequences converging to the boundary of the moduli space. See Appendix B.

First, let R_1 and R_2 be two closed Riemann surfaces, and $\{A_j, B_j\}_{j=1}^{g_1}$ and $\{A_{g_1+k}, B_{g_1+k}\}_{k=1}^{g_2}$ be sets of simple closed curves on R_1 and R_2, respectively, which give a canonical homology basis on R_1 and R_2, respectively (see Fig. A.3). Here, g_j $(j = 1, 2)$ is the genus of R_j, which we assume to be positive.

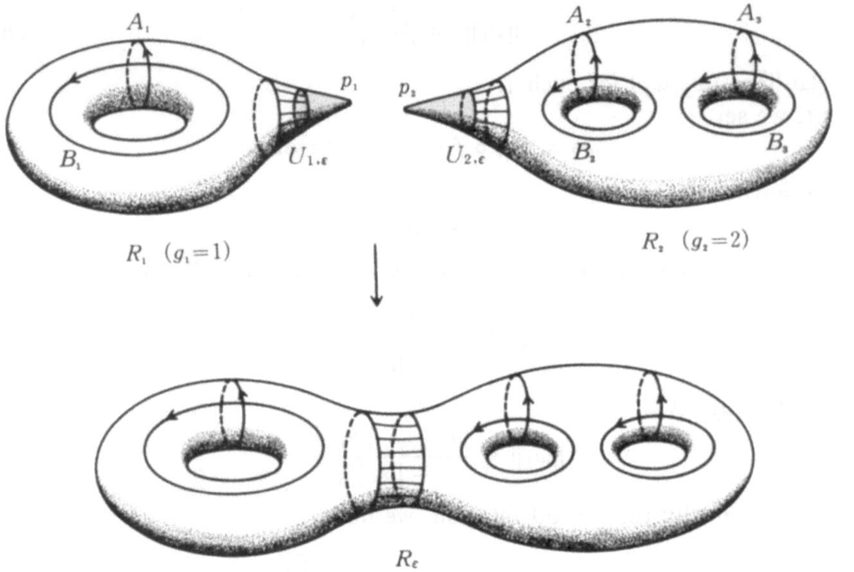

Fig. A.3.

For each j, fix a point $p_j \in R_j$, and a coordinate neighborhood (U_j, z_j) around p_j such that $z_j(p_j) = 0$ and $z_j(U_j) = B_j = \{ z \in \mathbf{C} \mid |z_j| < 1 \}$. For every complex ϵ with $0 < |\epsilon| < 1$, we set

$$U_{j,\epsilon} = U_j - \{ z \in \mathbf{C} \mid |z_j| \leq |\epsilon| \}, \quad j = 1, 2.$$

Then, identifying $U_{1,\epsilon}$ and $U_{2,\epsilon}$ by the mapping

$$z_1 \cdot z_2 = \epsilon,$$

we obtain a Riemann surface R_ϵ of genus $g = g_1 + g_2$ (see Fig. A.3). Note that $g \geq 2$.

When $\epsilon = 0$, we take as R_0 the closed Riemann surface with a node (which comes from the identification of $p_1 \in R_1$ and $p_2 \in R_2$. Also cf. Appendix B). Thus we have constructed a family $\{R_\epsilon \mid |\epsilon| < 1\}$, which we call a *degeneration* to R_0 with respect to (U_1, U_2). On every R_ϵ, we consider the canonical base $\{\theta_{j,\epsilon}\}_{j=1}^{g}$ of holomorphic Abelian differentials with respect to $\{A_j, B_j\}_{j=1}^{g}$, and define the canonical period matrix $\Pi(\epsilon) = (\pi_{jk}(\epsilon))$. Furthermore, the following variational formula is known. (For the proof, see Fay [A-30], and Yamada [261].)

Theorem A.4.

$$\Pi(\epsilon) = \begin{bmatrix} \Pi_1 & 0 \\ 0 & \Pi_2 \end{bmatrix} - 2\pi i\epsilon \begin{bmatrix} 0 & {}^tX_1X_2 \\ {}^tX_2X_1 & 0 \end{bmatrix} + O(\epsilon^2) \quad (|\epsilon| \to 0).$$

Here, for each $j = 1, 2$, Π_j is the canonical period matrix on R_j, and

$$X_j = (a_{j,1}(0), \cdots, a_{j,g_j}(0))$$

with $\theta_{j,k} = a_{j,k}(z_j)dz_j$ on U_j ($k = 1, \cdots, g_j$), where $\{\theta_{j,k}\}_{k=1}^{g_j}$ is the canonical base of holomorphic Abelian differentials on R_j.

Next, we can treat the case that $R_1 = R_2$ in a similar way. More precisely, let R be a closed Riemann surface of genus $g-1$ (≥ 1), and $\{A_j, B_j\}_{j=1}^{g-1}$ be a set of simple closed curves on R which induces a canonical homology base. Fix two distinct points p_1 and p_2 on R, and coordinate neighborhoods (U_j, z_j) around p_j ($j = 1, 2$) such that $z_j(p_j) = 0$ and $z_j(U_j) = \{ z \in \mathbf{C} \mid |z_j| < 1 \}$ for each j, and that $\overline{U_1} \cap \overline{U_2} = \phi$. Then, in the same way as before, we can construct a family $\{R_\epsilon \mid |\epsilon| < 1\}$, which we also call a *degeneration* to R_0 with respect to (U_1, U_2).

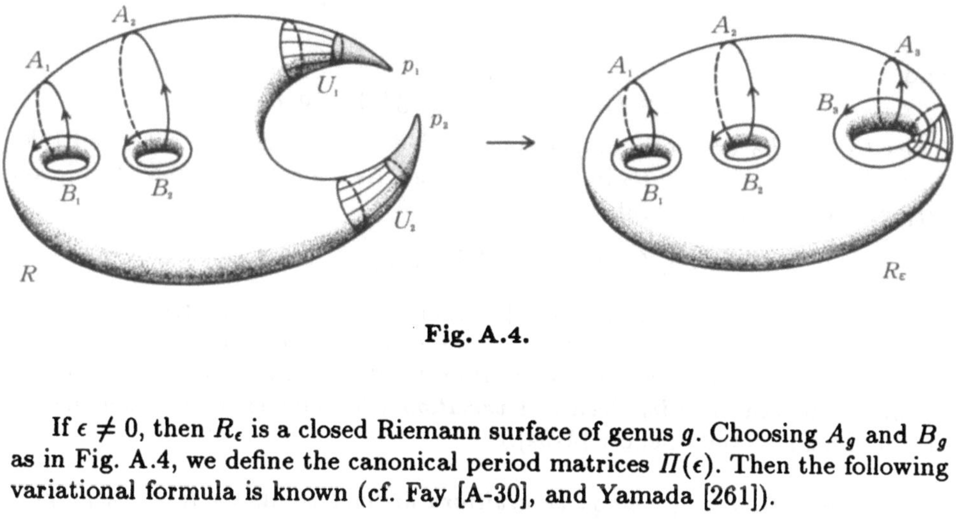

Fig. A.4.

If $\epsilon \neq 0$, then R_ϵ is a closed Riemann surface of genus g. Choosing A_g and B_g as in Fig. A.4, we define the canonical period matrices $\Pi(\epsilon)$. Then the following variational formula is known (cf. Fay [A-30], and Yamada [261]).

Theorem A.5.

$$\Pi(\epsilon) = \begin{bmatrix} \Pi_0 & {}^tX \\ X & \frac{1}{2\pi i}\log\epsilon + c_0 \end{bmatrix} + O(\epsilon) \quad (|\epsilon| \to 0).$$

Here, Π_0 is the canonical period matrix on R, X is a $(g-1)$-dimensional vector, and c_0 is a suitable constant.

We could write down explicitly the values of X and c_0, and the first order term with respect to ϵ. However, we do not delve into the details. See Corollary 6 in Yamada [261].

In the described degeneration to R_0, if we take a special (U_1, U_2), and restrict ϵ to the set $\{\eta t^2 \mid 0 \le t < 1\}$, where η is a constant with $|\eta| = 1$, then we obtain the so-called *Schiffer-Spencer's variation* (cf. Schiffer and Spencer [A-94]).

A variation corresponding to the second case is called the *variation by attaching a handle*, which we shall explain more closely.

Let R be as in the second case. Fix two distinct points p_1 and p_2 on R. Fix also a point $p_0 \in R - \{p_1, p_2\}$. Then there exists uniquely a harmonic function $G(p)$ on $R - \{p_1, p_2\}$ such that:

(i) $G(p_0) = 0$,

(ii) $G(p) - \log(1/|z_1(p)|)$ is extended to a harmonic function in a neighborhood of p_1, and

(iii) $G(p) - \log|z_2(p)|$ is extended to a harmonic function in a neighborhood of p_2,

where z_j is a local coordinate with $z_j(p_j) = 0$ at p_j for each j. We may call this $G(p)$ the *Green function* on R normalized at p_0 with positive pole at p_1 and with negative pole at p_2.

Now, for a sufficiently small positive t_0, we put

$$U_1 = \left\{ p \in R \,\middle|\, G(p) > \log \frac{1}{t_0} \right\} \cup \{p_1\},$$

and

$$U_2 = \{p \in R \mid G(p) < \log t_0\} \cup \{p_2\}.$$

Then U_1 and U_2 are both simply connected domains. Fix such a t_0, and take these domains as U_1 and U_2 in the second case. As a local coordinate, say ζ_j, on U_j, we choose one such that

$$\operatorname{Re} \zeta_j(p) = \exp\left((-1)^j G(p)\right), \quad j = 1, 2.$$

Fix a complex constant η with $|\eta| = 1$, and consider $\epsilon = \eta t^2$ for every t with $0 < t < t_0$. In the same way as before, we can construct a family $\{R_{\eta t^2} \mid 0 \le t < t_0\}$. This is *Schiffer-Spencer's variation by attaching a handle*. We know the following classical *Schiffer-Spencer's variational formula*. (For the proof, see for instance Schiffer and Spencer [A-94].)

Theorem A.6. *Let q_1 and q_2 be arbitrary distinct points on $R - \{p_0, p_1, p_2\}$. Let $g_t(p)$ be the Green function on $R_{\eta t^2}$ normalized at p_0 with positive pole at q_1 and with negative pole at q_2. Then*

$$g_t(p) - g_0(p) = \frac{1}{2\log t} G(p)\left(G(q_1) - G(q_2)\right) + t^2 \operatorname{Re}(\eta c) + o(t^2) \quad (t \to 0).$$

Here, c is a constant independent of t, and the convergence is locally uniform on $R - \{p_1, p_2, q_1, q_2\}$.

Notes

In Proposition A.3, for the sake of simplicity, we consider only a smooth Beltrami differential μ. However, the result still holds for a general μ.

Even for an arbitrary Riemann surface R, we can consider the "canonical period matrix" by imposing a suitable condition on the boundary behavior of holomorphic Abelian differentials. These "canonical period matrices" move holomorphically on the Teichmüller space $T(R)$. See for instance Kusunoki and Maitani [129]. As for variations in the case of general Riemann surfaces, we also cite Maitani [134], Shiba [197], [198], Taniguchi [226], [227], and Yamaguchi [262].

When the Teichmüller space $T(R)$ of an open Riemann surface R is infinitely dimensional, such an assertion as Theorem A.1 does not necessarily hold. For example, let E be a subset of R. Then, in order to conclude that

$$\{[S, f] \in T(R) \mid f \text{ is conformal on } R - E\}$$

covers a neighborhood of the base point of $T(R)$, we should assume that E is "sufficiently large" (cf. Ohtake [171] and Taniguchi [228]).

For arbitrary Riemann surfaces, we can consider variations by amalgamating quasiconformal deformations and degenerations, and can obtain variational formulas similar to those derived here for the canonical period matrices. See Taniguchi [224] and [225]. Moreover, Schiffer-Spencer's formula is still valid for an arbitrary Riemann surface. See Taniguchi [223] and [224].

Degeneration of surfaces is a fundamental tool in the study of the compactified moduli spaces. See also Appendix B.

Appendix B

Compactification of the Moduli Space

Following Bers [32], [33], [34], and [40], we shall construct a compactification of the moduli space M_g of closed Riemann surfaces of genus g by adjoining to M_g the set of biholomorphic equivalence classes of closed Riemann surfaces of genus g with nodes.

B.1 Compactification of M_1

As an example, we construct a compactification $\hat{M}_1$ of the moduli space M_1 of tori.

As was seen in the remark in §2.1 of Chapter 1, M_1 is identified with the complex plane $\mathbf{C}$, and every point in M_1 is represented by the biholomorphic equivalence class $[S_\lambda]$ of the torus S_λ defined by the algebraic equation $w^2 = z(z-1)(z-\lambda)$ for a complex number λ with $\lambda \neq 0, 1$. A *compactification* $\hat{M}_1$ of M_1 is the Riemann sphere $\hat{\mathbf{C}} = \mathbf{C} \cup \{\infty\}$. Here the point ∞ in $\hat{\mathbf{C}}$ corresponds to an algebraic curve given by the equation $w^2 = z(z-1)(z-\lambda)$ for $\lambda = 0, 1$, or ∞.

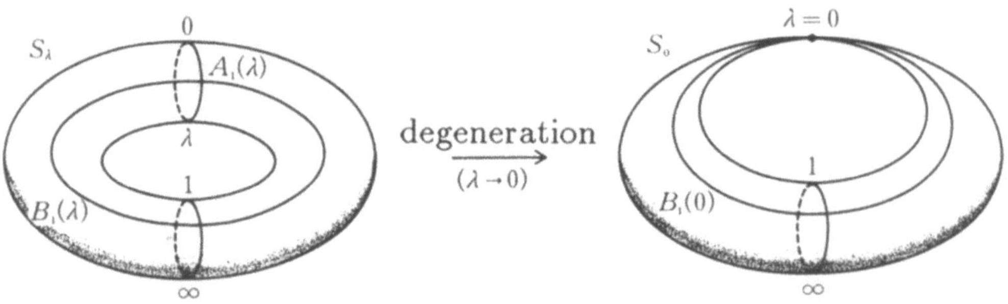

Fig. B.1.

For example, taking $\lambda = 0$, we see that the algebraic curve S_o given by the equation $w^2 = z^2(z-1)$ is the one which has a sole singular point at $p_o = (0, 0)$.

In a neighborhood of p_o, by the coordinate transformation $Z = w - z\sqrt{z-1}$ and $W = w + z\sqrt{z-1}$, the algebraic curve S_o is represented by

$$\{(Z,W) \in \mathbf{C} \times \mathbf{C} \mid ZW = 0,\ \ |Z| < r, |W| < r\},$$

where r is a positive constant. Thus, in a neighborhood of p_o, this S_o is regarded as the union of two disks $\{Z \in \mathbf{C} \mid |Z| < r\}$ and $\{W \in \mathbf{C} \mid |W| < r\}$, where two points $Z = 0$ and $W = 0$ are identified. Such a singular point p_o of S_o is called an *ordinary double point* or *node*. From this observation, we see that S_o looks like Fig. B.1, and it is obtained by the degeneration of a closed curve $A_1(\lambda)$ on S_λ into a node p_o as $\lambda \to 0$.

Next, we identify M_1 with the quotient space $H/PSL(2,\mathbf{Z})$ as in §2.1 of Chapter 1. Then the point $\infty \in \hat{\mathbf{C}}$ corresponds to the Riemann surface $\mathbf{C}/\Gamma_\infty$, where Γ_∞ is a subgroup of $Aut(\mathbf{C})$ generated by the translation $\delta(z) = z + 1$. This group Γ_∞ is obtained by the following process: let Γ_τ be the lattice group generated by two translations $\gamma_\tau(z) = z + \tau$, $\delta_\tau(z) = z + 1$ for some $\tau \in H$. Taking the shaded area in Fig. 1.9 as a fundamental domain F for $PSL(2,\mathbf{Z})$, we see that Γ_∞ is obtained by the degeneration of γ_τ as $\tau \to \infty$ through F. The Riemann surface $\mathbf{C}/\Gamma_\infty$ is biholomorphically equivalent to $\mathbf{C} - \{0\}$, and its one-point compactification is regarded as S_o.

B.2 Compactification of M_g for g ($\geqq 2$)

We construct a compactification of the moduli space M_g of closed Riemann surfaces of genus g ($\geqq 2$).

A compact connected Hausdorff space R is called a *closed Riemann surface of genus g with nodes* if R satisfies the following three conditions:

(i) Every $p \in R$ has a system of fundamental neighborhoods whose elements are homeomorphic to the unit disk $\{z \in \mathbf{C} \mid |z| < 1\}$ or the set $\{(z_1, z_2) \in \mathbf{C} \times \mathbf{C} \mid |z_1| < 1 \text{ and } |z_2| < 1\}$. (In the latter case, p is called a *node* of R. Since R is compact, the number m of nodes of R is finite. A connected component of R minus all its nodes is called a *part* of R.)

(ii) Every part R_j of R is a Riemann surface of analytically finite type (g_j, n_j), i.e., a Riemann surface which is obtained by removing n_j distinct points from a closed Riemann surface of genus g_j, and satisfies $2g_j - 2 + n_j > 0$; that is, R_j has the universal covering surface H.

(iii) If m and k are numbers of nodes and parts of R, respectively, then

$$g = \sum_{j=1}^{k} g_j + m + 1 - k.$$

Remark. Condition (iii) means that we get a closed Riemann surface R_o of genus g without nodes by opening each node on R (see Fig. B.2).

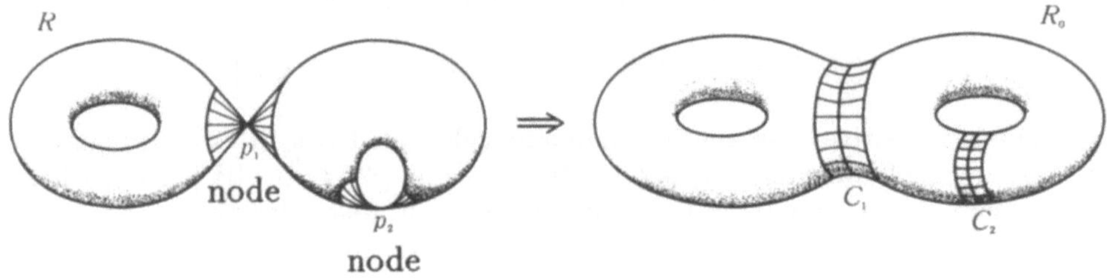

Fig. B.2.

A homeomorphism $f: R \to S$ between Riemann surfaces with nodes is said to be ***biholomorphic*** if f induces a biholomorphic mapping of R_j to a part of S for every part R_j of R. If there exists a biholomorphic mapping of R to S, then R and S are said to be ***biholomorphically equivalent***. We denote by $[R]$ the biholomorphic equivalence class of a closed Riemann surface R with nodes. As a *compactification* $\hat{M}_g$ of M_g with $g \geqq 2$, we take the union of M_g and the set of all biholomorphic equivalence classes of closed Riemann surfaces of genus g with at least one node.

Now, we define a topology on $\hat{M}_g$ by using the Fenchel-Nielsen coordinates as follows: let R be a closed Riemann surface of genus g with m nodes and k parts. As stated in the preceding remark, we can take a closed Riemann surface R_o of genus g without nodes and a system of decomposing curves $\mathcal{L} = \{C_1, \ldots, C_{3g-3}\}$ on R_o so that R is obtained from R_o by degenerating each element of a subset $\{C_{j_1}, \ldots, C_{j_m}\}$ of $\mathcal{L}$ into a point. Denote by $(\ell, \theta) = (\ell_1, \ldots, \ell_{3g-3}, \theta_1, \ldots, \theta_{3g-3})$ the Fenchel-Nielsen coordinates on the Teichmüller space T_g of genus g associated with $\mathcal{L}$ (see §2.1 of Chapter 3). From the proof of Theorem 3.10, for any point $(\ell, \theta) \in (\mathbf{R}^+)^{3g-3} \times \mathbf{R}^{3g-3}$, we can construct a closed Riemann surface $R_{\ell,\theta}$ of genus g such that $R_{\ell,\theta}$ induces a point in T_g whose Fenchel-Nielsen coordinates are (ℓ, θ).

Here, we admit the case where some $\ell_{j_1}, \ldots, \ell_{j_n}$ in $\{\ell_1, \ldots, \ell_{3g-3}\}$ vanish. In this case, we get a closed Riemann surface $R_{\ell,\theta}$ of genus g with n nodes by a construction similar to that in the proof of Theorem 3.10. (We consider that each element $C_{j_1}, \ldots, C_{j_n}$ in $\mathcal{L}$ degenerates into a point on $R_{\ell,\theta}$.) Then R has the Fenchel-Nielsen coordinates $(\ell(R), \theta(R)) = (\ell_1(R), \ldots, \ell_{3g-3}(R), \theta_1(R), \ldots, \theta_{3g-3}(R))$ with $\ell_{j_1}(R) = \cdots = \ell_{j_n}(R) = 0$. For any $(\epsilon, \delta) = (\epsilon_1, \ldots, \epsilon_{3g-3}, \delta_1, \ldots, \delta_{3g-3})$ with positive ϵ_j and δ_j, the (ϵ, δ)-neighborhood of $[R]$ in $\hat{M}_g$ is given by the set of all biholomorphic equivalence class $[R_{\ell,\theta}]$ of closed Riemann surfaces $R_{\ell,\theta}$ with or without nodes satisfying the following two conditions:

(i) $\quad |\ell_j - \ell_j(R)| < \epsilon_j$ for all $j = 1, \ldots, 3g-3$,

(ii) $\quad |\theta_j - \theta_j(R)| < \delta_j$ for all j with $\ell_j(R) \neq 0$.

These (ϵ, δ)-neighborhoods give a system of fundamental neighborhoods of $[R]$ in $\hat{M}_g$, which induces a Hausdorff topology on $\hat{M}_g$. Now we have the following theorem.

Theorem B.1. *The space $\hat{M}_g$ $(g \geq 2)$, consuructed as described, is a compact Hausdorff space.*

An outline of a proof of this theorem is explained in the next section.

B.3 Strong Deformation Spaces

In order to prove Theorem B.1, we construct the strong deformation space of a Riemann surface with nodes, a generalization of Teichmüller space in which closed Riemann surfaces are replaced by Riemann surfaces with nodes.

Let R and S be two closed Riemann surfaces with nodes of genus g (≥ 2). A continuous surjection $f: S \to R$ is called a *strong deformation* of R if it satisfies the following three conditions:

(i) The image of a node of S is a node of R.
(ii) The inverse image of a node of R is a node of S or a simple closed curve on S avoiding all nodes.
(iii) The restriction of f to the complement of the inverse images of all nodes of R is an orientation-preserving homeomorphism.

We say R is *terminal* if it has the largest possible number of nodes, namely $3g-3$ nodes. Note that there are, up to biholomorphic mappings, only finitely many terminal closed Riemann surfaces with nodes of genus g for a given g.

Two strong deformations $f_1: S_1 \to R$ and $f_2: S_2 \to R$ are said to be *equivalent* if there is a commutative diagram

$$\begin{array}{ccc} S_1 & \xrightarrow{f_1} & R \\ {\scriptstyle h}\downarrow & & \downarrow{\scriptstyle h_o} \\ S_2 & \xrightarrow[f_2]{} & R, \end{array}$$

where h and h_o are bijective homeomorphisms homotopic to a biholomorphic mapping and to the identity, respectively. The *strong deformation space* $\mathcal{D}(R)$ of R is the set of all equivalence classes $[f: S \to R]$ of strong deformations $f: S \to R$. If R has no nodes, $\mathcal{D}(R)$ is identified with the Teichmüller space $T(R)$ of R.

Every strong deformation $f_0: R' \to R$ induces a mapping $[f_0]_*: \mathcal{D}(R') \to \mathcal{D}(R)$ which sends $[f: S \to R']$ to $[f_0 \circ f: S \to R]$. The mappings $[\omega]_*$, induced by all orientation-preserving homeomorphisms $\omega: R \to R$, form a group. We call this group the *Teichmüller modular transformation group* of $\mathcal{D}(R)$ and denote it by $Mod(R)_*$.

We define a Hausdorff topology on $\mathcal{D}(R)$ as follows. Let S be a closed Riemann surface with nodes, and C be a closed curve on a part S_i of S. We set

$$\ell_S[C] = \inf_{C'} \ell_{S_i}(C'),$$

where C' runs over all closed curves on S_i freely homotopic to C, and $\ell_{S_i}(C')$ is the length of C' with respect to the hyperbolic metric on S_i. We also set $\ell_S[P] = 0$ if P is a node of S.

Let $\mathcal{C} = \{C_1, \dots, C_r\}$ be a finite set of closed curves on parts of S and ε a positive number. A strong deformation $h\colon S' \to S$ is said to be $(\mathcal{C}, \varepsilon)$-*small* if it satisfies

(i) $\quad |\ell_{S'}[h^{-1}(C_j)] - \ell_S[C_j]| < \varepsilon \quad$ for $j = 1, \dots, r$,

(ii) $\quad |\ell_{S'}[h^{-1}(q)]| < \varepsilon \quad$ for all nodes q of S.

We say that a set U in $\mathcal{D}(R)$ is open, if for every $[f\colon S \to R] \in U$, there exists a finite set $\mathcal{C}$ of closed curves on parts of S, and a positive number ε such that, whenever $h\colon S' \to S$ is $(\mathcal{C}, \varepsilon)$-small, the point $[f \circ h\colon S' \to R] \in \mathcal{D}(R)$ belongs to U (compare with §3 of Chapter 3).

There is a canonical projection $\Pi_R\colon \mathcal{D}(R) \to \hat{M}_g$ which sends $[f\colon S \to R]$ to $[S]$. It is seen that the canonical projection Π_R is a continuous open mapping.

Let us introduce the Fenchel-Nielsen coordinates on a strong deformation space.

First, assume that R is a terminal Riemann surface of genus g with nodes $\{p_j\}_{j=1}^{3g-3}$. Take an arbitrary point $[f\colon S \to R] \in \mathcal{D}(R)$. If $f^{-1}(p_j)$ is a simple closed curve on a part S_k of S, then we can choose a unique simple closed geodesic L_j on S_k which is freely homotopic to $f^{-1}(p_j)$. If $f^{-1}(p_j)$ is a node, then we put $L_j = f^{-1}(P_j)$. In this way, we have a system $\mathcal{L} = \{L_j\}_{j=1}^{3g-3}$ consisting of all nodes of S and some simple closed geodesics on parts of S.

If $f^{-1}(p_j)$ is not a node, we set

$$z_j = \frac{1}{2\pi} \ell_{S_k}(L_j)\, e^{i\theta_j}, \tag{B.1}$$

where θ_j is the twisting parameter with respect to L_j such that $0 \le \theta_j < 2\pi$ (cf. §2 of Chapter 3), and $\ell_{S_k}(L_j)$ is the length of the geodesic L_j measured by the hyperbolic metric on S_k. If $f^{-1}(p_j)$ is a node, put $z_j = 0$.

It is shown that the numbers $(z_1, \dots, z_{3g-3})$ depend only on the equivalence class $[f\colon S \to R]$ and the mapping of $\mathcal{D}(R)$ to $\mathbf{C}^{3g-3}$ sending $[f\colon S \to R]$ to $(z_1, \dots, z_{3g-3})$ is a homeomorphism, which is called the *Fenchel-Nielsen coordinates* on $\mathcal{D}(R)$. It is also proved that for every strong deformation $f_0\colon R' \to R$, the induced mapping $[f_0]_*\colon \mathcal{D}(R') \to \mathcal{D}(R)$ is a universal covering map onto its image, the image being the set of those $[f\colon S \to R]$ for which $f^{-1}(p_j)$ is not a node whenever $f_0^{-1}(p_j)$ is not a node. Furthermore, $\mathcal{D}(R')$ is homeomorphic to $\mathbf{C}^{3g-3}$.

Next, the *Fenchel-Nielsen coordinates* on the strong deformation space $\mathcal{D}(R')$ of an arbitrary closed Riemann surface R' of genus g with nodes are defined as follows: for every point $[h\colon S \to R'] \in \mathcal{D}(R')$, its Fenchel-Nielsen coordinates $(w_1, \ldots, w_{3g-3})$ are given by

$$\begin{aligned} w_j &= z_j && \text{if } f_0^{-1}(p_j) \text{ is a node of } R', \\ w_j &= \log z_j && \text{if } f_0^{-1}(p_j) \text{ is not a node of } R', \end{aligned}$$

where $(z_1, \ldots, z_{3g-3})$ are the Fenchel-Nielsen coordinates of $[f_0 \circ h\colon S \to R]$ given by (B.1). Here, if $f_0^{-1}(p_j)$ is not a node of S, then the monodromy theorem implies that the function $\log z_j$ has a single-valued continuous branch on $\mathcal{D}(R')$, since $\mathcal{D}(R')$ is homeomorphic to $\mathbf{C}^{3g-3}$ (cf. Lemma 3.9 in Chapter 3). It is shown that the mapping of $\mathcal{D}(R')$ to $\mathbf{C}^{3g-3}$ sending $[h\colon S \to R']$ to $(w_1, \ldots, w_{3g-3})$ is a homeomorphism.

In order to show the compactness of $\hat{M}_g$, we need the following lemma.

Lemma B.2. *For arbitrary $g \geqq 2$, there exists a positive constant L depending only on g such that every closed Riemann surface of genus g has a system $\mathcal{L} = \{C_1, \ldots, C_{3g-3}\}$ of decomposing curves satisfying $\ell(C_j) \leqq L$ for all $j = 1, \ldots, 3g-3$.*

This lemma is proved by using the *collar lemma*. See, for example, Abikoff [A-1], Lemma 3 in Chapter II, p. 98; Furusawa [78], and Matelski [150]. A different proof of this lemma is given by Buser (cf. Bers [40]).

Now, take all non-equivalent terminal Riemann surfaces $R_1, \ldots, R_r$ with nodes of genus g. Then, using Fenchel-Nielsen coordinates, Lemma B.2 implies that there are compact sets $K_j \subset \mathcal{D}(R_j)$ such that the images of $K_1, \ldots, K_r$ under the canonical projections $\Pi_{R_j}\colon \mathcal{D}(R_j) \to \hat{M}_g$ cover $\hat{M}_g$. Thus we conclude that $\hat{M}_g$ is compact.

B.4 Complex Structure of $\hat{M}_g$

We describe briefly how to introduce the complex structure of $\hat{M}_g$ in two ways.

The first introduction of the complex structure of $\hat{M}_g$ is the following. Let R_o be a closed Riemann surface of genus g without nodes, and let R be any other closed Riemann surface of the same genus with or without nodes. As stated in the previous section, $\mathcal{D}(R_o)$ is identified with the Teichmüller space $T(R_o)$, and hence $\mathcal{D}(R_o)$ has a complex structure. Take a strong deformation $f_0\colon R_o \to R$. Then the image $[f_0]_*(\mathcal{D}(R_o))$ of the induced mapping $[f_0]_*\colon \mathcal{D}(R_o) \to \mathcal{D}(R)$ is a domain in $\mathcal{D}(R)$ and $\mathcal{D}(R) - [f_0]_*(\mathcal{D}(R_o))$ is nowhere dense. Moreover, $[f_0]_*\colon \mathcal{D}(R_o) \to [f_0]_*(\mathcal{D}(R_o))$ is a universal covering map. Hence, $[f_0]_*(\mathcal{D}(R_o))$ has a natural complex structure induced from the complex structure of $\mathcal{D}(R_o)$. Then $\mathcal{D}(R)$ has a ringed structure, i.e., a continuous function on an open set U in $\mathcal{D}(R)$ is called ***holomorphic*** if its restriction to $U \cap [f_0]_*(\mathcal{D}(R_o))$ is holomorphic. This ringed structure induces a complex structure on $\mathcal{D}(R)$ and the induced mapping $[f_0]_*$ becomes holomorphic.

Using the Klein-Maskit combination theorems (Maskit [A-71]), we have the following theorem (cf. Bers [40]).

Theorem B.3. *$\mathcal{D}(R)$ is a complex manifold and is realized as a bounded domain in $\mathbf{C}^{3g-3}$.*

Now, we get the following two results.

Theorem B.4. *The Teichmüller modular transformation group $Mod(R)_*$ is a discrete subgroup of the analytic automorphism group of $\mathcal{D}(R)$. Moreover, the subroup $Mod_o(R)_*$ induced by the biholomorphic mappings of R onto itself is finite and is the stabilizer of $[id\colon R \to R]$ in $Mod(R)_*$.*

Theorem B.5. *There exists a neighborhood N of $[id\colon R \to R]$ in $\mathcal{D}(R)$, invariant under $Mod_o(R)_*$, such that the quotient space $N/Mod_o(R)_*$ is homeomorphic to a neighborhood of $[R]$ in $\hat{M}_g$.*

By H. Cartan's theorem, the quotient space $N/Mod_o(R)_*$ has a normal complex space structure. Thus $\hat{M}_g$ becomes a normal complex space, and it leads to the following, the main theorem in this Appendix.

Theorem B. 6. *The compactification $\hat{M}_g$ of the moduli space of closed Riemann surfaces of genus g ($\geqq 2$) has a normal complex space structure of dimension $3g-3$.*

The second introduction of the complex structure of $\hat{M}_g$ is the following. For a given closed Riemann surface R with nodes of genus g, considering both quasiconformal deformations and degenerations, we construct Riemann surfaces which represent a neighborhood of $[R]$ in $\hat{M}_g$ as follows.

Assume that R has m nodes $p_1, \dots, p_m$ and k parts $R_1, \dots, R_k$ such that each R_j is of type (g_j, n_j). Then it follows that

$$\sum_{j=1}^{k} n_j = 2m, \quad \sum_{j=1}^{k} g_j + m - k = g.$$

For each node p_α on R, suppose that p_α corresponds to a point obtained by identifying a point a_α in $\overline{R_{\alpha_1}} - R_{\alpha_1}$ with a point b_α in $\overline{R_{\alpha_2}} - R_{\alpha_2}$ for some α_1 and α_2 in $\{1, \dots, k\}$. Here, $\overline{R_{\alpha_j}}$ denotes the natural compactification of R_{α_j}. Take a local coordinate neighborhood (U_α^1, z_α) at a_α on $\overline{R_{\alpha_1}}$ such that $z_\alpha(a_\alpha) = 0$ and $z_\alpha(U_\alpha^1) = \Delta$, the unit disk. Similarly, choose a local coordinate neighborhood (U_α^2, w_α) at b_α on $\overline{R_{\alpha_2}}$ such that $w_\alpha(b_\alpha) = 0$ and $w_\alpha(U_\alpha^2) = \Delta$. Further, take a relatively compact open set V_j in R_j for every $j = 1, \dots, k$ so that $V = \bigcup_{j=1}^{k} V_j$ meets neither U_α^1 nor U_α^2 for all $\alpha = 1, \dots, m$.

Setting $N = 3g - 3 + m$, we can find Beltrami differentials $\mu_1, \dots, \mu_N$ on $R' = R - \{p_1, \dots, p_m\}$ such that μ_j vanishes identically outside V for all $j =$

$1, \ldots, N$, and such that they induce a basis of the tangent space of the product space $\prod_{j=1}^{k} T(R_j)$ of Teichmüller spaces $T(R_j)$ at the base point $[R', id]$ (cf. Theorem A.1 in Appendix A).

Let D be a sufficiently small open neighborhood of the origin in $\mathbf{C}^N$. Then, for any $s = (s_1, \ldots, s_N) \in D$, there exists a quasiconformal mapping $f_s : R' \to R'_s$ with Beltrami coefficient

$$\mu_s = \sum_{j=1}^{N} s_j \, \mu_j .$$

We may assume that the mapping of D into $\prod_{j=1}^{k} T(R_j)$ sending s to $[R'_s, f_s]$ is a biholomorphic mapping of D onto a neighborhood of $[R', id]$ in $\prod_{j=1}^{k} T(R_j)$. The image $R_{j,s} = f_s(R_j)$ of a part R_j under f_s is a Riemann surface of analytically finite type (g_j, n_j), and the restriction of f_s to R_j extends to a quasiconformal mapping of $\overline{R_j}$ to $\overline{R_{j,s}}$. Since f_s is conformal on U_α^1, we see that $(U_\alpha^1, z_{\alpha,s}) = (U_\alpha^1, z_\alpha \circ f_s^{-1})$ is a coordinate neighborhood of $f_s(a_\alpha)$ on $\overline{R_{\alpha_1,s}}$. Similarly, we have a coordinate neighborhood $(U_\alpha^2, w_{\alpha,s}) = (U_\alpha^2, w_\alpha \circ f_s^{-1})$ of $f_s(b_s)$ on $\overline{R_{\alpha_2,s}}$. From $R_{1,s}, \ldots, R_{k,s}$, identifying $f_s(a_\alpha)$ with $f_s(b_\alpha)$ for each $\alpha = 1, \ldots, m$, we obtain a closed Riemann surface R_s of genus g with m nodes and k parts. Denote by $p_{\alpha,s}$ the node on R_s corresponding to the node p_α on R.

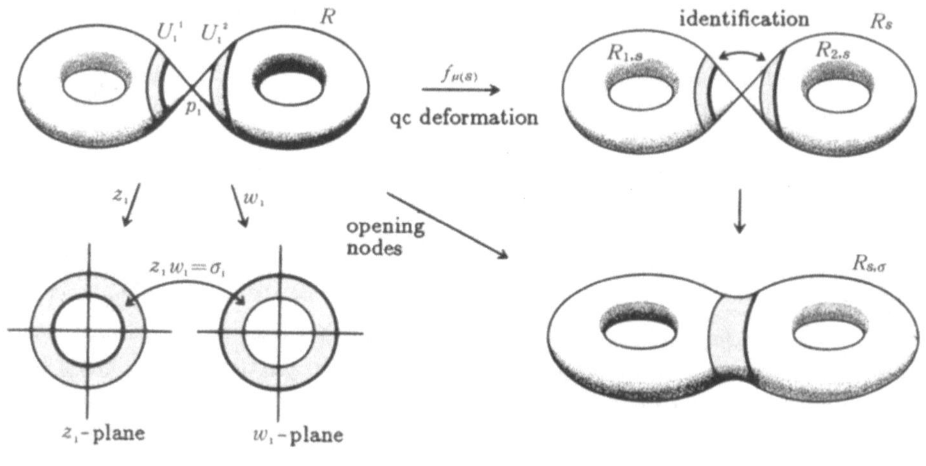

Fig. B.3.

Let $\sigma = (\sigma_1, \ldots, \sigma_m)$ be a point in $\mathbf{C}^m$ such that $|\sigma_\alpha| < 1$ for all $\alpha = 1 \ldots, m$. We put

$$U_{\alpha,s,\sigma}^1 = \{ p \in U_{\alpha,s}^1 \mid |z_{\alpha,s}(p)| \le |\sigma_\alpha| \},$$
$$U_{\alpha,s,\sigma}^2 = \{ p \in U_{\alpha,s}^2 \mid |w_{\alpha,s}(p)| \le |\sigma_\alpha| \},$$
$$R_{s,\sigma}^* = R_s - \left(\bigcup_{\alpha=1}^{m} U_{\alpha,s,\sigma}^1 \right) \bigcup \left(\bigcup_{\alpha=1}^{m} U_{\alpha,s,\sigma}^2 \right).$$

Identify any two points a and b in $R^*_{s,\sigma}$ if a and b are contained in $U^1_{\alpha,s,\sigma}$ and $U^2_{\alpha,s,\sigma}$, respectively, for some α, and if they satisfy $z_{\alpha,s}\, w_{\alpha,s} = \sigma_\alpha$. By this identification, we obtain a closed Riemann surface $R_{s,\sigma}$ of genus g with n nodes, where $m-n$ is the number of α's with $\sigma_\alpha \neq 0$ (see Fig. B.3).

Now, we set

$$\tilde{\mathcal{D}} = \{\, (s,\sigma) \in D \times \mathbf{C}^m \mid |\sigma_j| < 1, \quad j = 1,\ldots,m \,\},$$
$$\mathcal{D} = \{\, [R_{s,\sigma}] \in \hat{M}_g \mid (s,\sigma) \in \tilde{D} \,\}.$$

Then, $\mathcal{D}$ is a neighborhood of $[R]$ in $\hat{M}_g$. However, the mapping of $\tilde{\mathcal{D}}$ onto $\mathcal{D}$ sending (s,σ) to $[R_{s,\sigma}]$ is not always injective. By changing $\tilde{\mathcal{D}}$ suitably, we may assume that the biholomorphic automorphism group $Aut(R)$ of R induces a finite group G, consisting of analytic automorphisms of $\tilde{\mathcal{D}}$, such that the quotient space $\tilde{\mathcal{D}}/G$ is homeomorphic to a neighborhood of $[R]$ in $\hat{M}_g$. For details, we refer to Bers [33], XIII in §7, and Masur [143], §2, and Wolpert [249], §4. As before, H. Cartan's theorem implies that the quotient space $\tilde{\mathcal{D}}/G$ has a normal complex space structure. Thus, $\hat{M}_g$ becomes a normal complex space of dimension $3g-3$. Note that this complex space structure on $\hat{M}_g$ is equivalent to the one given in the first introduction.

B.5 Weil-Petersson Kähler Form on the Moduli Space

Since the Weil-Petersson Kähler form ω_{WP} on the Teichmüller space $T(R)$ of genus g ($\geqq 2$) is invariant under the action of the Teichmüller modular group $Mod(R)$, it is regarded as a form on the moduli space M_g. This form is denoted by the same notation ω_{WP}. We are interested in the behavior of ω_{WP} near the boundary of M_g in $\hat{M}_g$.

From the construction of $\hat{M}_g$ and Wolpert's formula (Theorem 8.6), ω_{WP} extends smoothly to the boundary with respect to the coordinates $(\ell,\tau) = (\ell_1,\ldots,\ell_{3g-3},\tau_1,\ldots,\tau_{3g-3})$, where $\tau_j = \ell_j\theta_j/2\pi$ for $j = 1,\ldots,3g-3$ and $(\ell,\theta) = (\ell_1,\ldots,\ell_{3g-3},\theta_1,\ldots,\theta_{3g-3})$ are the Fenchel-Nielsen coordinates associated with a system of decomposing curves on R. In particular, $\hat{M}_g$ has a finite volume with respect to the Weil-Petersson metric.

On the other hand, the boundary behavior of ω_{WP} with respect to the coordinates (s,σ) given in the previous section is studied by Masur [143], and Wolpert [251].

Next, ω_{WP} induces a cohomology class $[\omega_{WP}]$ on $\hat{M}_g$ such that $[\omega_{WP}]/\pi^2$ is a rational class (see Wolpert [249]). Thus multiplying $[\omega_{WP}]/\pi^2$ by some integer, we get a line bundle over $\hat{M}_g$. Wolpert [252] proved that this line bundle is positive, and consequently $\hat{M}_g$ is embedded in a complex projective space. Hence, we have the following result, which was first proved by Knudsen and Mumford [117] by using algebraic geometry.

Theorem B.7. *The compactification $\hat{M}_g$ of the moduli space M_g of genus g ($\geqq 2$) is a projective algebraic variety.*

Notes

The compactification of the moduli space of closed Riemann surfaces given in this appendix was first introduced by Deligne and Mumford [51]. Our treatment is due to Bers [32], [33], [34], [40], Masur [143], and Wolpert [249], [252]. We also refer to Kra [126]. Furthermore, see Mabuchi [133], Marden [136], and Namikawa [A-83].

For the geometry of the moduli space, there are many papers such as Arbarello and Cornalba [18], Bowditch and Epstein [46], Eisenbud and Harris [68], Giddings and Wolpert [86], Harer [90], [91], [92], Harer and Zagier [93], Harris [94], [95], Harris and Mumford [96], Maclachlan [132], Mumford [162], [163], Penner [175], and Wolpert [250], [255].

Recently, on Jacobi's problem on Jacobian varieties, there has been made a great progress. For this subject see, for example, Arbarello [17], Beauville [21], Mulase [160], and Shiota [204] and [205].

In string theory of particle physics, the theory of Teichmüller spaces and moduli spaces play an important role. For this interesting topic, we refer to Yau [A-112], Alvarez [15], Alvarez-Gaume and Nelson [16], Friedan [75], Nelson [167], and Polyakov [176]. There it is necessary to know the behavior of the Selberg zeta function and the Green functions near the boundary of the moduli space. These are studied in Hejhal [99], Taniguchi [226], and Wolpert [257]. In connection with super string theory, Teichmüller spaces and moduli spaces of super Riemann surfaces have been studied: for example, see Batchelor and Bryant [20], Crane and Rabin [50], LeBrun and Rothstein [131], Nelson [168], and Rabin and Freund [177].

References

Books and Proceedings

[A-1] Abikoff, W. : *The Real Analytic Theory of Teichmüller Space*, Lecture Notes in Math., Vol. 820, Springer-Verlag, Berlin and New York, 1980.

[A-2] Ahlfors, L. V. : *Lectures on Quasiconformal Mappings*, D. Van Nostrand, Princeton, New Jersey, 1966.

[A-3] Ahlfors, L. V. : *Conformal Invariants*, McGraw-Hill, New York, 1973.

[A-4] Ahlfors, L. V. : *Complex Analysis*, 3rd ed. McGraw-Hill, New York, 1979.

[A-5] Ahlfors, L. V. : *Collected Papers,* Vols. 1, 2, Birkhäuser, Boston, 1982.

[A-6] Ahlfors, L. V. and Sario, L. : *Riemann Surfaces*, Princeton University Press, Princeton, New Jersey, 1960.

[A-7] Ahlfors, L. V. et al. (eds.) : *Advances in the Theory of Riemann Surfaces, 1969 Stony Brook Conference*, Ann. Math. Studies, No. 66, Princeton University Press, Princeton, New Jersey, 1971.

[A-8] Ahlfors, L. V. et al. (eds.) : *Contributions to Analysis, A Collection of Papers Dedicated to L. Bers*, Academic Press, London, 1974.

[A-9] Arbarello, E., Cornalba, M., Griffiths, P. A. and Harris, J. : *Geometry of Algebraic Curves*, Vol. I, Springer-Verlag, Berlin and New York, 1984.

[A-10] Baily, W. L., Jr. : *Introductory Lectures on Automorphic Forms*, Iwanami-Shoten, Tokyo, and Princeton University Press, Princeton, New Jersey, 1973.

[A-11] Beardon, A. F. : *The Geometry of Discrete Groups*, Springer-Verlag, Berlin and New York, 1983.

[A-12] Bers, L. : *Topology,* Courant Institute of Mathematical Science, New York University Press, New York, 1956-1957.

[A-13] Bers, L. : *Riemann Surfaces*, Courant Institute of Mathematical Science, New York University Press, New York, 1957-1958.

[A-14] Bers, L. : *Introduction to Several Complex Variables*, Courant Institute of Mathematical Sciences, New York University Press, New York, 1964.

[A-15] Bers, L. et al. (eds.) : *A Crash Course on Kleinian Groups*, Lecture Notes in Math., Vol. 400, Springer-Verlag, Berlin and New York, 1974.

[A-16] Bers, L., John, F. and Schechter, M. : *Partial Differential Equations*, American Mathematical Society, Providence, Rhode Island, Providence, Rhode Island, Providence, Rhode Island, 1964.

[A-17] Besse, A. L. : *Einstein Manifolds*, Springer-Verlag, Berlin and New York, 1987.

[A-18] Birman, J. S. : *Braids, Links, and Mapping Class Groups*, Ann. Math. Studies, No. 82, Princeton University Press, Princeton, New Jersey, 1975.

[A-19] Casson, A. J. and Bleiler, S. A. : *Automorphisms of Surfaces after Nielsen and Thurston*, Cambridge Univ. Press, Cambridge, 1988.

[A-20] Chavel, I. and Farkas, H. M. (eds.) : *Differential Geometry and Complex Analysis* (*H. E. Rauch Memorial Volumes*), Springer-Verlag, Berlin and New York, 1985.

[A-21] Clemens, C. H. : *A Scrapbook of Complex Curve Theory*, Plenum Press, New York and London, 1980.

[A-22] Cohn, H.: *Conformal Mapping on Riemann Surfaces*, Dover, New York, 1967.

[A-23] Courant, R. : *Dirichlet's Principle, Conformal Mappings, and Minimal Surfaces*, Interscience, New York, 1950 (reprint, Springer-Verlag, Berlin and New York, 1977).

[A-24] Drasin, D. et al. (eds.) : *Holomorphic Functions and Moduli*, Vols. I and II, 1986 MSRI Conference, Springer-Verlag MSRI series, Springer-Verlag, Berlin and New York, 1988.

[A-25] Epstein, D. B. (ed.) : *Low-dimensional Topology and Kleinian Groups*, Cambridge Univ. Press, Cambridge, 1986.

[A-26] Epstein, D. B. (ed.) : *Analytical and Geometric Aspects of Hyperbolic Space*, Cambridge Univ. Press, Cambridge, 1987.

[A-27] Ezawa, H. and Kojima, I. (eds.) : *Developments of Mathematical Physics*, Tokyo-Tosho, Tokyo, 1988 (Japanese).

[A-28] Farkas, H. M. and Kra, I. : *Riemann Surfaces*, Springer-Verlag, Berlin and New York, 1980.

[A-29] Fathi, A., Laudenbach, F. and Poénaru, V. : *Travaux de Thurston sur les Surfaces*, 1976-1977 Orsay Seminaire, Astérisque, Vols. 66-67, Société Mathématique de France, Paris, 1979.

[A-30] Fay, J. D. : *Theta Functions on Riemann Surfaces*, Lecture Notes in Math., Vol. 352, Springer-Verlag, Berlin and New York, 1973.

[A-31] Ford, L. R. : *Automorphic Functions*, 2nd Ed., Chelsea, New York, 1951.

[A-32] Forster, O. : *Lectures on Riemann Surfaces*, Springer-Verlag, Berlin and New York, 1981.

[A-33] Fricke, R. and Klein, F. : *Vorlesungen über die Theori der Automorpfen Functionen*, Vol.I, Vol.II (1, 2), B. G. Teubner, Stuttbart, 1897, 1901, 1912.

[A-34] Gardiner, F. P. : *Teichmüller Theory and Quadratic Differentials*, Wiley, New York, 1987.

[A-35] Gehring, F. W. : *Characteristic Properties of Quasidisks*, Les Presses de L'université de Montreal, Montreal, 1982.

[A-36] Goldman, W. M. and Magid, A. R. (eds.) : *Geometry of Group Representations*, American Mathematical Society, Providence, Rhode Island, 1988.

[A-37] Greenberg, L. : *Discontinuous Groups and Riemann Surfaces, 1973 Maryland Conference*, Ann. Math. Studies No. 79, Princeton University Press, Princeton, New Jersey, 1974.

[A-38] Griffiths, P. A. : *Introduction to Algebraic Curves*, Translations of Mathematical Monographs, Vol.76, American Mathematical Society, Providence, Rhode Island, 1989.

[A-39] Griffiths, P. A. and Harris, J. : *Principles of Algebraic Geometry*, Wiley, New York, 1978.

[A-40] Gunning, R. C. : *Lectures on Riemann Surfaces*, Princeton University Press, Princeton, New Jersey, 1966.

[A-41] Harvey, W. J. (ed.) : *Discrete Groups and Automorphic Functions, 1975 Cambridge Institutional Conference*, Academic Press, London, 1977.

[A-42] Hirsh, M. W. : *Differential Topology*, Springer-Verlag, Berlin and New York, 1976.

[A-43] Hitotsumatsu, S. : *Theory of Analytic Functions of Several Complex Variables*, Baihukan, Tokyo, 1960 (Japanese).

[A-44] Hubbard, J. H. : *Sur les Sections Analytiques de la Courbe Universelle de Teichmüller*, Memoires of the American Mathematical Society Vol. 4, Number 166, American Mathematical Society, Providence, 1976.

[A-45] Iitaka, S., Ueno, K. and Namikawa, Y. : *Spirit of Descartes and Algebraic Geometry*, Nihonhyôron-sha, Tokyo, 1980 (Japanese).

[A-46] Ito, S. : *Introduction to Lebesgue Integral*, Shôkabo, Tokyo, 1963 (Japanese).

[A-47] Jenkins, J. A. : *Univalent Functions and Conformal Mappings*, Springer-Verlag, Berlin and New York, 1958.

[A-48] Jones, G. A. and Singerman, D. : *Complex Functions: An Algebraic and Geometric Viewpoint*, Cambridge University Press, Cambridge, 1987.

[A-49] Jost, J. : *Two-Dimensional Geometric Variational Problems*, Wiley, New York, 1991.

[A-50] Kato, M. : *Topology*, Science-sha, Tokyo, 1978 (Japanese).

[A-51] Kawada, Y. : *Theory of Automorphic Functions of One Variable*, Department of Mathematics, Tokyo University, Tokyo, 1963, 1964 (Japanese).

[A-52] Kawai, S. : *Algebraic Geometry*, Baihûkan, Tokyo, 1979 (Japanese).

[A-53] Klein, F. : *On Riemann's Theory of Algebraic Functions and their Integrals*, Dover, New York, 1963.

[A-54] Kobayashi, S. : *Hyperbolic Manifolds and Holomorphic Mappings*, Marcel Dekker, New York, 1970.

[A-55] Kobayashi, S. and Nomizu, K. : *Foundations of Differential Geometry*, Vols. 1, 2, Wiley, New York, 1963, 1969.

[A-56] Kodaira, K. : *Complex Analysis, I, II, III*, Iwanami-Shoten, Tokyo, 1978 (Japanese).

[A-57] Kodaira, K. : *Complex Manifolds and Deformation of Complex Structures*, Springer-Verlag, Berlin and New York, 1986.

[A-58] Kra, I. : *Automorphic Forms and Kleinian Groups*, W. A. Benjamin, Menlo Park, California, 1972.

[A-59] Kra, I. and Maskit, B. (eds.) : *Riemann Surfaces and Related Topics, 1978 Stony Brook Conference*, Ann. Math. Studies, No. 97, Princeton University Press, Princeton, New Jersey, 1981.

[A-60] Krushkal', S. L. : *Quasiconformal Mappings and Riemann Surfaces*, Winston, Washington, D. C., 1979.

[A-61] Krushkal', S. L., Apanasov, B. N. and Gusevkiĭ, N. A. : *Kleinian Groups and Uniformization in Examples and Problems*, Translations of Mathematical Monographs, Vol. 62, American Mathematical Society, Providence, Rhode Island, 1986.

[A-62] Kusunoki, Y. : *Theory of Analytic Functions*, Hirokawa-Shoten, Tokyo, 1962 (Japanese).

[A-63] Kusunoki, Y. : *Theory of Functions (Riemann Surfaces and Conformal Mappings)*, Asakura-Shoten, Tokyo, 1973 (Japanese).

[A-64] Lang, S. : *Introduction to Complex Hyperbolic Spaces*, Springer-Verlag, Berlin and New York, 1987.

[A-65] Lawrynowicż, J. and Krzyz, J. : *Quasiconformal Mappings in the Plane*, Lecture Notes in Math., Vol. 978, Springer-Verlag, Berlin and New York, 1983.

[A-66] Lehner, J. : *Discontinuous Groups and Automorphic Functions*, American Mathematical Society, Providence, Rhode Island, 1964.

[A-67] Lehner, J. : *A Short Course in Automorphic Functions*, Holt, Rinehart and Winston, New York, 1966.

[A-68] Lehto, O. : *Univalent Functions and Teichmüller Spaces*, Springer-Verlag, Berlin and New York, 1987.

[A-69] Lehto, O. and Virtanen, K. I. : *Quasiconformal Mappings in the Plane*, 2nd Ed., Springer-Verlag, Berlin and New York, 1973.

[A-70] Magnus, W. : *Noneuclidean Tesselations and their Groups*, Academic Press, London, 1974.

[A-71] Maskit, B. : *Kleinian Groups*, Springer-Verlag, Berlin and New York, 1987.

[A-72] Matsushima, Y. : *Differentiable Manifolds*, Marcel Dekker, New York, 1972.

[A-73] Mizohata, S. : *Theory of Partial Differential Equations*, Iwanami-Shoten, Tokyo, 1965 (Japanese).

[A-74] Mizohata, S. : *Lebesgue Integral*, Iwanami-Shoten, Tokyo, 1966 (Japanese).

[A-75] Moise, E. E. : *Geometric Topology in Dimensions 2 and 3*, Springer-Verlag, Berlin and New York, 1977.

[A-76] Morgan, J. W. and Bass, H. (eds.) : *The Smith Conjecture*, Academic Press, London, 1984.

[A-77] Morrow, J. and Kodaira, K. : *Complex Manifolds*, Holt, Rinehart and Winston, New York, 1971.

[A-78] Mumford, D. : *Curves and their Jacobians*, The University of Michigan Press, Ann Arbor, 1975.

[A-79] Murakami, S. : *Manifolds*, Kyôritsu-Shuppan, Tokyo, 1969 (Japanese).

[A-80] Nag, S. : *The Complex Analytic Theory of Teichmüller Spaces*, Wiley, New York, 1988.

[A-81] Nakai, M. : *Theory of Riemann Surfaces*, Morikita-Shuppan, Tokyo, 1980 (Japanese).

[A-82] Namba, M. : *Geometry of Projective Algebraic Curves*, Marcel Dekker, New York, 1984.

[A-83] Namikawa, Y. : *Troidal Compactification of Siegel Spaces*, Lecture Notes in Math., Vol. 812, Springer-Verlag, Berlin and New York, 1980.

[A-84] Nevanlinna, R. et al. (eds.) : *Analytic Functions, Princeton Conference*, Princeton University Press, Princeton, New Jersey, 1960.

[A-85] Newman, M. H. A. : *Elements of the Topology of Plane Sets of Points*, 2nd Ed., Cambridge University Press, Cambridge, 1951.

[A-86] Nicholls, P. J. : *The Ergodic Theory of Discrete Groups*, Cambridge University Press, Cambridge, 1989.

[A-87] Nielsen, J. : *Collected Mathematical Papers*, Vols. 1 and 2, Birkhäuser, Boston, 1986.

[A-88] Noguchi, J. and Ochiai, T. : *Geometric Function Theory in Several Complex Variables*, Translations of Mathematical Monographs, Vol. 80, American Mathematical Society, Providence, Rhode Island, 1990.

[A-89] Oikawa, K. : *Riemann Surfaces*, Kyôritsu-Shuppan, Tokyo, 1987 (Japanese).

[A-90] Poincaré, H. : *Papers on Fuchsian Functions*, Springer-Verlag, Berlin and New York, 1985.

[A-91] Reimann, H. M. and Rychener, T. : *Funktionen beschränkter mittlerer Oszillation*, Lecture Notes in Math., Vol. 487, Springer-Verlag, Berlin and New York, 1975.

[A-92] Riemann, B. : *Collected Works*, Dover, New York, 1953.

[A-93] Sario, L. and Nakai, M. : *Classification Theory of Riemann Surfaces*, Springer-Verlag, Berlin and New York, 1970.

[A-94] Schiffer, M. and Spencer, D. C. : *Functionals of Finite Riemann Surfaces*, Princeton University Press, Princeton, New Jersey, 1954.

[A-95] Schlichenmaier, M. : *An Introduction to Riemann Surfaces, Algebraic Curves and Moduli Spaces*, Lecture Notes in Physics Vol. 322, Springer-Verlag, Berlin and New York, 1989.

[A-96] Schober, G. : *Univalent Functions-Selected Topics*, Lecture Notes in Math., Vol. 478, Springer-Verlag, Berlin and New York, 1975.

[A-97] Shafarevich, I. R. *Basic Algebraic Geometry*, Springer-Verlag, Berlin and New York, 1974.

[A-98] Siegel, C. L. : *Topics in Complex Function Theory*, Vols. I, II and III, Wiley, New York, 1969, 1971, 1973.

[A-99] Springer, G. : *Introduction to Riemann Surfaces*, Addison-Wesley, Reading, Massachusetts, 1957.

[A-100] Stein, E. M. : *Singular Integrals and Differentiability Properties of Functions*, Princeton University Press, Princeton, New Jersey, 1970.

[A-101] Stillwell, J. : *Classical Topology and Combinatorial Group Theory*, Springer-Verlag, Berlin and New York, 1980.

[A-102] Strebel, K. : *Quadratic Differentials*, Springer-Verlag, Berlin and New York, 1984.

[A-103] Suita, N. : *Modern Function Theory II (Theory of Conformal Mappings)*, Morikita-Shuppan, Tokyo, 1977 (Japanese).

[A-104] Sunada, T. : *Fundamental Groups and Laplacian*, Kinokuniya-Shoten, Tokyo, 1988 (Japanese).

[A-105] Suzuki, S. : *PL Topology of Surfaces I, II*, Maki-Shoten, Tokyo, 1986, 1987 (Japanese).

[A-106] Teichmüller, O. : *Collected Papers*, Springer-Verlag, Berlin and New York, 1982.

[A-107] Toda, N. : *Riemann Surfaces*, Science-Sha, Tokyo, 1976 (Japanese).

[A-108] Tsuji, M. : *Potential Theory in Modern Function Theory*, Chelsea, New York, 1959.

[A-109] Väisälä, J. : *Lectures on n-Dimensional Quasiconformal Mappings*, Lecture Notes in Math., Vol. 229, Springer-Verlag, Berlin and New York, 1971.

[A-110] Venkov, A. : *Spectral Theory of Automorphic Functions and its Applications*, Kluwer Academic Publishers, Dordrecht, The Netherlands, 1990.

[A-111] Weyl, H. : *Die Idee der Riemannschen Fläche*, B. G. Teubner, Stuttgart, 1913.

[A-112] Yau, S. T. : *Mathematical Aspects of String Theory*, World Scientific, Singapore, New York and Hong Kong, 1987.

[A-113] Yoshida, M. : *Fuchsian Differential Equations*, Friedr. Vieweg & Shon, Braunschweig, Germany, 1987.

[A-114] Zieschang, H., Vogt, E. and Coldewey, H.-D. : *Surfaces and Planar Discontinuous Groups*, Lecture Notes in Math., Vol. 835, Springer-Verlag, Berlin and New York, 1980.

Papers

[1] Abikoff, W. : On boundaries of Teichmüller spaces and Kleinian groups, III, *Acta Math.* **134** (1975), 212-237.

[2] Abikoff, W. : The uniformization theorem, *Amer. Math. Monthly* **88** (1981), 574-592.

[3] Agard, S. : A geometric proof of Mostow's rigidity theorem for groups of divergence type, *Acta Math.* **151** (1983), 231-252.

[4] Ahlfors, L. V. : Development of the theory of conformal mappings and Riemann surfaces through a century, in *Contributions to the Theory of Riemann Surfaces*, Ann. Math. Studies No. 30, Ahlfors, L. V. et al. (eds.), pp. 3-13, Princeton University Press, Princeton, New Jersey, 1953.

[5] Ahlfors, L. V. : The complex analytic structure of the space of closed Riemann surfaces, in **[A-84]**, pp. 45-66, 1960.

[6] Ahlfors, L. V. : Some remarks on Teichmüller space of Riemann surfaces, *Ann. Math.* **74** (1961), 171-191.

[7] Ahlfors, L. V. : Curvature properties of Teichmüller space, *J. d'Analyse Math.* **9** (1961), 161-176.

[8] Ahlfors, L. V. : Teichmüller spaces, *Proc. Internat. Congr. Math.*, pp. 3-9, Inst. Mittag-Leffler, 1962.

[9] Ahlfors, L. V. : Quasiconformal reflections, *Acta Math.* **109** (1963), 291-301.

[10] Ahlfors, L. V. : Quasiconformal Mappings and their Applications, in *Lectures on Modern Mathematics*, Vol. II, pp. 151-164, Saaty, T. L. (ed.), Wiley, New York, 1964.

[11] Ahlfors, L. V. : Quasiconformal mappings, Teichmüller spaces and Kleinian groups, *Proc. Internat. Congr. Math.*, pp. 71-84, Helsinki, 1978.

[12] Ahlfors, L. V. : Ergodic properties of groups of Möbius transformations, in *Analytic Functions, Kozubnik 1979*, Lawrynowicż, J. (ed.), Lecture Notes in Mathematics, Vol. 798, pp. 1-9, Springer-Verlag, Berlin and New York, 1980.

[13] Ahlfors, L. V. and Bers, L. : Riemann's mapping theorem for variable metrics, *Ann. Math.* **72** (1960), 385-404.

[14] Ahlfors, L. V. and Weill, G. : A uniqueness theorem for Beltrami equations, *Proc. A. M. S.* **13** (1962), 975-978.

[15] Alvarez, O. : Theory of string with boundaries, *Nucl. Phys.* **B. 216** (1983), 125-184.

[16] Alvarez-Gaume, L. and Nelson, P. : Riemann surfaces and string theories, in *Supersymmetry, Supergravity, Superstrings '86, Proceedings of the Trieste Spring School, 7-15 April 1986*, de Wit, B., Fayet, P. and Grisaru, M. (eds.), pp. 419-510, World Scientific, Singapore, New York and Hong Kong, 1986.

[17] Arbarello, E. : Periods of Abelian integrals, theta functions, and differential equations of KdV type, in *Proc. Internat. Congr. Math.*, pp. 623-627, Berkeley, 1986.

[18] Arbarello, E. and Cornalba, M. : The Picard groups of the moduli spaces of curves, *Topology* **26** (1987), 153-171.

[19] Astala, K. : Selfsimilar zippers, in **[A-24]**, Vol. I. pp. 61-73, 1988.

[20] Batchelor, M. and Bryant, P. : Graded Riemann surfaces, *Commun. Math. Phys.* **114** (1988), 243-255.

[21] Beauville, A. : L'approche géométrique du problème de Schottky, *Proc. Internat. Congr. Math.*, pp. 628-633, Berkeley, 1986.

[22] Bers, L. : Spaces of Riemann surfaces, *Proc. Internat. Congr. Math.*, pp. 349-361, Edinburgh, 1958.

[23] Bers, L. : Quasiconformal mappings and Teichmüller's theorem, in **[A-84]**, pp. 89-119, 1960.

[24] Bers, L. : Simultaneous uniformization, *Bull. A. M. S.* **66** (1960), 94-97.

[25] Bers, L. : Correction to "Spaces of Riemann surfaces as bounded domains", *Bull. A. M. S.* **67** (1961), 465-466.

[26] Bers, L. : Uniformization by Beltrami equations, *Comm. Pure Appl. Math.* **14** (1961), 215-228.

[27] Bers, L. : A non-standard integral equation with applications to quasiconformal mappings, *Acta Math.* **116** (1966), 113-134.

[28] Bers, L. : On boundaries of Teichmüller spaces and on Kleinian groups, I, *Ann. Math.* **91** (1970), 570-600.

[29] Bers, L. : Uniformization, moduli, and Kleinian groups, *Bull. London Math. Soc.* **4** (1972), 257-300.

[30] Bers, L. : A remark on Mumford's compactness theorem, *Israel J. Math.* **12** (1972), 400-407.

[31] Bers, L. : Fiber spaces over Teichmüllar spaces, *Acta Math.* **130** (1973), 89-126.

[32] Bers, L. : On spaces of Riemann surfaces with nodes, *Bull. A. M. S.* **80** (1974), 1219-1222.

[33] Bers, L. : Spaces of degenerating Riemann surfaces, in **[A-37]**, pp. 43-55, 1974.

[34] Bers, L. : Deformations and moduli of Riemann surfaces with nodes and signatures, *Math. Scand.* **36** (1975), 12-16.

[35] Bers, L. : Automorphic forms for Schottky groups, *Adv. in Math.* **16** (1975), 332-361.

[36] Bers, L. : On Hilbert's 22nd problem, in *Mathematical Developments Arising from Hilbert Problems*, Browder, F. E. (ed.), Proc. Sympos. Pure Math., Vol. 28, American Mathematical Society, Providence, Rhode Island, pp. 559-609, 1976.

[37] Bers, L. : Quasiconformal mappings, with applications to differential equations, function theory and topology, *Bull. A. M. S.* **83** (1977), 1083-1100.

[38] Bers, L. : An extremal problem for quasiconformal mappings and a theorem by Thurston, *Acta Math.* **141** (1978), 73-98.

[39] Bers, L. : The action of the modular group on the complex boundary, in **[A-59]**, pp. 33-52, 1981.

[40] Bers, L. : Finite dimensional Teichmüller spaces and generalizations, *Bull. A. M. S. (New Series)* **5** (1981), 131-172.

[41] Bers, L and Ehrenpreis, L. : Holomorphic convexity of Teichmüller spaces, *Bull. A. M. S.* **70** (1964), 761-764.

[42] Bers, L and Gardiner, F. P. : Fricke spaces, *Adv. in Math.* **62** (1986), 249-284.

[43] Bers, L and Royden, H. : Holomorphic families of injections, *Acta Math.* **157** (1986), 259-286.

[44] Blanchard, P. : Complex analytic dynamics on the Riemann sphere, *Bull. A. M. S. (New Series)* **11** (1984), 85-141.

[45] Bonahon, F. : The geometry of Teichmüller space via geodesic currents, *Invent. math.* **92** (1988), 139-162.

[46] Bowditch, B. H. and Epstein, D. B. A. : Natural triangulations associated to a surface, *Topology* **27** (1988), 91-117.

[47] Bowen, R. and Series, C. : Markov maps associated with Fuchsian groups, *IHES Publ. Math.*, No. **50** (1979), 153-170.

[48] Cartan, H. : Quotient d'un espace analytique par un group d'automorphisms, in *Algebraic Geometry and Algebraic Topology, A Symposium in Honor of S. Lefschetz*, Fox, R. H., Spencer, D. C., and Tucker, A. W. (eds.), pp. 90-102. Princeton University Press, Princeton, New Jersey, 1957.

[49] Chu, T. : The Weil-Petersson metric in the moduli space, *Chinese J. Math.* **4** (1976), 29-51.

[50] Crane, L. and Rabin, J. M. : Super Riemann surfaces (uniformization and Teichmüller theory), *Commun. Math. Phys.* **113** (1988), 601-623.

[51] Deligne, P. and Mumford, D. : The irreducibility of the space of curves of given genus, *IHES Publ. Math.* No. **36** (1969), 75-109.

[52] Douady, A. : Chirugie sur les applications holomorphes, *Proc. Internat. Congr. Math.*, pp. 724-738, Berkeley, 1986.

[53] Douady, A. and Earle, C. J. : Conformally natural extension of homeomorphisms of the circle, *Acta Math.* **157** (1986), 23-48.

[54] Douady, A. and Hubbard, J. : On the density of Strebel differentials, *Invent. math.* **30** (1975), 175-179.

[55] Douady, A. and Hubbard, J. : On the dynamics of polynomial-like mappings, *Ann. Sci. École Norm. Sup.* **18** (1985), 287-343.

[56] Drasin, D. : Proof of a conjecture of F. Nevanlinna concerning functions which have deficiency sum two, *Acta Math.* **158** (1987), 1-94.

[57] Earle, C. J. : Reduced Teichmüller spaces, *Trans. A. M. S.* **126** (1967), 54-63.

[58] Earle, C. J. : On holomorphic families of pointed Riemann surfaces, *Bull. A. M. S.* **79** (1973), 163-166.

[59] Earle, C. J. : On the Carathéodory metric in Teichmüller spaces, in **[A-37]**, pp. 99-103, 1974.

[60] Earle, C. J. : The Teichmüller distance is differentiable, *Duke Math. J.* **44** (1977), 389-397.

[61] Earle, C. J. : Families of Riemann surfaces and Jacobi varieties, *Ann. Math.* **107** (1978), 255-286.

[62] Earle, C. J. and Eells, Jr., J. : On the differential geometry of Teichmüller spaces, *J. d'Analyse Math.* **19** (1967), 35-52.

[63] Earle, C. J. and Eells, Jr., J. : A fiber bundle description of Teichmüller theory, *J. Diff. Geom.* **3** (1969), 19-43.

[64] Earle, C. J. and Fowler, R. S. : Holomorphic families of open Riemann surfaces, *Math. Ann.* **270** (1985), 249-273.

[65] Earle, C. J. and Kra, I. : On isometries between Teichmüller spaces, *Duke Math. J.* **41** (1974), 583-591.

[66] Earle, C. J. and Kra, I. : On sections of some holomorphic families of closed Riemann surfaces, *Acta Math.* **137** (1976), 49-79.

[67] Eells, Jr., J. and Lemaire, L. : Another report on harmonic maps, *Bull. London Math. Soc.* **20** (1988), 385-524.

[68] Eisenbud, D. and Harris, J. : Progress in the theory of complex algebraic curves, *Bull. A. M. S. (New Series)* **21** (1989), 205-232.

[69] Epstein, D. B. A. : Curves on 2-manifolds and isotopics, *Acta Math.* **115** (1966), 83-107.

[70] Fehlmann, R. and Sakan, K. : On extremal quasiconformal mappings with varying dilatation bounds, *Osaka J. Math.* **23** (1986), 751-764.

[71] Fischer, A. E. and Tromba, A. J. : On a purely "Riemannian" proof of the structure and dimension of the unramified moduli space of a compact Riemann surface, *Math. Ann.* **267** (1984), 311-345.

[72] Fischer, A. E. and Tromba, A. J. : On the Weil-Petersson metric on Teichmüller space, *Trans. A. M. S.* **284** (1984), 319-335.

[73] Fischer, A. E. and Tromba, A. J. : Almost complex principal fiber bundles and the complex structure on Teichmüller space, *J. Reine Angew. Math.* **352** (1984), 151-160.

[74] Fischer, A. E. and Tromba, A. J. : A new proof that Teichmüller space is a cell, *Trans. A. M. S.* **303** (1987), 257-262.

[75] Friedan, D. : Introduction to Polyakov's string theory, in *1982 Les Houches Summer School*, Zuber, J. -B. and Strora, R. (eds.), North Holland, Amsterdam, pp. 839-867, 1984.

[76] Fujiki, A. : The Moduli space of polarized algebraic manifolds and Kähler metrics, *Sûgaku*, **42** (1990), 231-243 (Japanese).

[77] Fujiki, A. and Schumacher, G. : On the moduli space of extremal compact Kähler manifolds, *Publ. RIMS, Kyoto Univ.* **26** (1990), 101-183.

[78] Furusawa, H. : A remark on the hyperbolic collar lemma, *Tôhoku Math. J.* **39** (1987), 291-298.

[79] Gardiner, F. P. : Schiffer's interior variation and quasiconformal mappings, *Duke Math. J.* **42** (1975), 371-380.

[80] Gardiner, F. P. : On partially Teichmüller Beltrami differentials, *Michigan Math. J.* **29** (1982), 237-242.

[81] Gardiner, F. P. : Approximation of infinite dimensional Teichmüller spaces, *Trans. A. M. S.* **282** (1984), 367-383.

[82] Gardiner, F. P. and Kra, I. : Quasiconformal stability of Kleinian groups, *Indiana Univ. Math. J.* **21** (1972), 1037-1059.

[83] Gehring, F. W. : Univalent functions and the Schwarzian derivative, *Comment. Math. Helv.* **52** (1977), 561-572.

[84] Gehring, F. W. : Spirals and the universal Teichmüller space, *Acta Math.* **141** (1978), 99-113.

[85] Gehring, F. W. : Topics in quasiconformal mappings, *Proc. Internat. Congr. Math.*, pp. 62-80, Berkeley, 1986.

[86] Giddings, S. B. and Wolpert, S. A. : A triangulation of moduli space from light-cone string theory, *Commun. Math. Phys.* **109** (1987), 177-190.

[87] Griffiths, P. A. : Complex-analytic properties of certain Zariski open sets on algebraic varieties, *Ann. Math.* **94** (1971), 21-51.

[88] Gunning, R. C. : Affine and projective structures on Riemann surfaces, in **[A-59]**, pp. 225-244, 1981.

[89] Hamilton, R. S. : Extremal quasiconformal mappings with prescribed boundary values, *Trans. A. M. S.* **138** (1969), 399-406.

[90] Harer, J. : The second homology group of the mapping class group of an orientable surface, *Invent. math.* **72** (1983), 221-239.

[91] Harer, J. : Stability of the homology of the mapping class groups of orientable surfaces, *Ann. Math.* **121** (1985), 215-249.

[92] Harer, J. : The virtual cohomological dimension of the mapping class group of an orientable surface, *Invent. math.* **84** (1986), 157-176.

[93] Harer, J. and Zagier, D. : The Euler characteristic of the moduli space of curves, *Invent. math.* **85** (1986), 457-485.

[94] Harris, J. : On the Kodaira dimension of the moduli space of curves, II. The even genus case, *Invent. math.* **75** (1984), 437-466.

[95] Harris, J. : Curves and their moduli, in *Algebraic Geometry, Bowdoin 1985*, Bloch, S. J. (ed.) Proceedings of Symposia in Pure Mathematics, Vol. 46, Part 1, pp. 99-143, American Mathematical Society, Providence, Rhode Island, 1987.

[96] Harris, J. and Mumford, D. : On the Kodaira dimension of the moduli space of curves, *Invent. math.* **67** (1982), 23-86.

[97] Hayman, W. K. and Reich, E. : On Teichmüller mappings of the disk, *Complex Variables* **1** (1982), 1-12.

[98] Hejhal, D. A. : On Schottky and Teichmüller spaces, *Adv. in Math.* **15** (1975), 133-156.

[99] Hejhal, D. A. : Regular B-groups, degenerating Riemann surfaces, and spectral theory, *Chalmers Univ. Tech. Report*, **No. 19**, 1987.

[100] Hejhal, D. A. : Regular b-group and repeated Dehn-twists, in *Complex Analysis, Joensuu 1987*, Laine, I., Rickman, S., and Sorvali, T. (eds.), Springer Lecture Notes in Mathematics, Vol. 1351, pp.169-192, 1988.

[101] Hubbard, J. and Masur, H. : Quadratic differentials and foliations, *Acta Math.* **142** (1979), 221-274.

[102] Imayoshi, Y. : Holomorphic families of Riemann surfaces and Teichmüller spaces, in **[A-59]**, pp. 277-300, 1981.

[103] Imayoshi, Y. : Universal covering spaces of certain quasi-projective algebraic surfaces, *Osaka J. Math.* **20** (1983), 581-598.

[104] Imayoshi, Y : Generalizations of de Franchis theorem, *Duke Math. J.* **50** (1983), 393-408.

[105] Imayoshi, Y. : An analytic proof of Severi's theorem, *Complex Variables* **2** (1983), 151-155.

[106] Imayoshi, Y. : Holomorphic maps of compact Riemann surfaces into two dimensional compact C-hyperbolic manifolds, *Math. Ann.* **270** (1985), 403-416.

[107] Imayoshi, Y. and Shiga, H. : A finiteness theorem of holomorphic families of Riemann surfaces, in **[A-24]**, Vol. II, pp. 207-219, 1988.

[108] Ivanov, N. V. : Complexes of curves and the Teichmüller modular group, *Russian Math. Surveys*, **42** (1987), 55-107.

[109] Jones, P. : Extension theorems for BMO, *Indiana Univ. Math. J.* **29** (1980), 41-66.

[110] Keen, L. : On Fricke moduli, in **[A-7]**, pp. 205-224, 1971.

[111] Kerckhoff, S. P. : The asymptotic geometry of Teichmüller spaces, *Topology* **19** (1980), 23-41.

[112] Kerckhoff, S. P. : The Nielsen realization problem, *Ann. Math.* **117** (1983), 235-265.

[113] Kerckhoff, S. P. : The geometry of Teichmüller space, *Proc. Internat. Congr. Math.*, pp. 665-678, Warszawa, 1983.

[114] Kerckhoff, S. P. : Earthquakes are analytic, *Comment. Math. Helv.* **60** (1985), 17-30.

[115] Kerckhoff, S. P., Masur, H. and Smillie, J. : Ergodicity of billiard flows and quadratic differentials, *Ann. Math.* **124** (1986), 293-311.

[116] Kerckhoff, S. P. and Thurston, W. P. : Non-continuity of the action of the modular group at Bers' boundary of Teichmüller space, *Invent. math.* **100** (1990), 25-47.

[117] Knudsen, F. and Mumford, D. : The projectivity of the moduli space of stable curves I, Preliminaries on "det" and "div", *Math. Scand.* **39** (1976), 19-55.

[118] Koiso, N. : Einstein metrics and complex structures, *Invent. math.* **73** (1983), 71-106.

[119] Koiso, N. : Yang-Mills connections and moduli space, *Osaka J. Math.* **24** (1987), 147-171.

[120] Kra, I. : Deformations of Fuchsian groups, I, II, *Duke Math. J.* **36** (1969), 537-546, *ibid.* **38** (1971), 499-508.

[121] Kra, I. : On the Nielsen-Thurston-Bers type of some self-maps of Riemann surfaces, *Acta Math.* **146** (1981), 231-270.

[122] Kra, I. : The Carathéodory metric on abelian Teichmüller disks, *J. d'Analyse Math.* **40** (1981), 129-143.

[123] Kra, I. : Canonical mappings between Teichmuüller spaces, *Bull. A. M. S. (New Series)* **4** (1981), 143-179.

[124] Kra, I. : On the vanishing of and spanning sets for Poincaré series for cusp forms, *Acta Math.* **153** (1984), 47-116.

[125] Kra, I. : Non-variational global coordinates for Teichmüller spaces, in **[A-24]**, Vol. II, pp. 221-249, 1988.

[126] Kra, I. : Horocyclic coordinates for Riemann surfaces and moduli spaces I: Teichmüller and Riemann spaces of Kleinian groups, *J. A. M. S.* **3** (1990), 497-578.

[127] Kra, I and Maskit, B. : The deformation space of a Kleinian group, *Amer. J. Math.* **103** (1981), 1065-1102.

[128] Kravetz, S. : On the geometry of Teichmüller spaces and the structure of their modular groups, *Ann. Acad. Sci. Fenn.* AI**278** (1959), 1-35.

[129] Kusunoki, Y. and Maitani, F. : Variations of abelian differentials under quasiconformal deformations, *Math. Z.* **181** (1982), 435-450.

[130] Lewittes, J. : Invariant quadratic differentials, *Bull. A. M. S.* **68** (1962), 320-322.

[131] LeBrun, C. and Rothstein, M. : Moduli of super Riemann surfaces, *Commun. Math. Phys.* **117** (1988), 159-176.

[132] Maclachlan, C. : Modulus space is simply-connected, *Proc. A. M. S.* **29** (1971), 85-86.

[133] Mabuchi, T. : Compactification of the moduli space of Einstein-Kähler orbifolds, in *Recent Topics in Differential and Analytic Geometry*, Ochiai, T. (ed.) Adv. Studies in Pure Math., Vol. 18-I, pp.359-384, Kinokuniya-Shoten, Tokyo, and Academic Press, London, 1990.

[134] Maitani, F. : Variations of meromorphic differentials under quasiconformal deformations, *J. Math. Kyoto Univ.* **24** (1984), 49-66.

[135] Mañé, R., Sad, P. and Sullivan, D. P. : On the dynamics of rational maps, *Ann. Sci. École Norm. Sup.* **16** (1983), 193-217.

[136] Marden, A. : Geometric relations between homeomorphic Riemann surfaces, *Bull. A. M. S. (New Series)* **3** (1980), 1001-1017.

[137] Marden, A., and Strebel, K. : The heights theorem for quadratic differentials on Riemann surfaces, *Acta Math.* **153** (1984), 153-211.

[138] Maskit, B. : On boundaries of Teichmüller spaces and on Kleinian groups, II, *Ann. Math.* **91** (1970), 607-639.

[139] Maskit, B. : Moduli of marked Riemann surfaces, *Bull. A. M. S.* **80** (1974), 773-777.

[140] Masumoto, M. : A characterization of the kernel of the Poincaré series operator, *Trans. A. M. S.* **300** (1987), 695-704.

[141] Masumoto, M. : Adjoints of the Poincaré series operators, *J. Math. Kyoto Univ.* **29** (1989), 569-589.

[142] Masur, H. : On a class of geodesics in Teichmüller space, *Ann. Math.* **102** (1975), 205-221.

[143] Masur, H. : The extension of the Weil-Petersson metric to the boundary of Teichmüller space, *Duke Math. J.* **43** (1976), 623-635.

[144] Masur, H. : The Jenkins-Strebel differentials with one cylinder are dense, *Comment. Math. Helv.* **54** (1979), 179-184.

[145] Masur, H. : Transitivity properties of the horocyclic and geodesic flows on moduli space, *J. d'Analyse Math.* **39** (1981), 1-10.

[146] Masur, H : Two boundaries of Teichmüller space, *Duke Math. J.* **49** (1982), 183-190.

[147] Masur, H. : Interval exchange transformations and measured foliations, *Ann. Math.* **115** (1982), 169-200.

[148] Masur, H. : Ergodic actions of the mapping class group, *Proc. A. M. S.* **94** (1985), 455-459.

[149] Masur, H. : Closed trajectories for quadratic differentials with an application to billiards, *Duke Math. J.* **53** (1986), 307-314.

[150] Matelski, J. P. : A compactness theorem for Fuchsian groups of the second kind, *Duke Math. J.* **43** (1976), 829-840.

[151] Matsuzaki, K. : Geometric finiteness, quasiconformal stability and surjectivity of the Bers map for Kleinian groups, to appear in *Tôhoku Math. J.*

[152] McKean, H. P. : Selberg's trace formula as applied to a compact Riemann surface, *Commun. Pure Appl. Math.* **25** (1972), 225-246.

[153] McMullen, C. : Amenability, Poincaré series and quasiconformal maps, *Invent. math.* **97** (1989), 95-127.

[154] McMullen, C. : Iteration on Teichmüller space, *Invent. math.* **99** (1990), 425-454.

[155] McMullen, C. : Cusps are dense, *Ann. Math.* **133** (1991), 217-247.

[156] Mori, A. : On Quasiconformality and Pseudo-analyticity, *Sûgaku* **7** (1955), 75-89 (Japanese).

[157] Mori, A. : On an absolute constant in the theory of quasi-conformal mappings, *J. Math. Soc. Japan* **8** (1956), 156-166.

[158] Morosawa, S. : Invariant subsets of the limit set for a Fuchsian group, *Tôhoku Math. J.* **42** (1990), 429-437.

[159] Mostow, G. D. : Quasi-conformal mappings in n-space and the rigidity of hyperbolic space forms, *IHES Publ. Math.* **34** (1968), 53-104.

[160] Mulase, M. : Cohomological structure in soliton equations and Jacobian varieties, *J. Diff. Geom.* **19** (1984), 403-430.

[161] Mumford, D. : Abelian quotients of the Teichmüller modular group, *J. d'Analyse Math.* **18** (1967), 227-244.

[162] Mumford, D. : The structure of the moduli space of curves and Abelian varieties, *Proc. Internat. Congr. Math.*, Vol. 1, pp. 457-465, Nice, 1970.

[163] Mumford, D. : Towards an enumerative geometry of the moduli space of curves, in *Arithmetic and Geometry*, Artin, M. and Tate, J. (eds.), Vol. II, pp. 271-327, Birkhäuser, Boston, 1983.

[164] Nag, S. and Verjovsky, A. : Diff(S^1) and Teichmüller spaces, *Commun. Math. Phys.* **130** (1990), 123-138.

[165] Nakada, M. : Quasi-conformal stability of finitely generated function groups, *Tôhoku Math. J.* **30** (1978), 45-58.

[166] Nakanishi, T. : A theorem on the outradii of Teichmüller spaces, *J. Math. Soc. Japan* **40** (1988), 1-8.

[167] Nelson, P. : Lectures on strings and moduli space, *Physics Reports* **149**, **No. 6** (1987), 337-375.

[168] Nelson, P. : Holomorphic coordinates for supermoduli space, *Commun. Math. Phys.* **115** (1988), 167-175.

[169] O'Byrne, B. : On Finsler geometry and applications to Teichmüller spaces, in **[A-7]**, pp. 317-328, 1971.

[170] Ohtake, H. : Lifts of extremal quasiconformal mappings of arbitrary Riemann surfaces, *J. Math. Kyoto Univ.* **22** (1982), 191-200.

[171] Ohtake, H. : On the deformation of Fuchsian groups by quasiconformal mappings with partially vanishing Beltrami coefficients, *J. Math. Kyoto Univ.* **29** (1989), 69-90.

[172] Oikawa, K. : On moduli of Riemann surfaces, *Sûgaku* **12** (1960/61), 79-104 (Japanese).

[173] Okumura, Y. : On the global real analytic coordinates for Teichmüller spaces, *J. Math. Soc. Japan* **42** (1990), 91-101.

[174] Patt, C. : Variations of Teichmüller and Torelli surfaces, *J. d'Analyse Math.* **11** (1963), 221-247.

[175] Penner, R. C. : The moduli space of punctured surfaces, in **[A-112]**, pp. 313-340, 1987.

[176] Polyakov, A. M. : Quantum geometry of bosonic strings, *Phys. Lett.* **103 B** (1981), 207-210.

[177] Rabin, J. M. and Freund, P. G. O. : Supertori are algebraic curves, *Commun. Math. Phys.* **114** (1988), 131-145.

[178] Rauch, H. E. : The singularities of the modulus spaces, *Bull. A. M. S.* **68** (1962), 390-394.

[179] Rauch, H. E. : Transcendental view of the space of algebraic Riemann surfaces, *Bull. A. M. S.* **71** (1965), 1-39.

[180] Reich, E. and Strebel, K. : On the extremality of certain Teichmüller mappings, *Comm. Math. Helv.* **45** (1970), 353-362.

[181] Riemann, B. : Grundlagen für eine allgemeine Theorie der Functionen einer veränderlichen complexen Grosse, in **[A-92]**, pp. 3-48, 1851.

[182] Riemann, B. : Theorie der Abel'schen Functionen, in **[A-92]**, pp. 88-144, 1857.

[183] Riera, G. : Semi-direct products of Fuchsian groups and uniformization, *Duke Math. J.* **44** (1977), 291-304.

[184] Royden H. L. : Automorphisms and isometries of Teichmüller space, in **[A-7]**, pp. 369-383, 1971.

[185] Royden, H. L. : Invariant metrics on Teichmüller space, in **[A-8]**, pp. 393-399, 1974.

[186] Saito, K. : Moduli space for Fuchsian groups, in *Algebraic Analysis*, Kashiwara, M. and Kawai, T. (eds.) Vol. II, pp.735-786, Academic Press, London, 1988.

[187] Sakan, K. : Necessary and sufficient conditions for extremality in certain classes of quasiconformal mappings, *J. Math. Kyoto Univ.* **26** (1986), 31-37.

[188] Sato, H. : Introduction to new coordinates to the Schottky space-The general case, *J. Math. Soc. Japan* **35** (1983), 23-35.

[189] Sato, H. : Limits of sequences of Riemann surfaces represented by Schottky groups, *Tôhoku Math. J.* **36** (1984), 521-539.

[190] Schumacher, G. : On the geometry of moduli spaces, *Manuscripta math.* **50** (1985), 229-267.

[191] Segawa, S. : Martin boundaries of Denjoy domains and quasiconformal mappings, *J. Math. Kyoto Univ.* **30** (1990), 297-316.

[192] Sekigawa, H. : The outradius of the Teichmüller space, *Tôhoku Math. J.* **30** (1978), 607-612.

[193] Sekigawa, H. and Yamamoto, H. : Outradii of the Teichmüller spaces of Fuchsian groups of the second kind, *Tôhoku Math. J.* **38** (1986), 365-370.

[194] Seppälä, M. and Sorvali, T. : Parametrization of Teichmüller spaces by geodesic length functions, in **[A-24]**, Vol. II, pp. 267-284, 1988.

[195] Series, C. : Geometrical Markov coding of geodesics on surfaces of constant negative curvature, *Ergod. Th. and Dynam. Sys.* **6** (1986), 601-625.

[196] Sethares, G. C. : The extremal property of certain Teichmüller mappings, *Comment. Math. Helv.* **43** (1968), 98-119.

[197] Shiba, M. : Realization of open Riemann surfaces by streamfunctions, *Sûgaku* **36** (1984), 208-226 (Japanese).

[198] Shiba, M. : The moduli of compact continuations of an open Riemann surfaces of genus one, *Trans. A. M. S.* **301** (1987), 299-311.

[199] Shiga, H. : On analytic and geometric properties of Teichmüller spaces, *J. Math. Kyoto Univ.* **24** (1984), 441-452.

[200] Shiga, H. : Characterization of quasi-disks and Teichmüller spaces, *Tôhoku Math. J.* **37** (1985), 541-552.

[201] Shiga, H. : On Teichmüller spaces and modular transformations, *J. Math. Kyoto Univ.* **25** (1985), 619-626.

[202] Shiga, H. : Projective structure on Riemann surfaces and Kleinian groups, *J. Math. Kyoto Univ.* **27** (1987), 433-438.

[203] Shimizu, H. : On discontinuous groups operating on the product of upper half-planes, *Ann. of Math.* **77** (1959), 33-71.

[204] Shiota, T. : Characterization of Jacobian varieties in terms of soliton equations, *Invent. math.* **83** (1986), 333-382.

[205] Shiota, T. : The KP equation and The Schottky problem, *Sugaku Expositions* **3** (1990), 183-211.

[206] Shishikura, M. : On the quasiconformal surgery of rational functions, *Ann. Sci. École Norm. Sup.* **20** (1987), 1-29.

[207] Shishikura, M. : On complex dynamics on the Riemann sphere, *Sûgaku* **41** (1989), 34-48 (Japanese).

[208] Siu, S. T. : Curvature of the Weil-Petersson metric in the moduli space of compact Kähler-Einstein manifolds of negative first Chern class, in *Contributions to Several Complex Variables, In Honour of W. Stoll*, Howard, A. and Wong, P. (eds.), Aspects of Math., Vol. 9, pp. 261-298, Friedr. Vieweg & Sohn, Braunschweig, Germany, 1986.

[209] Slodkowski, Z. : Holomorphic motions and polynomial hulls, *Proc. A. M. S.* **111** (1991), 347-355.

[210] Strebel, K. : On lifts of extremal quasiconformal mappings, *J. d'Analyse Math.* **31** (1977), 191-203.

[211] Strebel, K. : On quasiconformal mappings of open Riemann surfaces, *Comment. Math. Helv.* **53** (1978), 301-321.

[212] Sugawa, T. : On the Bers conjecture for Fuchsian groups of the second kind, *J. Math. Kyoto Univ.*, to appear.

[213] Sullivan, D. P. : On the ergodic theory at infinity of an arbitrary discrete group of hyperbolic motions, in **[A-59]**, pp. 465-496, 1981.

[214] Sullivan, D. P. : Quasiconformal homeomorphisms and dynamics I, *Ann. Math.* **122** (1985), 401-418.

[215] Sullivan, D. P. : Quasiconformal homeomorphisms and dynamics II, *Acta Math.* **155** (1985), 243-260.

[216] Sullivan, D. : Quasiconformal homeomorphisms in dynamics, topology, and geometry, *Proc. Internat. Congr. Math.*, pp. 1216-1228, Berkeley, 1986.

[217] Sullivan, D. P. and Thurston, W. P. : Extending holomorphic motions, *Acta Math.* **157** (1986), 243-257.

[218] Sunada, T. : Riemannian coverings and isospectral manifolds, *Ann. Math.* **121** (1985), 169-186.

[219] Takhtadzhyan, L. A. : Uniformization, local index theorem, and geometry of the moduli spaces of Riemann surfaces and vector bundles, in *Theta Functions, Bowdoin 1987*, Ehrenpreis, L. and Gunning, R. C. (eds.), Proceedings of Symposia in Pure Math. Vol. 49, Part I, pp. 581-596, American Mathematical Society, Providence, Rhode Island, 1989.

[220] Takhtadzhyan, L. A. and Zograf, P. G. : A local index theorem for families of $\bar{\partial}$-operators on punctured Riemann surfaces and a new Kähler metric on their moduli spaces, *Commun. Math. Phys.* **137** (1991), 399-426.

[221] Tanigawa, H. : Orbits and their accumulation points of cyclic subgroups of modular groups, *Tôhoku Math J.* **43** (1991), 289-299.

[222] Taniguchi, M. : Abelian differentials whose squares have closed trajectories on compact Riemann surfaces, *Japan. J. Math.* **4** (1978), 417-443.

[223] Taniguchi, M. : Variational formulas on arbitrary Riemann surfaces under pinching deformation, *J. Math. Kyoto Univ.* **27** (1987), 507-530; Supplements to my previous paper, *ibid.* **28** (1988), 81-86.

[224] Taniguchi, M. : Pinching deformation of arbitrary Riemann surfaces and variational formulas for abelian differentials, in *Analytic Function Theory of One Complex Variable*, Komatu, Y., Niino, K., and Yang C. (eds.) Pitman Research Notes in Math., Ser. 212, 330-345, Longman Scientific & Technical, 1989.

[225] Taniguchi, M. : Abelian differentials with normal behavior and complex pinching deformation, *J. Math. Kyoto Univ.* **29** (1989), 45-56.

[226] Taniguchi, M. : On the first variation of Green's functions under quasiconformal deformations, *J. Math. Kyoto Univ.* **29** (1989), 591-600.

[227] Taniguchi, M. : A note on the second variational formulas of functionals on Riemann surfaces, *Kôdai Math. J.* **12** (1989), 283-295.

[228] Taniguchi, M. : On rigidity of the ideal boundary of an infinite Riemann surface, *Complex Variables* **14** (1990), 161-167.

[229] Teichmüller, O. : Extremale quasikonforme Abbildungen und quadratische Differentiale, in **[A-106]**, pp. 335-531, 1939.

[230] Teichmüller, O. : Bestimmung der extremalen quasikonformen Abbildungen bei geschlossenen orientierten Riemannschen Flächen, in **[A-106]**, pp. 635-676, 1943.

[231] Thurston, W. P. : Three dimensional manifolds, Kleinian groups and hyperbolic geometry, *Bull. A. M. S.* (*New Series*) **6** (1982), 357-381.

[232] Thurston, W. P. : Zippers and univalent functions, in *The Bieberbach Conjecture*, Baernstein, A. et al. (eds.), pp. 185-197, American Mathematical Society, Providence, Rhode Island, 1986.

[233] Thurston, W. P. : Earthquakes in two-dimensional hyperbolic geometry, in **[A-25]**, pp. 91-112, 1986.

[234] Thurston, W. P. : On the geometry and dynamics of diffeomorphisms of surfaces, *Bull. A. M. S.* (*New Series*) **19** (1988), 417-431.

[235] Tromba, A. J. : On a natural algebraic affine connection on the space of almost complex structures and the curvature of Teichmüller space with respect to the Weil-Petersson metric, *Manuscripta math.* **56** (1986), 475-497.

[236] Tromba, A. J. : On an energy function for the Weil-Petersson metric on Teichmüller space, *Manuscripta math.* **59** (1987), 249-260.

[237] Tromba, A. J. : A classical variational approach to Teichmüller theory, in *Topics in Calculus of Variation*, Giaquinta, M. (ed.), Lecture Notes in Mathematics, Vol. 1365, pp. 155-185, Springer-Verlag, Berlin and New York, 1988?.

[238] Tukia, P. : Quasiconformal extension of quasisymmetric mappings compatible with a Möbius group, *Acta Math.* **154** (1985), 153-193.

[239] Veech, W. A. : The Teichmüller geodesic flow, *Ann. Math.* **124** (1986), 441-530.

[240] Veech, W. A. : Dynamics over Teichmüller space, *Bull. A. M. S.* **14** (1986), 103-106.

[241] Velling, J. A. and Matsuzaki K. : The action at infinity of conservative groups of hyperbolic motions need not have atoms, to appear in *Ergod. Th. and Dynam. Sys.*

[242] Vignéras, M. F. : Variétés riemnanniennes isospectrales et non isométriques, *Ann. Math.* **112** (1980), 21-32.

[243] Weil, A. : Sur les modules des surfaces de Riemann, *Sem. Bourbaki* **10** (1957/58).

[244] Wolf, M. : The Teichmüller theory of harmonic maps, *J. Diff. Geom.* **29** (1989), 449-479.

[245] Wolpert, S. A. : Non-completeness of the Weil-Petersson metric for Teichmüller space, *Pacific J. Math.* **61** (1975), 573-577.

[246] Wolpert, S. A. : The length spectra as moduli for compact Riemann surfaces, *Ann. Math.* **109** (1979), 323-351.

[247] Wolpert, S. A. : The Fenchel-Nielsen deformation, *Ann. Math.* **115** (1982), 501-528.

[248] Wolpert, S. A. : On the symplectic geometry of deformations of a hyperbolic surface, *Ann. Math.* **117** (1983), 207-234.

[249] Wolpert, S. A. : On the homology of the moduli space of stable curves, *Ann. of Math.* **118** (1983), 491-523.

[250] Wolpert, S. A. : On the Kähler form of the moduli space of once punctured tori, *Comment. Math. Helvetici* **58** (1983), 246-256.

[251] Wolpert, S. A. : On the Weil-Petersson geometry of the moduli space of curves, *Amer. J. Math.* **107** (1985), 969-997.

[252] Wolpert, S. A. : On obtaining a positive line bundle from the Weil-Petersson class, *Amer. J. Math.* **107** (1985), 1485-1507.

[253] Wolpert, S. A. : Chern forms and the Riemann tensor for the moduli space of curves, *Invent. math.* **85** (1986), 119-145.

[254] Wolpert, S. A. : Thurston's Riemannian metric for Teichmüller space, *J. Diff. Geom.* **23** (1986), 143-174.

[255] Wolpert, S. A. : The topology of the moduli space of Riemann surfaces, in *Arbeitstagung Bonn 1984*, Hirzebruch, F., Schwermer, J., and Suter, S. (eds.) Lecture Notes in Mathetatics, Vol. 1111, pp. 431-451, Springer-Verlag, Berlin and New York, 1986.

[256] Wolpert, S. A. : Geodesic length functions and the Nielsen problem, *J. Diff. Geom.* **25** (1987), 275-296.

[257] Wolpert, S. A. : Asymptotics of the spectrum and the Selberg zeta function on the space of Riemann surfaces, *Commun. Math. Phys.* **112** (1987), 283-315.

[258] Wolpert, S. A. : Cut-paste deformations of Riemann surfaces, *Ann. Acad. Sci. Fenn. A. I. Math.* **13** (1988), 401-412.

[259] Wolpert, S. A. : The hyperbolic metric and the geometry of the universal curve, *J. Diff. Geom.* **31** (1990), 417-472.

[260] Wolpert, S. A. : The Bers embedding and the Weil-Petersson metric, *Duke Math. J.* **60** (1990), 497-508.

[261] Yamada, A. : Precise variational formulas for abelian differentials, *Kôdai Math. J.* **3** (1980), 114-143.

[262] Yamaguchi H. : Calcul des variations analytiques, *Japan. J. Math.* **7** (1981), 319-377.

[263] Yoshida, M. : The Schwarz program, *Sûgaku* **40** (1988), 36-46 (Japanese).

[264] Zograf, P. G. and Takhtadzhyan, L. A. : On Liouville's equation, accessory parameters, and the geometry of Teichmüller space for Riemann surfaces of genus 0, *Math. USSR Sbornik* **60** (1988), 143-161.

[265] Zograf, P. G. and Takhtadzhyan, L. A. : On uniformization of Riemann surfaces and the Weil-Petersson metric on Teichmüller and Schottky spaces, *Math. USSR Sbornik* **60** (1988), 297-313.

[266] Zograf, P. G. and Takhtadzhyan, L. A. : On the geometry of moduli spaces of vector bundles over a Riemann surface, *Math. USSR Izvestiya* **35** (1990), 83-100.

List of Symbols

$\mathbf{N}$	natural numbers
$\mathbf{Z}$	integers
$\mathbf{Q}$	rational numbers
$\mathbf{R}$	real numbers, real axis
$\hat{\mathbf{R}}$	$\mathbf{R} \cup \{\infty\}$
$\mathbf{R}^+$	positive real numbers
$\overline{\mathbf{R}^+}$	nonnegative real numbers
$\mathbf{C}$	complex numbers, complex plane
$\hat{\mathbf{C}}$	Riemann sphere
H	upper half-plane
H^*	lower half-plane
Δ	unit disk
Δ^*	$\hat{\mathbf{C}} - \bar{\Delta}$
$\mathbf{R}^n$	n-dimensional real Euclidean space
$\mathbf{C}^n$	n-dimensional complex Euclidean space
Re	real part
Im	imaginary part
$w = u + iv$	complex variable w with $\operatorname{Re} w = u$, $\operatorname{Im} w = v$
$z = x + iy$	complex variable z with $\operatorname{Re} z = x$, $\operatorname{Im} z = y$
$\zeta = \xi + i\eta$	complex variable ζ with $\operatorname{Re} \zeta = \xi$, $\operatorname{Im} \zeta = \eta$
$\partial/\partial z$	$\{\partial/\partial x - i\partial/\partial y\}/2$
$\partial/\partial \bar{z}$	$\{\partial/\partial x + i\partial/\partial y\}/2$
f_z	$\partial f/\partial z$
$f_{\bar{z}}$	$\partial f/\partial \bar{z}$
id	identity mapping
$\lvert\cdot\rvert$	absolute value
$[\cdot]$	equivalence class

Index

N

O

P

Q

R

U

V

W

MIX
Papier aus verantwortungsvollen Quellen
Paper from responsible sources
FSC® C105338

If you have any concerns about our products,
you can contact us on
ProductSafety@springernature.com

In case Publisher is established outside the EU,
the EU authorized representative is:
Springer Nature Customer Service Center GmbH
Europaplatz 3, 69115 Heidelberg, Germany

Printed by Libri Plureos GmbH
in Hamburg, Germany